# Lecture Notes in Mathematics 1788

Editors:
J.-M. Morel, Cachan
F. Takens, Groningen
B. Teissier, Paris

T0183498

**Springer**
*Berlin*
*Heidelberg*
*New York*
*Barcelona*
*Hong Kong*
*London*
*Milan*
*Paris*
*Tokyo*

Alexander Vasil'ev

# Moduli of Families of Curves for Conformal and Quasiconformal Mappings

Springer

Author

Alexander Vasil'ev
Departamento de Matemática
Universidad Técnica Federico Santa María
Casilla 110-V, Valparaíso, Chile

E-mail: alexander.vasiliev@mat.utfsm.cl

Cataloging-in-Publication Data applied for

Die Deutsche Bibliothek - CIP-Einheitsaufnahme

Vasil'ev, Aleksandr:
Moduli of families of curves for conformal and quasiconformal mappings /
Alexander Vasil'ev. - Berlin ; Heidelberg ; New York ; Barcelona ; Hong Kong
; London ; Milan ; Paris ; Tokyo : Springer, 2002
 (Lecture notes in mathematics ; 1788)
 ISBN 3-540-43846-7

Mathematics Subject Classification (2000):
30C35, 30C55, 30C62, 30C75, 30F10, 30F60

ISSN 0075-8434
ISBN 3-540-43846-7 Springer-Verlag Berlin Heidelberg New York

Springer-Verlag Berlin Heidelberg New York a member of BertelsmannSpringer
Science + Business Media GmbH

http://www.springer.de

© Springer-Verlag Berlin Heidelberg 2002
Printed in Germany

Typesetting: Camera-ready TeX output by the author

SPIN: 10878683      41/3142/du-543210 - Printed on acid-free paper

# Preface

In the present monograph, we consider the extremal length method in its form of *the method of moduli of families of curves* in applications to the problems of conformal, quasiconformal mapping, and Teichmüller spaces. This method going back to H. Grötzsch, A. Beurling, L. V. Ahlfors, J. Jenkins is now one of the basic methods in various parts of Analysis. Several surveys and monographs, e.g., [30], [64], [78], [107], [139] are devoted to the development of this method and applications. However, we want to give here a useful guide: how one can start to solve extremal problems of conformal mapping beginning with simple but famous classical theorems and ending at difficult new results. Some more non-traditional applications we consider in the quasiconformal case. The modulus method permits us to consider the problems in question from a single point of view.

At the mid-century it was established that the classical methods of the geometric function theory could be extended to complex hyperbolic manifolds. The Teichmüller spaces turned out to be the most important of them. Recently, it has become clear that some forms of the extremal length method could be applied to examine different properties of Teichmüller spaces (see e.g. [43], [44]).

Thus, we are concerned with *the modulus method and its applications to extremal problems for conformal, quasiconformal mappings, the extension of moduli onto Teichmüller spaces*. The book is intended for different groups of readers:

(1) Non-experts who want to know about how one can use the modulus technique to solve extremal problems of Complex Analysis. One can find proofs of classical theorems of conformal and quasiconformal mapping by means of the modulus method as well as many examples of symmetrization and polarization. Graduate students will find here some useful exercises to check their understanding.

(2) Experts who will find new results about solution of difficult extremal problems for conformal and quasiconformal mappings and about the extension of the modulus onto Teichmüller spaces.

For the most part of this book we assume the background provided by the usual graduate courses in complex analysis, in particular, the theory of conformal mappings.

This book is not an exhaustive survey of quasiconformal or conformal mapping. Here we mostly consider applications of the modulus method to extremal problems and to the Teichmüller theory. Some of results are known but we present them from the modern point of view and some of them appear here for the first time. One can find either the results in the proper development of the modulus method or its applications.

We omit some difficult proofs of theorems which one can find in already existing monographs, however, we prove various introductory theorems in Chapter 2 to give the reader the flavour of the modulus method. To facilitate matters along this line, we present some exercises (marked by $\star$) which are either simple examples or else theorems that we suggest to prove independently.

First of all, this book reflects the scientific interests and results of the author and does not pretend to be an exhaustive treatment of the field. In particular, we have deliberately omitted a number of results by R. Kühnau and his alumni. Most of them are covered in depth in [75]. We only mention here the work by F. Gardiner and H. Masur [44] on the relationship between a special embedding of a Teichmüller space by extremal lengths and the Thurston embedding that turns out to be the starting point of a new interesting direction. A thorough treatment of the proper development of the modulus method in connection with the extremal partitions one can find in the already mentioned books by G. Kuz'mina [78], M. Ohtsuka [107], K. Strebel [141], and of symmetrization and polarization in a series of articles by A. Solynin [133]–[136] and V. Dubinin [30]. So this work is neither a complete exposition of the classical theory nor a complete survey of the latest results. But we hope one can find here a step to new ways of investigation to make progress in modern and classical problems.

**Acknowledgements.** There are groups of mathematicians in Saratov, Kazan, Novosibirsk, Kiev, Tomsk, and St.-Petersburg who initiated powerful and informative discussions with the author. The author particularly wants to acknowledge many conversations with I. Alexandrov, V. Aseev, D. Prokhorov, S. Fedorov, V. Gutlyanskiǐ, S. Nasyrov, A. Solynin, A. Sychëv, and P. Tamrazov. The author especially wants to thank Professor Christian Pommerenke for his useful suggestions and the referees for their remarks and observations. The author would like to express his gratitude to Professor Dmitri Prokhorov for his constant warm attention to this work and for his great participation in the author's scientific and human aspirations. Special thanks go to my wife and colleague Irina who inspired my work on this book. The project has been partially supported by FONDECYT (Chile), grants # 1010093, 1020067.

Bogotá–Valparaíso, 2001–2002                                   *Alexander Vasil'ev*

# Contents

# 1. Introduction

An important part of complex analysis is extremal problems, first of all, estimation of functionals and description of the ranges of system of functionals in various compact classes of conformal or quasiconformal maps. Many powerful methods have been developed to solve the problems for conformal maps. We mention here the area principle [48], [88], the variational method of Goluzin–Schiffer [48], [118], [120], the Löwner–Kufarev parametric method [4], [56], [94], [112]. The latter has allowed, finally, to solve the Bieberbach Conjecture (see the history, e.g., in [21], [111]). One of the crucial problems is the uniqueness of the extremal functions for a given functional. The complete solution of such a problem is given by the modulus method that uses either variational principles or topological-geometric ones.

The method of extremal metrics (extremal length method) goes back to the strip method by H. Grötzsch [53], [54]. "H. Grötzsch first used it as a method in the theory of univalent functions. He states that its use was suggested to him by the work of Faber" (see [64], page 7). Later, L. V. Ahlfors and A. Beurling [9], [12] introduced the notion of *the extremal length* (the reciprocal of the *modulus*) that gave rise to the active development of the method. The great contribution to the subject has been made by J. Jenkins [63], [64] and K. Strebel [141] who connected the modulus problem with the problem of the extremal partitioning of a Riemann surface and proved the existence of the extremal metric by Schiffer's variations [119]. Further development and generalization of the modulus were made by G. V. Kuz'mina [78]. Now on, the modulus method (the extremal length method in Ahlfors' terminology) is a powerful tool for the solution of problems of various types. It turns out to be rather universal and provides easily and naturally the uniqueness of solution in many extremal problems of conformal mapping what is complicate on applying other methods. It is important, that the modulus became a source of the definition of quasiconformal maps.

The theory of quasiconformal mapping emerged at the beginning of the twentieth century. At that time, these maps arose by geometric reasons based on the works of H. Grötzsch [53], [54] (who introduced so-called regular quasiconformal maps) and the notion of the extremal length suggested by L. V. Ahlfors and A. Beurling and, on the other hand, as solutions of a special type of elliptic systems of differential equations in the works by

M. A. Lavrentiev (see e.g. [86]). Important applications to various fields of mathematics such as the discrete group theory, mathematical physics, complex differential geometry, have caused a great development of the theory of quasiconformal mappings that, nowadays, is an important branch of Complex Analysis. A great contribution to this theory has been made by M. Lavrentiev, H. Grötzsch, L. Ahlfors (who became one of the first Fields laureates (1936)), L. Bers, O. Teichmüller, P. Belinskiĭ, and L. Volkovyskiĭ in the past, and C. Earle, I. Kra, M. Ohtsuka, V. Zorich, S. Krushkal, Yu. Reshetnyak, A. Sychëv, F. Gehring, O. Lehto, J. Väisälä, C. Andreian Cazacu, R. Kühnau, V. Aseev, V. Gutlyanskiĭ, V. Sheretov, etc., recently.

Various methods such as the area method, the variational method, the method of parametric representations, the length-area principle, the extremal length method were developed for solution of the problems for quasiconformal maps. Many of them became more or less transformations of the ideas from classical methods of conformal analysis. However, not all of them provide similar effect solving concrete problems. So, one can compare the Löwner-Kufarev method [4], [94], [112] for univalent functions in the conformal case and the parametric method by Shah Dao-Shing [126] and Gehring-Reich [46] (see also [87]) in the quasiconformal case. Only few problems have been solved by the latter but the main extremal problems like the Bieberbach Conjecture in the conformal case have used the Löwner method. Thus, the problems in the quasiconformal case are, generally, much more difficult.

At the mid-century it was established that the classical methods of the geometric function theory could be extended to complex hyperbolic manifolds. The Teichmüller spaces became the most important of them. In 1939 O. Teichmüller [150] proposed and partially realized an adventurous program of investigation in the moduli problem for Riemann surfaces. In the present work we often use the term "modulus" with a different meaning. Só, we should clarify that the classical modulus problem starts with the work by B. Riemann [115]. It is based on the fact that the conformal structure of a compact Riemann surface of genus $g > 1$ depends on $3g - 3$ complex parameters (moduli). The problem proceeds with a study of the nature of these parameters for general Riemann surfaces and with the induction of a real or a complex structure into the corresponding space of Riemann surfaces. O. Teichmüller has brought together the moduli problem, extremal quasiconformal maps, and relevant quadratic differentials on a Riemann surface. This led him to the well known theory of the Teichmüller spaces. Later on, Teichmüller's ideas were thoroughly substantiated by L. Ahlfors, L. Bers [11] and other specialists. Investigations in Teichmüller spaces were carried out in various directions: the topological one which is connected with homotopy classes of diffeomorphisms of surfaces of finite conformal type (W. Abikoff, R. Fricke, J. Nielsen, W. Thurston), mathematical physics which is connected with applications to the conformal-gauge string theory (L. Takhtadzhyan, P. Zograf), the theory of discontinuous groups (B. Maskit, H. Zieschang, E. Fogt,

H.-D. Coldewey, I. Kra, S. Krushkal, B. Apanasov, N. Gusevskiĭ), the theory of dynamical systems (R. Devaney, A. Douady, J. Hubbard, L. Keen, D. Sullivan, C. McMullen, A. Eremenko, M. Lyubich), complex analysis and metric theory (C. Earle, I. Kra, S. Nag, H. Royden, F. Gardiner, S. Krushkal, V. Sheretov), etc.

However, returning to the appearance of quasiconformal maps one can consider the notion of the modulus of families of curves as the basis of the notion of quasiconformality (surface strips in Grötzsch's terminology, extremal length in Ahlfors' terminology). We refer the reader to a simple and elegant formulation of the problem (known as the Grötzsch problem (see [9]; [43], Section 1.10). It is known, that the simplest and most important conformal invariant is the modulus of a quadrilateral, i.e., the ratio of the lengths of the sides of a relevant conformally equivalent rectangle. Suppose that two topological quadrilaterals $D$ and $D'$ are defined by a fixed order of their vertices. We intend to construct a homeomorphism $f : D \to D'$ with a natural agreement of the vertices which is the most nearly conformal map. The deviation of this homeomorphism from a conformal one can be measured by $K(f)$ that maximizes the ratio of the maximal and minimal diameters of the images of infinitesimal disks from $D$. Then, the extremal homeomorphism $f^*$ is an affine map up to the conformal maps of $D$ and $D'$ onto rectangles. If $M(D)$ is the modulus of $D$ and $M(D') \geq M(D)$, then $K(f^*) = M(D')/M(D)$.

A sense preserving homeomorphism $f$ of a domain $\Omega \subset \overline{\mathbb{C}}$ into the Riemann sphere $\overline{\mathbb{C}}$ is said to be *a quasiconformal map* if for any quadrilateral $D \subset \Omega$ the ratio $M(f(D))/M(D)$ is finite; *a K–quasiconformal map* if $M(D)/K \leq M(f(D)) \leq K M(D)$. So, the homeomorphism $f^*$ is extremal for the latter inequality.

The conformal invariant $M(D)$ of a quadrilateral $D$ is equal to the modulus of the family of curves that join two opposite sides of $D$. Here a link appears between the notion of quasiconformality and the modulus of a family of curves. The modulus is a realization of a so-called length-area principle where the lengths of curves in some metric and the reduced area that they sweep out are taken into account together. Let us give a simple definition. Let $\Gamma$ be a family of curves in $\Omega$ and $P$ be a family of differential metrics $\rho(z)|dz|$ on $\Omega$ with a non-negative, real, square integrable density $\rho(z)$ on $\Omega$, and

$$\int_{\gamma} \rho(z)|dz| \geq 1, \quad \text{for any} \quad \gamma \in \Gamma. \tag{1.1}$$

If $P \neq \emptyset$, then one can say that for $\Gamma$ the modulus problem is formulated and the modulus in this problem is defined as

$$m(\Gamma) := \inf_{\rho \in P} \iint_{\Omega} \rho^2(z) \, dxdy, \quad z = x + iy. \tag{1.2}$$

In the classical definition (see e.g. [9]; [64], Definition 2.6) the Lebesgue integrals are considered in (1.1), (1.2) as well as locally rectifiable curves. Here and further on, we refer the reader to Tamrazov's [146], [149] approach to conditions of admissibility, to definitions of metrics, families of curves, linear integrals, and moduli where one can abandon rectifiability of curves and Borelian metrics and can consider only measured metrics and arbitrary (not necessary rectifiable) curves. It is important when we consider the quasiconformal images of rectifiable curves (see an example by P. Belinskiĭ [16] of the non-rectifiable quasiconformal image of a rectifiable curve). To make this possible we consider in (1.1) the lower Darboux integral instead of the usual Lebesgue one, and keep the Lebesgue integral in (1.2). Under these conditions all the results about the moduli and the extremal partitions remain true. This will help us to avoid difficulties when we will consider quasiconformal images of curves. We remark here that there are other approaches to generalize the definition of the modulus. For example, J. Hersch, A. Pfluger [59] suggested to use the upper Darboux integral in (1.2); B. Fuglede [42] considered Borelian metrics $\rho(\zeta)|d\zeta|$. But in a view of later applications, Tamrazov's approach seems to be the best one. There exists another way to avoid the problem of non-rectifiability. Considering the quasiconformal image of a homotopy class of curves one observes that the set of non-rectifiable images sweeps out a plane null set. Therefore, they does not contribute to the value of (1.2) (see C. Anreian Cazacu [7]).

The main and simple properties of the modulus one can learn from the advanced monograph by J. Jenkins [64]. They are the conformal invariance of the modulus and the uniqueness (essentially) of the extremal metric in (1.2). But the existence and the form of such a metric turns out to be a much more difficult problem. Simple cases of families of curves on a quadrilateral and on a doubly connected hyperbolic domain still have been only known for a long time. The works by J. Jenkins [63], [64], K. Strebel [141], and G. Kuz'mina [78] are concerned with the proofs of the existence of the extremal metrics in non-trivial cases. Important results have been obtained by J. Jenkins in his famous article [63] by the application of Schiffer's special variations [119], and then, generalized by G. Kuz'mina [78]–[85], E. Emel'yanov [34], [36], and A. Solynin [138], [139] . They established the connection between the modulus problem and the problem of the extremal partition of a Riemann surface by domains of special shape. These results reflect the geometric and variational-analytic nature of the modulus and stipulate further development of the modulus method. Nowadays, it has become a powerful tool of investigation in the theory of conformal and quasiconformal mappings. Many difficult problems have been solved by this method, e.g., geometric problems posed by N. Chebotarev, M. Lavrentiev, O. Teichmüller (see G. Kuz'mina [78], S. Fedorov [40]); problems of the potential theory (see M. Ohtsuka [107], P. Tamrazov [147]–[149], N. Zoriĭ [180], [181], V. Shlyk [128], [129]); variational and isoperimetric problems (see J. Jenkins [62], [64], G. Kuz'mina [78]–[80], I. Mityuk [101], A. Solynin [132], E. Emel'yanov [34], S. Fedorov [41], [163], A. Vasil'ev [163]); problems for spatial mappings and mappings on

Riemann manifolds (see J. Väisälä [171], P. Tamrazov [149], F. Gehring [45], M. Vuorinen [173], [174], V. Zorich [178], [179], A. Sychëv [142], V. Aseev, B. Sultanov [13], and V. Shlyck [128]).

There are many problems of conformal mapping solved by the modulus method that could not be solved by other methods or there were insuperable technical obstacles surveyed, e.g., in [64], [78], [30]. Among these problems we meet covering theorems in different classes of conformal mappings [40], [78], [79]; estimations of growth, distortion, and their mutual variation in these classes [34], [41], [62], [79], [163]; estimations of the product of conformal radii of non-overlapping domains [29], [78], problems connected with the harmonic measure [28], [133]–[136], etc.

This monograph covers applications of the modulus method in connection with the extremal partitions of Riemann surfaces to a number of extremal problems of conformal mapping (Chapter 3) and quasiconformal mapping (Chapter 4). In Chapter 5 we extend the modulus onto Teichmüller spaces and derive various properties of the modulus dependent on the conformal structure of the underlying Riemann surface. There we also deal with metric problems of the Teichmüller theory.

# 2. Moduli of Families of Curves and Extremal Partitions

This chapter contains an introductory course of the extremal length method in the form of *the modulus of a family of curves* in connection with *the extremal partition of Riemann surfaces*. For convenience, we introduce in Section 2.4 a catalogue of the most frequently used moduli of rings and quadrilaterals, reduced moduli of circular domains and digons, and summarize in Section 2.3 the information on some elliptic functions and integrals.

Throughout the book we use the following notations:

$\mathbb{C} = \{z = x + iy : |z| < \infty\}$ the finite complex plane,

$\overline{\mathbb{C}} = \mathbb{C} \cup \{\infty\}$ the compactificated complex plane or the Riemann sphere,

$U = \{z \in \mathbb{C} : |z| < 1\}$ the unit disk,

$U(a, r) = \{z \in \mathbb{C} : |z - a| < r\}, \quad a \in \mathbb{C}, \quad r > 0, \quad U(0,1) = U,$

if $D \subset \overline{\mathbb{C}}$ is a domain, then $\partial D$ is the usual boundary operator of $D$, $\overline{D}$ is the closure of $D$,

$L^p$ is the class of Lebesgue $p$-integrable functions.

If $f(z)$ is a complex valued function, $z \in D$, then $f(D) := \{w \in \overline{\mathbb{C}} : w = f(z), z \in D\}$.

## 2.1 Simple definition and properties of the modulus

Basic definitions and theorems of this section in some form one can find in [9], [64], [78], [107], [141].

### 2.1.1 Definition

Let $D$ be a domain in $\mathbb{C}$ and a function $\rho(z)$ be real-valued, measurable, almost everywhere non-negative, and from $L^2(D)$. Let this function define a differential metric $\rho$ on $D$ by $\rho := \rho(z)|dz|$.

Moreover, we define a conformal invariant metric $\rho$ which is represented in $D$ as above. If $\widetilde{D} = f(D)$, where $w = f(z)$ is a conformal map, then $\rho(f^{-1}(w))|f'(w)|^{-1}|dw| \equiv \widetilde{\rho}(w)|dw|$ is the metric $\widetilde{\rho}(w)|dw|$ that represents $\rho$ in $\widetilde{D}$. Thus, we construct a metric which is defined on the complete collection of conformally equivalent domains. This refers to the definition for an abstract Riemann surface where the map $f$ means the change of charts.

Let $\gamma$ be a curve in $D$. The lower Darboux integral

$$\int_\gamma \rho(z)|dz| =: l_\rho(\gamma) \tag{2.1}$$

is said to be the $\rho$-*length* of $\gamma$. If $\rho(z) \equiv 1$ almost everywhere in $D$, then the 1-length of any rectifiable $\gamma \subset D$ coincides with its Euclidian length. In the non-rectifiable case, nevertheless, the integral exists (in the sense of Darboux) and the definition is valid. The integral

$$\iint_D \rho^2(z)d\sigma_z =: A_\rho(D), \quad d\sigma_z = dx \cdot dy, \tag{2.2}$$

is called the $\rho$-*area* of $D$.

Let $\Gamma$ be a family of curves $\gamma$ in $D$. Denote by

$$L_\rho(\Gamma) := \inf_{\gamma \in \Gamma} l_\rho(\gamma)$$

the $\rho$-length of the family $\Gamma$. Then, the quantity

$$m(D, \Gamma) = \inf_\rho \frac{A_\rho(D)}{L_\rho^2(\Gamma)}$$

is said to be the *modulus* of the family $\Gamma$ in $D$ where the infimum is taken over all metrics $\rho$ in $D$.

Another equivalent and suitable (in a view of further applications) definition of the modulus can be formulated as follows. Denote by $P$ the family of all *admissible* (for $\Gamma$) metrics in $D$, that is, a metric $\rho \in P$ that satisfies the additional condition $l_\rho(\gamma) \geq 1$ for all $\gamma \in \Gamma$. If $P \neq \emptyset$, then we can define the modulus as

$$m(D, \Gamma) = \inf_{\rho \in P} A_\rho(D).$$

If there is such metric $\rho^*$ that $m(D, \Gamma) = A_{\rho^*}(D)$, then this metric is called *extremal*. We follow Tamrazov's approach as it has been mentioned in Introduction. See there also the discussion about generalizations of this definition.

## 2.1.2 Properties

Now we establish simple but important properties of the modulus $m(D, \Gamma)$.

**Theorem 2.1.1** (conformal invariance). *Let $\Gamma$ be a family of curves in a domain $D \in \overline{\mathbb{C}}$, and $w = f(z)$ be a conformal map of $D$ onto $\tilde{D} \in \overline{\mathbb{C}}$. If $\tilde{\Gamma} := f(\Gamma)$, then*

$$m(D, \Gamma) = m(\tilde{D}, \tilde{\Gamma}). \tag{2.3}$$

*In other words, the modulus is well defined in the whole class of conformally invariant domains.*

*Proof.* Let $\widetilde{P}$ be the family of all admissible metrics for $\widetilde{\Gamma}$ and $P$ be the family of admissible metrics for $\Gamma$. We set $\rho(z)|dz| = \widetilde{\rho}(f(z))|f'(z)||dz|$ for $\widetilde{\rho} \in \widetilde{P}$. Since

$$\int_{\gamma \in \Gamma} \rho(z)|dz| = \int_{\gamma} \widetilde{\rho}(f(z))|f'(z)||dz| = \int_{f(\gamma) \in \widetilde{\Gamma}} \widetilde{\rho}(w)|dw| \geq 1,$$

for any $\gamma \in \Gamma$, we have $\rho \in P$. Making use of change of variables under a conformal map in (2.2) we have $A_\rho(D) = A_{\widetilde{\rho}}(\widetilde{D})$. Hence,

$$m(D, \Gamma) = \inf_P A_\rho(D) \leq \inf_{\widetilde{\rho} \in P} A_{\widetilde{\rho}}(\widetilde{D}) = m(\widetilde{D}, \widetilde{\Gamma}).$$

Considering the inverse map $z = f^{-1}(w)$ we obtain the reverse inequality and finish the proof. □

**Theorem 2.1.2** (properties of the extremal metric). *(i) Let $\rho_1$ and $\rho_2$ be two extremal metrics for the modulus $m(D, \Gamma)$. Then, $\rho^* := \rho_1 = \rho_2$ almost everywhere.*

*(ii) Moreover, $L_{\rho^*}(\Gamma) = 1$.*

*Proof.* (i) Since $\rho_1$ and $\rho_2$ are the extremal metrics for the modulus $m(D, \Gamma)$, they are admissible. Then $\frac{1}{2}(\rho_1(z) + \rho_2(z))|dz|$ is an admissible metric, and

$$\iint_D \left( \frac{\rho_1(z) + \rho_2(z)}{2} \right)^2 d\sigma_z \geq m(D, \Gamma).$$

We have the chain of inequalities

$$0 \leq \iint_D \left( \frac{\rho_1(z) - \rho_2(z)}{2} \right)^2 d\sigma_z$$

$$= \iint_D \frac{\rho_1^2(z) + \rho_2^2(z)}{2} d\sigma_z - \iint_D \left( \frac{\rho_1(z) + \rho_2(z)}{2} \right)^2 d\sigma_z$$

$$= m(D, \Gamma) - \iint_D \left( \frac{\rho_1(z) + \rho_2(z)}{2} \right)^2 d\sigma_z \leq 0,$$

which is valid only if $\rho_1(z) = \rho_2(z)$ almost everywhere.

(ii) We denote by $\rho^*(z)|dz|$ this (essentially unique) extremal metric. Let $L_{\rho^*}(\Gamma) = c > 1$. Then the metric $\frac{1}{c}\rho^*(z)|dz|$ is admissible and

$$m(D, \Gamma) \leq \frac{1}{c^2} \iint_D (\rho^*(z))^2 d\sigma_z = \frac{1}{c^2} m(D, \Gamma) < m(D, \Gamma).$$

This contradiction proves (ii). □

**Theorem 2.1.3** (monotonicity). *If $\Gamma_1 \subset \Gamma_2$ in $D$, then $m(D, \Gamma_1) \leq m(D, \Gamma_2)$.*

The *proof* immediately follows from the inequality $L_\rho(\Gamma_1) \geq L_\rho(\Gamma_2)$.

Thus, we have established the properties of the conformal invariance of the modulus and the essential uniqueness of the extremal metric if it exists. But the most difficult question in the theory is exactly about the existence of this metric and about its form.

### 2.1.3 Examples

Now, we give simple examples of domains and families of curves that sheds light on the general principle of the theory *to surmise which metric must be thought of as an extremal one.*

*Example 2.1.1.* Let $D$ be a rectangle $\{z = x + iy : 0 < x < l, 0 < y < 1\}$ and $\Gamma$ be the family of curves in $D$ that connect the opposite horizontal sides of $D$. Then, $m(D, \Gamma) = l$.

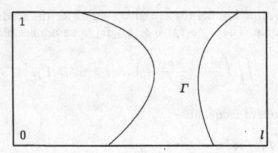

**Fig. 2.1.** The family of curves in the rectangle for Example 2.1.1

*Proof.* The Euclidean metric $\rho^* = |dz|$ is obviously admissible, therefore, the family $P$ of admissible metrics is non-empty. Moreover, $A_{\rho^*}(D) = l$ and, consequently, $m(D, \Gamma) \leq l$.

Now, let $\rho$ be an admissible metric from $P$. Then,

$$1 \leq \int_0^1 \rho(x + iy) dy$$

and, with the Fubini formula, we derive that

$$l \leq \int_0^l \left( \int_0^1 \rho(x + iy) dy \right) dx = \iint_D \rho(z) d\sigma_z.$$

The following chain

$$0 \leq \iint\limits_{D} (1 - \rho(z))^2 d\sigma_z = l - 2 \iint\limits_{D} \rho(z) d\sigma_z + \iint\limits_{D} \rho^2(z) d\sigma_z \leq \iint\limits_{D} \rho^2(z) d\sigma_z - l$$

leads to the inequality

$$\iint\limits_{D} \rho^2(z) d\sigma_z \geq l$$

for any admissible $\rho$ and, taking the infimum over all $\rho$, we have $m(D, \Gamma) \geq l$. This means that $m(D, \Gamma) = l$ and, by Theorem 2.1.2, only the Euclidean metric is extremal. $\qquad\square$

*Example 2.1.2.* Let $D$ be an annulus $\{z = re^{i\theta} : 1 < r < R, 0 < \theta \leq 2\pi\}$ and $\Gamma$ be the family of curves in $D$ that separate the opposite boundary components of $D$. Then, $m(D, \Gamma) = \frac{1}{2\pi} \log R$.

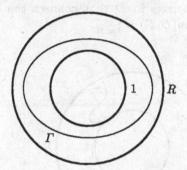

**Fig. 2.2.** The family of curves in the annulus for Example 2.1.2

*Proof.* The metric $\rho^* = \frac{|dz|}{2\pi|z|}$ is obviously admissible, therefore, the family $P$ of admissible metrics is non-empty. Moreover, $A_{\rho^*}(D) = \frac{1}{2\pi} \log R$ and $m(D, \Gamma) \leq \frac{1}{2\pi} \log R$.

Now, let $\rho$ be an admissible metric from $P$. Then,

$$\frac{1}{r} \leq \int\limits_0^{2\pi} \rho(re^{i\theta}) d\theta$$

and, with the Fubini formula, taking into account the polar coordinates, we have

$$\log R \leq \int\limits_0^R \left( \int\limits_0^{2\pi} \rho(re^{i\theta}) d\theta \right) dr = \iint\limits_{D} \frac{\rho(z)}{|z|} d\sigma_z.$$

The following chain

$$0 \leq \iint\limits_D (\frac{1}{2\pi|z|} - \rho(z))^2 d\sigma_z = \frac{1}{2\pi} \log R - \frac{1}{\pi} \iint\limits_D \frac{\rho(z)}{|z|} d\sigma_z + \iint\limits_D \rho^2(z) d\sigma_z \leq$$

$$\leq \iint\limits_D \rho^2(z) d\sigma_z - \frac{1}{2\pi} \log R,$$

leads to the inequality

$$\iint\limits_D \rho^2(z) d\sigma_z \geq \frac{1}{2\pi} \log R$$

for any admissible $\rho$ and, taking the infimum over all $\rho$, we have $m(D, \Gamma) \geq \frac{1}{2\pi} \log R$. This means that $m(D, \Gamma) = \frac{1}{2\pi} \log R$ and, by Theorem 2.1.2, only the metric $\rho^*$ is extremal. $\qquad \square$

*Example 2.1.3.* Let $D$ be an annulus $\{z = re^{i\theta} : 1 < r < R, 0 < \theta \leq 2\pi\}$ and $\Gamma$ be the family of curves in $D$ that connect the opposite boundary components of $D$. Then, $m(D, \Gamma) = \frac{2\pi}{\log R}$.

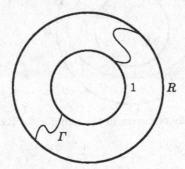

**Fig. 2.3.** The family of curves in the annulus for Example 2.1.3

★ We will not repeat the *proof* and ask the reader to do this independently as an exercise.

From, e.g., [97] one can learn that any hyperbolic doubly connected domain $D$ can be conformally mapped onto an annulus $\{w : 1 < |w| < R\}$ and $R$ is a conformal invariant of $D$. This means that if two doubly connected domains $D_1$ and $D_2$ happen to have the same moduli $m(D_1, \Gamma_1) = m(D_2, \Gamma_1)$ with respect to the families of curves that separate their boundary components, then there must be a conformal map of $D_1$ onto $D_2$. This special modulus we denote by $M(D)$. In order to simplify further notations, the analogous conformal invariant for a quadrilateral $D$ is designated also by $M(D)$.

*Remark 2.1.1.* Generally, one could not deduce the modulus of Example 2.1.2 directly as a corollary from Example 2.1.1 using the logarithmic map.

Namely, we can map the rectangle $D_z = \{z : 0 < x < \frac{1}{2\pi} \log R,\ 0 < y \leq 1\}$ onto the annulus $\{w : 1 < |w| < R\}$ by $w = g(z) := \exp(2\pi z)$. If $\Gamma$ is the family of curves that connect the horizontal sides of the rectangle, then $m(D_z, \Gamma) = \frac{1}{2\pi} \log R$. But $g(\Gamma)$ does not coincide with the family of curves that separate the boundary components of the annulus.

By Example 2.1.2 we know now that if $D = \{z : 1 < |z| < R\}$ and $\Gamma$ is the family of curves that separate its boundary components; if $D' = D \setminus (1, R)$ is a quadrilateral with the vertices with the order $1, 1, R, R$ and $\Gamma'$ is the family of curves that connect the horizontal sides of $D'$, then $\Gamma \subset \Gamma'$ and $m(D', \Gamma') = m(D, \Gamma)$.

### 2.1.4 Grötzsch lemmas

**Theorem 2.1.4** (1-st Grötzsch lemma). *Let $D$ be an annulus $D := \{z : r < |z| < R\}$ and $D_1, \ldots, D_n$ be non-overlapping doubly connected domains that separate the boundary components of $\partial D$. If $M(D)$ stands for the modulus of $D$ with respect to the family of curves that separate its boundary components, then*

$$M(D) \geq \sum_{j=1}^{n} M(D_j). \qquad (2.4)$$

*The equality occurs only in the case when $\cup_{j=1}^{n} \overline{D_j} = D$ and $\partial D_j$ are the concentric circles for any $j$.*

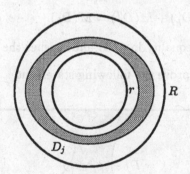

**Fig. 2.4.** Illustration of the 1-st Grötzsch Lemma

*Proof.* The metric $\rho^* = \frac{|dz|}{2\pi|z|}$ is extremal (see Example 2.1.2) for $M(D)$ and, simultaneously, admissible for $M(D_j)$. Then,

$$M(D) = \frac{1}{2\pi} \log \frac{R}{r} = \iint_{D} (\rho^*(z))^2 d\sigma_z \geq \sum_{j=1}^{n} \iint_{D_j} (\rho^*(z))^2 d\sigma_z \geq \sum_{j=1}^{n} M(D_j).$$

Obviously, the equality sign appears in (2.4) when $\cup_{j=1}^{n} \overline{D_j} = D$ and $\partial D_j$ are concentric circles for any $j$.

In order to prove the uniqueness, we suppose first that there is a domain $D_k$, $1 \leq k \leq n$, such that there is no circle centered at the origin which lies in $\overline{D_k}$. Hence, if $\Gamma_k$ is the family of curves that separate the boundary components of $D_k$, then $L_{\rho^*}(\Gamma) > 1$, and by Theorem 2.1.2 this metric is not extremal for $M(D_k)$ and the inequality sign in (2.4) is substituted by the strict inequality sign.

Now, we assume that there is a domain $D_k$, $1 \leq k \leq n$ such that one of the boundary components $\delta_k$ of $D_k$ is not a concentric circle. There exists a conformal map $w = f(z)$ of $D_k$ onto the annulus $1 < |w| < \exp(2\pi M(D_k))$. Then, the metric

$$\frac{|f'(z)|}{2\pi|f(z)|}|dz|, \ z \in D_k$$

is extremal for $M(D_k)$. Without loss of generality, we suppose that $\delta_k \subset \partial D'_k := \partial(f^{-1}(1 < |w| < 1 + \varepsilon))$. Denote by $D''_k := D_k \setminus D'_k$. Then $M(D_k) = M(D'_k) + M(D''_k)$. But for a sufficiently small $\varepsilon$ the domain $D'_k$ does not contain a concentric circle. Therefore, the previous case yields that the metric $\rho^*$ is not extremal for $M(D'_k)$ and

$$\iint\limits_{D_k} (\rho^*(z))^2 d\sigma_z =$$

$$= \iint\limits_{D'_k} (\rho^*(z))^2 d\sigma_z + \iint\limits_{D''_k} (\rho^*(z))^2 d\sigma_z >$$

$$> M(D'_k) + M(D''_k) = M(D_k).$$

This implies the strict inequality in (2.4) and finishes the whole proof.    □

Analogously, one can prove the following statement.

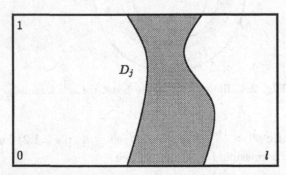

**Fig. 2.5.** Illustration of the 2-nd Grötzsch Lemma

**Theorem 2.1.5** (2-nd Grötzsch lemma). *Let $D$ be a rectangle $\{z = x + iy : 0 < x < l, \ 0 < y < 1\}$ and $D_1, \ldots, D_n$ be non-overlapping quadrilaterals in $D$ with horizontal opposite sides on those of $D$. If $M(D)$ is the modulus of $D$ with respect of the family of curves that connect its horizontal sides, then*

$$M(D) \geq \sum_{j=1}^{n} M(D_j).$$

*The equality occurs only when $\cup_{j=1}^{n} \overline{D_j} = D$ and when $D_j$ are rectangles.*

### 2.1.5 Exercises

★ Let $D$ be a parallelogram based on the vectors $\tau \in \mathbb{C}$ and $l > 0$; $\Gamma$ be the family of curves in $D$ that connect the points $z$ on the segment $[0, \tau]$ and $z + l$. Prove that $m(D, \Gamma) = \text{Im} \ (\tau/l)$.

★ Let $D$ be a doubly connected domain, $\partial D = \{|z| = R\} \cup \{|z - a| = r\}$, $a > 0$, $a + r < R$; $\Gamma$ be the family of curves in $D$ that separate its boundary components. Find the modulus $m(D, \Gamma)$ (for a particular case see Section 2.4).

★ Let $D$ be an annulus $D := \{z : r < |z| < R\}$ and $D_1, \ldots, D_n$ be non-overlapping quadrilaterals in $D$ with the opposite sides on the circles $|z| = r$ and $|z| = R$. If $M(D)$ is the modulus of $D$ with respect to the family of curves that connect these circular sides, then

$$M(D) \geq \sum_{j=1}^{n} M(D_j).$$

The equality sign appears only in the case when $\cup_{j=1}^{n} \overline{D_j} = D$ and when $D_j$ are quadrilaterals with one pair of circular opposite sides and the other of radial ones.

## 2.2 Reduced moduli and capacity

### 2.2.1 Reduced modulus

Let $D \subset \overline{\mathbb{C}}$ be a simply connected hyperbolic domain, $a \in D$, $|a| < \infty$. We construct a doubly connected domain $D_\varepsilon = D \setminus U(a, \varepsilon)$ for a sufficiently small $\varepsilon$. The quantity

$$m(D, a) := \lim_{\varepsilon \to 0} \left( M(D_\varepsilon) + \frac{1}{2\pi} \log \varepsilon \right)$$

is said to be the *reduced modulus* of the circular domain $D$ with respect to the point $a$ where $M(D_\varepsilon)$ is the modulus of the doubly connected domain $D_\varepsilon$ with respect to the family of curves that separate its boundary components.

By the Riemann mapping theorem there is a unique conformal map $w = f(z)$ of $D$ onto a disk $|w| < R < \infty$, such that $f(a) = 0$, $f'(a) = 1$. The number $R$ is called the conformal radius of $D$ with respect to the point $a$. We denote it by $R(D, a)$.

In the case $a = \infty$ the function $f$ has the expansion about $\infty$ as $f(z) = z + a_0 + a_1/z + \ldots$ and maps $D$ onto the exterior part of the disk $|w| > R := R(D, \infty)$.

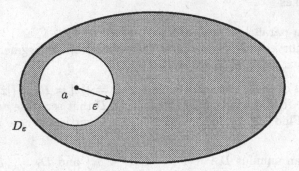

**Fig. 2.6.** To the definition of the reduced modulus

**Theorem 2.2.1.** *Let a simply connected hyperbolic domain $D$ have the conformal radius $R(D, a)$ with respect to a fixed point $a \in D$. Then, the quantity $m(D, a)$ exists, finite, and equal to $\frac{1}{2\pi} \log R(D, a)$.*

*Proof.* Let $w = f(z) = z - a + b_2(z - a)^2 + \ldots$ be the Riemann mapping. Then,

$$\{\varepsilon(1 + |b_2|\varepsilon + o(\varepsilon)) < |w| < R(D, a)\} \subset f(D_\varepsilon) \subset$$

$$\subset \{\varepsilon(1 - |b_2|\varepsilon - o(\varepsilon)) < |w| < R(D, a)\}.$$

By Theorem 2.1.3 and Example 2.1.2 we have the inequality

$$\frac{1}{2\pi} \log \frac{R(D, a)}{\varepsilon(1 + |b_2|\varepsilon + o(\varepsilon))} \leq M(D_\varepsilon) = M(f(D_\varepsilon)) \leq \frac{1}{2\pi} \log \frac{R(D, a)}{\varepsilon(1 - |b_2|\varepsilon - o(\varepsilon))}.$$

Then, we add $\frac{1}{2\pi} \log \varepsilon$ to all parts of the latter inequality and take the limit as $\varepsilon \to 0$. This leads to the statement of the theorem. □

**Corollary 2.2.1.** *Let $D$ be a simply connected hyperbolic domain, $a \in D$, $|a| < \infty$. If $f(z)$ is a conformal map of $D$ such that $|f(a)| < \infty$, then $m(f(D), f(a)) = m(D, a) + \frac{1}{2\pi} \log |f'(a)|$*

Now, we define the reduced modulus $m(D, \infty)$ of a simply connected domain $D$, $\infty \in D$ with respect to infinity as the reduced modulus of the image of $D$ under the map $1/z$ with respect to the origin,

$$m(D, \infty) = -\frac{1}{2\pi} \log R(D, \infty).$$

So, if $D$ is a simply connected hyperbolic domain, $a \in D$, $|a| < \infty$, and $f(z) = a_{-1}/(z - a) + a_0 + a_1(z - a) + \ldots$ is a conformal map from $D$, then $m(f(D), \infty) = m(D, a) - \frac{1}{2\pi} \log |a_{-1}|$.

## 2.2.2 Capacity and transfinite diameter

We give all definitions only for compact sets, however, there are generalizations to the Borel sets as well. Denote by $\mathrm{Lip}\,(D)$ the class of functions $u(z) : D \to \mathbb{R}$ satisfying the Lipschitz condition in $D$, i.e., for every function $u \in \mathrm{Lip}\,(D)$ there is a constant $K$ such that for any two points $z_1, z_2 \in D$ the inequality $|u(z_1) - u(z_2)| \leq K|z_1 - z_2|$ holds. In the case $\infty \in D$ the continuity of $u(z)$ at $\infty$ is required. Functions from $\mathrm{Lip}\,(\overline{\mathbb{C}})$ are absolutely continuous on lines which are parallel to the axes and the integral

$$I(u) := \iint\limits_{\overline{\mathbb{C}}} |\nabla u(z)|^2 d\sigma_z$$

exists.

An ordered pair of non-overlapping compact sets $D_1$, $D_2$ is called the *condenser* $C = \{D_1, D_2\}$ with the field $\overline{\mathbb{C}} \setminus \{D_1 \cup D_2\}$. The *capacity of a condenser* $C$ is the quantity $\mathrm{cap}\,C := \inf I(u)$ as $u$ ranges over the class $\mathrm{Lip}\,(\overline{\mathbb{C}})$ and $0 \leq u(z) \leq 1$ whenever $z \in \overline{\mathbb{C}}$, $u(z) \equiv 0$ whenever $z \in D_1$, $u(z) \equiv 1$ whenever $z \in D_2$.

A condenser $C$ is said to be *admissible* if there exists a continuous real-valued in $\overline{\mathbb{C}}$ function $\omega(z)$, $0 \leq \omega(z) \leq 1$ which is harmonic in $\overline{\mathbb{C}} \setminus \{D_1 \cup D_2\}$ and $\omega(z) \equiv 0$ for $z \in D_1$, $\omega(z) \equiv 1$ for $z \in D_2$. This function is said to be a *potential*. The Dirichlet principle yields that in the definition of capacity the equality appears only in the case of an admissible condenser and $u(z) \equiv \omega(z)$ almost everywhere for the potential function $\omega$.

Obviously, the capacity is a conformal invariant, that is, if $C_f$ is a condenser $\overline{\mathbb{C}} \setminus f(\overline{\mathbb{C}} \setminus \{D_1 \cup D_2\})$ for a conformal map $f$ in $\overline{\mathbb{C}} \setminus \{D_1 \cup D_2\}$, then $\mathrm{cap}\,C = \mathrm{cap}\,C_f$.

If $D_1$ and $D_2$ are simply connected domains, then we can construct the conformal map $w = f(z)$ of $\overline{\mathbb{C}} \setminus \{D_1 \cup D_2\}$ onto an annulus $1 < |w| < R$ and the potential function for the condenser $C = \{D_1, D_2\}$ is

$$\omega(z) = \frac{\log \frac{R}{|f(z)|}}{\log R}, \quad z \in \overline{\mathbb{C}} \setminus \{D_1 \cup D_2\},$$

$\omega(z) \equiv 0$ for $z \in D_1$, $\omega(z) \equiv 1$ for $z \in D_2$. Therefore, cap $C = 2\pi/\log R$.

Let $C = \{D_1, D_2\}$ and $C_k = \{D_1^k, D_2^k\}$, $k = 1, \ldots, n$ be such condensers that all $C_k$ have non-intersected fields and

$$D_1 \subset \bigcap_{k=1}^{n} D_1^k, \quad D_2 \subset \bigcap_{k=1}^{n} D_2^k.$$

From the definition of capacity and from the Dirichlet principle one can derive the inequality which is analogous to the first Grötzsch Lemma

$$\frac{1}{\text{cap } C} \geq \sum_{k=1}^{n} \frac{1}{\text{cap } C_k}. \tag{2.5}$$

(possibly with the equality sign see, e.g., [30]).

Let $D$ be a compact bounded set. We consider a condenser of special type $C_R = \{|z| \geq R, D\}$. If $C_{R_1, R_2} = \{|z| \leq R_1, |z| \geq R_2\}$, $R_1 < R_2$, then the inequality (2.5) implies

$$\frac{1}{\text{cap } C_{R_2}} \geq \frac{1}{\text{cap } C_{R_1}} + \frac{1}{2\pi} \log \frac{R_2}{R_1}.$$

Therefore, the function $\frac{1}{\text{cap } C_R} - \frac{1}{2\pi} \log R$ increases with increasing $R$ and the limit

$$\text{cap } D = \lim_{R \to \infty} R \exp\left(-\frac{2\pi}{\text{cap } C_R}\right) \tag{2.6}$$

exists and is said to be the *logarithmic capacity* of a compact set $D \subset \mathbb{C}$. The limit (2.6) is also known as Pfluger's theorem (see e.g. [112], Theorem 9.17).

Now we briefly summarize the definition and some properties of the logarithmic capacity of a compact bounded set $D \subset \mathbb{C}$ following Fekete. For $n = 2, 3, \ldots$ we consider

$$\Delta_n(D) = \max_{z_1, \ldots, z_n \in D} \prod_{1 \leq k < j \leq n} |z_k - z_j|.$$

The maximum exists and holds for so-called Fekete points $z_{nk} \in \partial D$, $k = 1, \ldots, n$. The value $\Delta_n$ is equal to the Vandermonde determinant

$$\Delta_n(D) = \left| \det_{k=1,\ldots,n} (1 \, z_{nk} \, \ldots \, z_{nk}^{n-1}) \right|.$$

Then, the limit

$$\text{cap } D = \lim_{n \to \infty} (\Delta_n(D))^{\frac{2}{n(n-1)}}$$

exists (see [112]) and gives the same quantity of the logarithmic capacity or the *transfinite diameter* (see also [109]).

Let $D$ be a continuum in $\overline{\mathbb{C}}$ and $D' = \overline{\mathbb{C}} \setminus D$. Then, from the definition of the logarithmic capacity and the reduced modulus it is clear that cap $D =$ cap $\partial D = \exp(-2\pi\, m(D', \infty))$.

It is well known that for $D = [0, 1]$ its capacity is given by cap $D = 1/4$.

If we have a condenser $C^{(h)} = \{D_1, D_2\}$ of special type $D_1 \subset U, D_2 = \overline{\mathbb{C}} \setminus U$, then cap $C^{(h)}$ is said to be the hyperbolic capacity of $D_1$ and cap $^{(h)}D_1 =$ cap $C^{(h)}$. One can define also cap $^{(h)}D$ by means of the *hyperbolic transfinite diameter*. Denote by

$$\Delta_n^{(h)}(D) = \max_{z_1, \ldots, z_n \in D} \prod_{1 \le k < j \le n} \left| \frac{z_k - z_j}{1 - z_k \overline{z_j}} \right|.$$

Then,

$$\mathrm{cap}\,^{(h)}D = \lim_{n \to \infty} (\Delta_n^{(h)}(D))^{\frac{2}{n(n-1)}}.$$

Now, if $D$ is a continuum in $U$ and $D' = U \setminus D$, then

$$\mathrm{cap}\,^{(h)}D = \mathrm{cap}\,^{(h)}\partial D = \exp(-2\pi M(D')),$$

where $M(D')$ is the modulus of the doubly connected domain $D'$ with respect to the family of separating curves.

### 2.2.3 Digons, triangles and their reduced moduli

We are also concerned with another quantity that has appeared rather recently in [36], [81], [139]. It is called the *reduced modulus of a digon*. We will use this reduced modulus as well as the reduced modulus of triangles to solve problems on conformal maps (see also [162], [164], [166], [168], [169]). The existence of the reduced moduli of digons or triangles is closely connected with the existence of the angular derivative of a conformal map at the vertex of the figure in question. Therefore, we are particularly interested in these quantities to apply them further to study extremal problems for the angular derivatives.

We define the reduced modulus of digons. For details we refer to the papers by E. G. Emel'yanov [36], G. V. Kuz'mina [81], some recent particulars and the connection with the angular derivatives one can see in [135], [139], [166], [168].

An important notion in the problems of the boundary behaviour of conformal maps (see [112]) is the *Stolz angle* at a point $\zeta \in \partial U$ that is of the form

$$\Delta_\zeta = \{z \in U : |\arg(1 - \overline{\zeta}z)| < \theta, |z - \zeta| < \eta\},$$

with $\theta \in (0, \frac{\pi}{2})$, $\eta \in (0, 2\cos\theta)$. We say that $f$ has the *angular limit* $a \in \overline{\mathbb{C}}$ at $\zeta \in \partial U$ if $f(\zeta) \to a$ as $z \in \Delta_\zeta$, $z \to \zeta$ for any Stolz angle $\Delta_\zeta$ at $\zeta$. We denote this angular limit by $f(\zeta)$. If the limit $f(z) \to a$ exists for all $z \in U, z \to \zeta$, then $f$ turns out to be continuous at $\zeta$ as a function on $U \cup \{\zeta\}$. As one can see [112], univalent functions in $U$ have the angular limits at almost all points in $\partial U$ but are continuous there only in some restricted cases. Let us consider an analytic function $f$ that maps the disk $U$ into itself, $f(0) = 0$. Then the angular limit $f(\zeta)$ exists for almost all $\zeta \in \partial U$. Moreover, the exceptional set in $\partial U$ has zero capacity.

We say that $f$ has the *angular derivative* $f'(\zeta)$ at $\zeta \in \partial U$ if the finite angular limit $f(\zeta)$ exists and if

$$\lim_{z \to \zeta, z \in \Delta_\zeta} \frac{f(z) - f(\zeta)}{z - \zeta} = f'(\zeta).$$

The angular derivative $f'(\zeta)$ exists if and only if the analytic function $f'(z)$ has the angular limit $f'(\zeta)$ (see [112], Proposition 4.7). Generally, very little can be said about the existence of the angular derivative. However, for self-maps of $U$ the Julia-Wolff lemma (see [112], Proposition 4.13) implies that the angular derivative $f'(\zeta)$ exists for all points $\zeta$ whenever the angular limit $f(\zeta)$ exists and $|f(\zeta)| = 1$ even without assumption of univalence. Furthermore, for the case of univalent functions the McMillan Twist Theorem [98] says that the angular derivative is finite at almost all such points. Thus, the points $\zeta$ such that $|f(\zeta)| = 1$ are of particular interest.

Let $D$ be a hyperbolic simply connected domain in $\mathbb{C}$ with two finite fixed boundary points $a$, $b$ (maybe with the same support) on its piecewise smooth boundary. It is called a *digon*. Set the region $S(a, \varepsilon)$ which is the connected component of $D \cap \{|z - a| < \varepsilon\}$ with the point $a$ in its border. Denote by $D_\varepsilon$ the domain $D \setminus \{S(a, \varepsilon_1) \cup S(b, \varepsilon_2)\}$ for sufficiently small $\varepsilon_{1,2}$ such that there is a curve connecting the opposite sides on $S(a, \varepsilon_1)$ and $S(b, \varepsilon_2)$. Let $M(D_\varepsilon)$ be the modulus of the family of arcs in $D_\varepsilon$ that connect the boundary arcs of $S(a, \varepsilon_1)$ and $S(b, \varepsilon_2)$ when lie in the circumferences $|z - a| = \varepsilon_1$ and $|z - b| = \varepsilon_2$ (we choose a single arc in each circle so that both arcs can be connected in $D_\varepsilon$). If the limit

$$m(D, a, b) = \lim_{\varepsilon_{1,2} \to 0} \left( \frac{1}{M(D_\varepsilon)} + \frac{1}{\varphi_a} \log \varepsilon_1 + \frac{1}{\varphi_b} \log \varepsilon_2 \right), \qquad (2.7)$$

exists, where $\varphi_a = \sup \Delta_a$ and $\varphi_b = \sup \Delta_b$ are the inner angles and $\Delta_{a,b}$ is the Stolz angle inscribed in $D$ at $a$ or $b$ respectively, then it is called the *reduced modulus of the digon* $D$. Various conditions guarantee the existence of this modulus, whereas even in the case of piecewise analytic boundary there are examples [139] which show that it is not always the case. The existence of the limit (2.7) is a local characteristic of the domain $D$ (see [139], Theorem 1.2). If the domain $D$ is conformal (see the definition in [112], page 80) at the points $a$ and $b$, then ([139], Theorem 1.3) the limit (2.7) exists. More generally, suppose that there is a conformal map $f(z)$ of the domain $S(a, \varepsilon_1) \subset D$ onto a circular sector, so that the angular limit $f(a)$ exists which is thought of as a vertex of this sector of angle $\varphi_a$. If the function

$f$ has the finite non-zero angular derivative $f'(a)$ we say that the domain $D$ is also *conformal at the point a* (compare [112], page 80). If the digon $D$ is conformal at the points $a$, $b$ then the limit (2.7) exists ([139], Theorem 1.3). It is noteworthy that J. Jenkins and K. Oikawa [68] in 1977 applied extremal length techniques to study the behaviour of a regular univalent map at a boundary point. Necessary and sufficient conditions were given for the existence of a finite non-zero angular derivative. Independently a similar results have been obtained by B. Rodin and S. E. Warschawski [116].

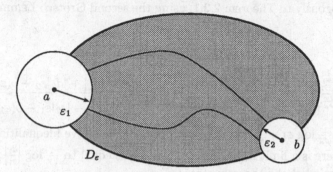

**Fig. 2.7.** To the definition of the reduced modulus of a digon

**Theorem 2.2.2.** *Let the digon $D$ with the vertices at $a$ and $b$ be so that the limit (2.7) exists and the Stolz angles are $\varphi_a$ and $\varphi_b$ . Suppose that there is a conformal map $f(z)$ of the digon $D$ (which is conformal at $a$, $b$) onto a digon $D'$, so that there exist the angular limits $f(a)$, $f(b)$ with the inner angles $\psi_a$ and $\psi_b$ at the vertices $f(a)$ and $f(b)$ which we also understand as the supremum over all Stolz angles inscribed in $D'$ with the vertices at $f(a)$ or $f(b)$ respectively. If the function $f$ has the finite non-zero angular derivatives $f'(a)$ and $f'(b)$, then $\varphi_a = \psi_a$, $\varphi_b = \psi_b$, and the reduced modulus (2.7) of $D'$ exists and changes [36], [81], [139] according to the rule*

$$m(f(D), f(a), f(b)) = m(D, a, b) + \frac{1}{\psi_a}\log|f'(a)| + \frac{1}{\psi_b}\log|f'(b)| \qquad (2.8)$$

*Proof.* First we consider a simple case of the slit plane $D_0 = \mathbb{C} \setminus [0, \infty)$ as a digon with the vertices at $0$, $\infty$. Following the definition of the reduced modulus, denote by $D_\varepsilon$ the domain $\{\varepsilon_1 < |z| < 1/\varepsilon_2\} \setminus (\varepsilon_1, 1/\varepsilon_2)$. The modulus of the quadrilateral $D_\varepsilon$ with respect to the family of curves that connect its circular sides is equal to $M(D_\varepsilon) = -\frac{2\pi}{\log(\varepsilon_1\varepsilon_2)}$. In this case $\varphi_0 = \varphi_\infty = 2\pi$. Therefore,

$$m(D_0, 0, \infty) = \lim_{\varepsilon_{1,2} \to 0} \left( \frac{1}{M(D_\varepsilon)} + \frac{1}{2\pi}\log\varepsilon_1 + \frac{1}{2\pi}\log\varepsilon_2 \right) = 0.$$

Now we consider a digon $D$ such that the conditions of the theorem hold with the vertices at $a$ and $b$. There is a conformal univalent map $f(z)$ that maps $D$ onto $D_0$. Suppose that $\lim_{z \to a} f(z)(z-a)^{-\frac{2\pi}{\varphi_a}} = c_1$ and $\lim_{z \to b} f(z)/(z-b)^{\frac{2\pi}{\varphi_b}} = d_1$. The image of the arc of $S(a, \varepsilon_1)$ lies in the ring

$$\varepsilon_1^{\frac{2\pi}{\varphi_a}}(|c_1| - |c_2|\varepsilon_1 - o(\varepsilon_1)) < |w| < \varepsilon_1^{\frac{2\pi}{\varphi_a}}(|c_1| + |c_2|\varepsilon_1 + o(\varepsilon_1))$$

(see e.g. [112], Proposition 11.8). The same inequality is easily verified for the point $b$. Analogously to Theorem 2.2.1, using the second Grötsch Lemma, we obtain that

$$\frac{1}{M(D_\varepsilon)} = \frac{1}{M(f(D_\varepsilon))} \in$$

$$\in \left( \frac{1}{2\pi} \log \frac{\varepsilon_2^{-\frac{2\pi}{\varphi_b}}(|d_1| - |d_2|\varepsilon_2 - o(\varepsilon_2))}{\varepsilon_1^{\frac{2\pi}{\varphi_a}}(|c_1| + |c_2|\varepsilon_1 + o(\varepsilon_1))}, \frac{1}{2\pi} \log \frac{\varepsilon_2^{-\frac{2\pi}{\varphi_b}}(|d_1| + |d_2|\varepsilon_2 + o(\varepsilon_2))}{\varepsilon_1^{\frac{2\pi}{\varphi_a}}(|c_1| - |c_2|\varepsilon_1 - o(\varepsilon_1))} \right).$$

Then we add $\frac{1}{\varphi_a} \log \varepsilon_1 + \frac{1}{\varphi_b} \log \varepsilon_2$ to each part of the above inequalities and derive that there is a limit in the theorem, and it is equal to $\frac{1}{2\pi} \log \left|\frac{d_1}{c_1}\right|$. This is equivalent to what the theorem states. $\quad\square$

If we suppose, moreover, that $f$ has the expansion

$$f(z) = w_1 + (z-a)^{\psi_a/\varphi_a}(c_1 + c_2(z-a) + \dots)$$

about the point $a$, and the expansion

$$f(z) = w_2 + (z-b)^{\psi_b/\varphi_b}(d_1 + d_2(z-a) + \dots)$$

about the point $b$, then the reduced modulus of $D$ changes according to the rule

$$m(f(D), f(a), f(b)) = m(D, a, b) + \frac{1}{\psi_a} \log |c_1| + \frac{1}{\psi_b} \log |d_1|. \tag{2.9}$$

Obviously, one can extend this definition to the case of vertices with infinite support.

We also consider a quantity which is called the *reduced modulus of a triangle*. It is closely connected with the reduced modulus of a digon. For details we refer to the paper by A. Yu. Solynin [139].

Let $D$ be a hyperbolic simply connected domain in $\mathbb{C}$ with three finite fixed boundary points $z_1, z_2$, and $a$ on its piecewise smooth boundary. Denote by $D_\varepsilon$ the domain $D \setminus S(a, \varepsilon)$ for a sufficiently small $\varepsilon$. Denote by $M(D_\varepsilon)$ the modulus of the family of arcs in $D_\varepsilon$ joining the boundary arc of $S(a, \varepsilon)$ that lies in the circumference $|z - a| = \varepsilon$ with the leg of the triangle $D$ which is opposite to $a$ (we again choose a single arc in the circle so that both arcs can be connected in $D_\varepsilon$). If the limit

$$m_\Delta(D, a) = \lim_{\varepsilon \to 0} \left( \frac{1}{M(D_\varepsilon)} + \frac{1}{\varphi_a} \log \varepsilon \right)$$

exists, where $\varphi_a$ is the inner angle defined as before, then it is called the *reduced modulus of the triangle* $D$. The conditions for the reduced modulus to exist are similar to those for the reduced modulus of digons. It turns out that the reduced modulus exists when $D$ is conformal at $a$. If there is a conformal map $f(z)$ of the triangle $D$ onto a triangle $D'$, so that there exists the angular limit $f(a)$ with the inner angle $\psi_a$ at the vertex $f(a)$, if the function $f$ has the finite non-zero angular derivative $f'(a)$, then $\varphi_a = \psi_{f(a)}$ and the reduced modulus of $D$ exists and changes [139] according to the rule

$$m_\Delta(f(D), f(a)) = m_\Delta(D, a) + \frac{1}{\psi_a} \log |f'(a)|.$$

If we suppose, moreover, that $f$ has the expansion

$$f(z) = w_1 + (z - a)^{\psi_a/\varphi_a}(c_1 + c_2(z - a) + \dots)$$

in a neighbourhood of the point $a$, then the reduced modulus of $D$ is changed by the rule

$$m_\Delta(f(D), f(a)) = m_\Delta(D, a) + \frac{1}{\psi_a} \log |c_1|.$$

## 2.3 Elliptic functions and integrals

In this section we present some useful information and formulae about elliptic functions and integrals that we will use throughout the following sections. A major part of this section can be found in e.g. [3], [22], [57].

### 2.3.1 Elliptic functions

A meromophic function in the complex plane with two periods $\omega_1$, $\omega_2$, such that $\operatorname{Im} \frac{\omega_2}{\omega_1} > 0$, is called an *elliptic function*. If $(\omega_1, \omega_2)$ is a pair of primitive (or reduced) periods of a non-constant meromorphic function $f(z)$, they form a basis of the set of all periods of $f$. Each period $\omega$ is of the form $\omega = m\omega_1 + n\omega_2$, where $m, n$ are some integers. The points $\omega$ form a lattice in $\mathbb{C}$, therefore, the set $\{m\omega_1 + n\omega_2\}$ is called the *period-lattice* with $(\omega_1, \omega_2)$ as the basis. Any other basic periods $(\omega_1^*, \omega_2^*)$ of the same elliptic function $f(z)$ can be described by a *unimodular transform* $\tau \to \tau^*$. That is, denote by $\tau = \frac{\omega_2}{\omega_1}$ and $\tau^* = \frac{\omega_2^*}{\omega_1^*}$. Let $\omega_1^* = m\omega_1 + n\omega_2$ and $\omega_2^* = p\omega_1 + p\omega_2$. Then,

$$\tau^* = \frac{q\tau + p}{m\tau + n}.$$

A pair of basic periods $(\omega_1, \omega_2)$ of a given elliptic function $f(z)$ is a pair of primitive periods if and only if the following conditions are satisfied:

$$|\tau| > 1, \quad \operatorname{Im} \tau > 0, \quad -\frac{1}{2} \leq \operatorname{Re} \tau \leq \frac{1}{2}.$$

A parallelogram constructed by basic periods is called a *period-parallelogram*. The parallelogram constructed by the basic primitive periods is called a *fundamental period parallelogram*. A non-constant elliptic function has at least one pole in any period-parallelogram.

Define a sum

$$\sideset{}{'}\sum_{m,n} = \sum_{m=-\infty}^{\infty} \sum_{\substack{m=-\infty \\ (m,n) \neq (0,0)}}^{\infty}.$$

Let $\omega_1$, $\omega_2$ be two complex numbers, both of them different from zero, and let $\tau = \omega_2/\omega_1$, with $\operatorname{Im} \tau > 0$. The series

$$\sideset{}{'}\sum_{m,n} \left( \frac{1}{(z - m\omega_1 - n\omega_2)^2} - \frac{1}{(m\omega_1 + n\omega_2)^2} \right)$$

converges absolutely for all $z$ different from the points of the period-lattice. For every finite $R > 0$, the series converges uniformly in the closed disk $|z| \leq R$, after the omission of a sufficient number of initial terms.

An important elliptic function is the Weierstrass function $\wp(z) \equiv \wp(z; \omega_1, \omega_2)$ defined by

$$\wp(z; \omega_1, \omega_2) = \frac{1}{z^2} + \sideset{}{'}\sum_{m,n} \left( \frac{1}{(z - m\omega_1 - n\omega_2)^2} - \frac{1}{(m\omega_1 + n\omega_2)^2} \right).$$

The function $\wp(z)$ is an elliptic function with the primitive periods $\omega_1, \omega_2$. Its poles are given by $z = m\omega_1 + n\omega_2$. The principal part of $\wp(z)$ at $z = 0$ is $1/z^2$.

$$\lim_{z \to 0} \left( \wp(z) - \frac{1}{z^2} \right) = 0, \quad \wp(z) = \wp(-z), \quad \wp'(-z) = -\wp'(z).$$

The elliptic function $\wp(z)$ satisfies the differential equation

$$(\wp'(z))^2 = 4\wp^3(z) - g_2\wp(z) - g_3,$$

where

$$g_2 = 60 \sideset{}{'}\sum_{m,n} \frac{1}{(m\omega_1 + n\omega_2)^4}, \quad g_3 = 140 \sideset{}{'}\sum_{m,n} \frac{1}{(m\omega_1 + n\omega_2)^6}.$$

Another traditional form of this equation is

$$(\wp'(z))^2 = 4(\wp(z) - e_1)(\wp(z) - e_2)(\wp(z) - e_3),$$

where $e_1 + e_2 + e_3 = 0$, $g_2 = -4(e_2 e_3 + e_3 e_1 + e_1 e_2)$, $g_3 = 4 e_1 e_2 e_3$.

The Laurent expansion of $\wp(z)$ at $z = 0$ is given by

$$\wp(z) = \frac{1}{z^2} + \sum_{n=1}^{\infty} b_n z^{2n},$$

where

$$b_1 = \frac{g_2}{20}, \quad b_2 = \frac{g_3}{28}, \ldots$$

The following relation holds

$$\wp(z_1 + z_2) = \frac{1}{4} \left( \frac{\wp'(z_1) - \wp'(z_2)}{\wp(z_1) - \wp(z_2)} \right)^2 - \wp(z_1) - \wp(z_2).$$

### 2.3.2 Elliptic integrals and Jacobi's functions

Integrals of the form $\int R(x, \sqrt{P(x)}) dx$, where $P(x)$ is a polynomial of the third or fourth degree and $R$ is a rational function are known to be the *elliptic integrals*. They were first treated systematically by Legendre, who showed that any elliptic integral can be represented by means of three fundamental integrals $\mathbf{F}(\varphi, k)$, $\mathbf{E}(\varphi, k)$, and $\Pi(\varphi, n, k)$. They are the Legendre canonical elliptic integrals of the first, second, and third kind respectively. The elliptic functions of Abel, Jacobi, and Weierstrass are obtained by the inversion of elliptic integrals of the first kind. For a real number $0 < k < 1$, that is called the *modulus*, let us define

$$\mathbf{F}(\varphi, k) := \int_0^{\varphi} \frac{d\varphi}{\sqrt{1 - k^2 \sin^2 \varphi}},$$

$$\mathbf{E}(\varphi, k) := \int_0^{\varphi} \sqrt{1 - k^2 \sin^2 \varphi} \, d\varphi,$$

$$\Pi(\varphi, n, k) := \int_0^{\varphi} \frac{d\varphi}{(1 + n \sin^2 \varphi) \sqrt{1 - k^2 \sin^2 \varphi}}.$$

The number $k' = \sqrt{1 - k^2}$ is called the *complementary modulus*. In another notation these integrals are rewritten as

$$\mathbf{F}(x, k) = \int_0^x \frac{dx}{\sqrt{(1 - x^2)(1 - k^2 x^2)}},$$

$$\mathbf{E}(x, k) := \int_0^x \sqrt{\frac{1 - k^2 x^2}{1 - x^2}} \, dx,$$

$$\Pi(x, n, k) := \int_0^x \frac{dx}{(1 + nx^2)\sqrt{(1 - x^2)(1 - k^2 x^2)}},$$

and $x = \sin \varphi$.

When $x = 1$ (or $\varphi = \pi/2$), the above integrals are said to be complete. In that case, one writes $\mathbf{K}(k) := \mathbf{F}(\pi/2, k)$, $\mathbf{E}(k) = \mathbf{E}(\pi/2, k)$, $\Pi(k) = \Pi(\pi/2, n, k)$. The *associated complete elliptic integrals* are given by $\mathbf{K}'(k) \equiv \mathbf{K}(k')$, $\mathbf{E}'(k) \equiv \mathbf{E}(k')$. They satisfy the Legendre relation $\mathbf{E}\mathbf{K}' + \mathbf{E}'\mathbf{K} - \mathbf{K}\mathbf{K}' = \pi/2$.

Some special values of the elliptic integrals and functional equations are presented as

$$\mathbf{K}(0) = \mathbf{K}'(1) = \mathbf{E}(0) = \mathbf{E}(1) = \frac{\pi}{2},$$

$$\frac{\mathbf{K}'}{4\mathbf{K}}\left(\frac{1-r}{1+r}\right) = \frac{\mathbf{K}}{2\mathbf{K}'}(r),$$

$$\mathbf{K}\left(\frac{2\sqrt{r}}{1+r}\right) = (1+r)\mathbf{K}(r), \quad \mathbf{K}'\left(\frac{2\sqrt{r}}{1+r}\right) = \frac{1}{2}(1+r)\mathbf{K}'(r),$$

$$\frac{\mathbf{K}'}{4\mathbf{K}}\left(\frac{1}{R}\right) \le \frac{1}{2\pi} \log 4R, \quad R > 1,$$

$$\mathbf{K}(\frac{1}{\sqrt{2}}) = \mathbf{K}'(\frac{1}{\sqrt{2}}) = \frac{[\Gamma(1/4)]^2}{4\sqrt{\pi}},$$

$$\mathbf{K}'(\sqrt{2} - 1) = \sqrt{2}\mathbf{K}(\sqrt{2} - 1), \quad \mathbf{K}'\left(\frac{\sqrt{2}-1}{\sqrt{2}+1}\right) = 2\mathbf{K}\left(\frac{\sqrt{2}-1}{\sqrt{2}+1}\right),$$

$$\mathbf{K}\left(\frac{1-k'}{1+k'}\right) = \frac{1+k'}{2}\mathbf{K}(k),$$

$$\log 4 \le \mathbf{K}(k) + \log k' \le \frac{\pi}{2}.$$

Some limiting values:

$$\lim_{k \to 1}(\mathbf{K}(k) - \log \frac{4}{k'}) = \lim_{k \to 0}(\frac{\mathbf{K}'}{\mathbf{K}}(k) - \frac{2}{\pi}\log\frac{4}{k}) = \lim_{k \to 0}(\mathbf{E}(k) - \mathbf{K}(k))\mathbf{K}'(k) = 0.$$

$$\lim_{k \to 0}\frac{e^{-(\pi\mathbf{K}'/\mathbf{K})}}{k^2} = \frac{1}{16}.$$

Upon differentiating the elliptic integrals result

$$\frac{d}{dk}\mathbf{K}(k) = \frac{\mathbf{E}(k) - (k')^2\mathbf{K}(k)}{k(k')^2}, \quad \frac{d}{dk}\mathbf{E}(k) = \frac{\mathbf{E}(k) - \mathbf{K}(k)}{k},$$

$$\frac{d}{dk}(\mathbf{E}(k) - \mathbf{K}(k)) = -\frac{k\mathbf{E}(k)}{(k')^2}, \quad \frac{d}{dk}(\mathbf{E}(k) - (k')^2\mathbf{K}(k)) = k\mathbf{K}(k).$$

The inverse function to the integral $\omega = \mathbf{F}(x, k)$ is called Jacobi's sine elliptic function $x = \mathbf{sn}(\omega, k) \equiv \mathbf{sn}(\omega)$. Also we will use the elliptic cosine $\mathbf{cn}(\omega) = \sqrt{1 - \mathbf{sn}^2(\omega)}$ and the elliptic amplitude $\mathbf{dn}(\omega) = \sqrt{1 - k^2\mathbf{sn}^2(\omega)}$. These functions can be represented by the Weierstrass elliptic function. Set

$$k^2 = \frac{e_2 - e_3}{e_1 - e_3}, \quad \gamma^2 = e_1 - e_3.$$

Then,

$$\mathbf{sn}(\omega, \gamma) = \frac{\gamma}{\sqrt{\wp(\omega) - e_3}},$$

$$\mathbf{cn}(\omega, \gamma) = \sqrt{\frac{\wp(\omega) - e_1}{\wp(\omega) - e_3}},$$

$$\mathbf{dn}(\omega, \gamma) = \sqrt{\frac{\wp(\omega) - e_2}{\wp(\omega) - e_3}}.$$

The periods of the Weierstrass functions $\wp$ are given by

$$\omega_1 = 2\frac{\mathbf{K}(k)}{\gamma} = 2\int_{e_1}^{\infty} \frac{dt}{\sqrt{4t^3 - g_2 t - g_3}} = 2\int_{e_3}^{e_2} \frac{dt}{\sqrt{4t^3 - g_2 t - g_3}},$$

$$\omega_2 = 2i\frac{\mathbf{K}'(k)}{\gamma} = 2\int_{e_3}^{\infty} \frac{dt}{\sqrt{4t^3 - g_2 t - g_3}} = 2\int_{e_2}^{e_1} \frac{dt}{\sqrt{4t^3 - g_2 t - g_3}}.$$

Jacobi's elliptic functions satisfy the differential equations

$$\frac{d}{d\omega}\mathbf{sn}(\omega) = \mathbf{cn}(\omega)\mathbf{dn}(\omega),$$

$$\frac{d}{d\omega}\mathbf{cn}(\omega) = -\mathbf{sn}(\omega)\mathbf{dn}(\omega),$$

$$\frac{d}{d\omega}\mathbf{dn}(\omega) = -k^2\mathbf{sn}(\omega)\mathbf{cn}(\omega).$$

Differentiation of Jacobi's elliptic functions leads to the formulae

$$\frac{\partial}{\partial k}\mathbf{sn}(\omega, k) = \frac{\mathbf{cn}(\omega)\mathbf{dn}(\omega)}{k(k')^2}\left[-\mathbf{E}(\omega, k) + (k')^2\omega + k^2\mathbf{sn}(\omega)\mathbf{cd}(\omega)\right],$$

$$\frac{\partial}{\partial k}\mathbf{cn}(\omega, k) = -\frac{\mathbf{sn}(\omega)\mathbf{dn}(\omega)}{k(k')^2}\left[-\mathbf{E}(\omega, k) + (k')^2\omega + k^2\mathbf{sn}(\omega)\mathbf{cd}(\omega)\right],$$

$$\frac{\partial}{\partial k}\mathbf{dn}(\omega, k) = -\frac{k\,\mathbf{sn}(\omega)\mathbf{cn}(\omega)}{(k')^2}\left[-\mathbf{E}(\omega, k) + (k')^2\omega + \mathbf{dn}(\omega)\mathbf{tn}(\omega)\right],$$

where $\mathbf{cd}(\omega) \equiv \mathbf{cn}(\omega)/\mathbf{dn}(\omega)$.

It is evident that

$$\mathbf{sn}\,(-\omega) = -\mathbf{sn}\,(\omega), \quad \mathbf{cn}\,(-\omega) = \mathbf{cn}\,(\omega), \quad \mathbf{dn}\,(-\omega) = \mathbf{dn}\,(\omega).$$

Jacobi's functions are $4\mathbf{K}\,(k)$-periodic. More rigorously,

$$\mathbf{sn}\,(\omega \pm 2\mathbf{K}\,(k)) = -\mathbf{sn}\,(\omega),\, \mathbf{cn}\,(\omega \pm 2\mathbf{K}\,(k)) = -\mathbf{cn}\,(\omega),$$
$$\mathbf{dn}\,(\omega + 2\mathbf{K}\,(k)) = \mathbf{dn}\,(\omega).$$

It is seen that $2\mathbf{K}\,(k)$ is a period of $\mathbf{dn}\,(\omega)$ and $\mathbf{tn}\,(\omega) \equiv \frac{\mathbf{sn}\,(\omega)}{\mathbf{cn}\,(\omega)}$. Some special values are given as

$$\mathbf{sn}\,(2\mathbf{K}\,(k)) = 0,\, \mathbf{cn}\,(2\mathbf{K}\,(k)) = -1,\, \mathbf{dn}\,(2\mathbf{K}\,(k)) = 1,$$
$$\mathbf{sn}\,(4\mathbf{K}\,(k)) = 0,\, \mathbf{cn}\,(4\mathbf{K}\,(k)) = 1, \quad \mathbf{dn}\,(4\mathbf{K}\,(k)) = 1.$$

$$\mathbf{sn}\,(\omega \pm 2\mathbf{K}\,'(k), k') = -\mathbf{sn}\,(\omega, k'), \quad \mathbf{sn}\,(\omega \pm 4\mathbf{K}\,'(k), k') = \mathbf{sn}\,(\omega, k'), \quad etc.$$

The addition-theorem for $\mathbf{sn}$, $\mathbf{cn}$, $\mathbf{dn}$ asserts that

$$\mathbf{sn}\,(u+v) = \frac{\mathbf{sn}\,(u)\mathbf{cn}\,(v)\mathbf{dn}\,(v) + \mathbf{sn}\,(v)\mathbf{cn}\,(u)\mathbf{dn}\,(u)}{1 - k^2\mathbf{sn}^2(u)\mathbf{sn}^2(v)},$$

$$\mathbf{cn}\,(u+v) = \frac{\mathbf{cn}\,(u)\mathbf{cn}\,(v) - \mathbf{dn}\,(u)\mathbf{dn}\,(v)\mathbf{sn}\,(u)\mathbf{sn}\,(v)}{1 - k^2\mathbf{sn}^2(u)\mathbf{sn}^2(v)},$$

$$\mathbf{dn}\,(u+v) = \frac{\mathbf{dn}\,(u)\mathbf{dn}\,(v) - k^2\mathbf{cn}\,(u)\mathbf{cn}\,(v)\mathbf{sn}\,(u)\mathbf{sn}\,(v)}{1 - k^2\mathbf{sn}^2(u)\mathbf{sn}^2(v)}.$$

The imaginary Jacobi's transforms are defined as

$$\mathbf{sn}\,(i\omega, k) = i\,\mathbf{tn}\,(\omega, k'),$$

$$\mathbf{cn}\,(i\omega, k) = \frac{1}{\mathbf{cn}\,(\omega, k')},$$

$$\mathbf{dn}\,(i\omega, k) = \frac{\mathbf{dn}\,(\omega, k')}{\mathbf{cn}\,(\omega, k')}.$$

The continuation of the real Jacobi's transforms into the complex $\omega$-plane gives the elliptic functions with the primitive periods represented by $\mathbf{K}$ and $\mathbf{K}'$.

The periods of $\mathbf{sn}\,(\omega, k)$ are $4\mathbf{K}\,(k)$ and $2i\mathbf{K}\,'(k)$.
The periods of $\mathbf{cn}\,(\omega, k)$ are $4\mathbf{K}\,(k)$ and $2\mathbf{K}\,(k) + 2i\mathbf{K}\,'(k)$.
The periods of $\mathbf{dn}\,(\omega, k)$ are $\mathbf{K}\,(k)$ and $4i\mathbf{K}\,'(k)$.

The Theta Functions of Jacobi are auxiliary functions in many problems involving elliptic functions. They are defined by the Fourier series

$$\Theta(\omega) = \vartheta_0(u) = 1 + 2\sum_{n=1}^{\infty}(-1)^n q^{n^2}\cos(2nu),$$

$$H(\omega) = \vartheta_1(u) = 2\sum_{n=1}^{\infty}(-1)^{n-1}q^{n-1/2^2}\sin(2n-1)u,$$

$$H_1(\omega) = \vartheta_2(u) = 1 + 2\sum_{n=1}^{\infty}q^{n^2}\cos(2n-1)u,$$

$$\Theta_1(\omega) = \vartheta_3(u) = 1 + 2\sum_{n=1}^{\infty}q^{n^2}\cos(2nu),$$

where

$$q = e^{-\frac{\pi K'}{K}(k)}, \quad u = \frac{\pi\omega}{2K(k)}.$$

Some special values are

$$\Theta(0) = \Theta_1(K) = \vartheta_0(0) = \vartheta_3(\frac{\pi}{2}) = \sqrt{\frac{2k'K(k)}{\pi}},$$

$$H(0) = -H_1(K) = \vartheta_1(0) = \vartheta_2(\frac{\pi}{2}) = 0,$$

$$H_1(0) = H(K) = \vartheta_2(0) = \vartheta_1(\frac{\pi}{2}) = \sqrt{\frac{2kK(k)}{\pi}},$$

$$\Theta_1(0) = \Theta(K) = \vartheta_3(0) = \vartheta_0(\frac{\pi}{2}) = \sqrt{\frac{2K(k)}{\pi}}.$$

These functions satisfy the following relations

$$\Theta(-\omega) = \Theta(\omega),\ H(-\omega) = H(\omega),\ \Theta_1(-\omega) = \Theta(\omega),\ H_1(-\omega) = H(\omega).$$

The Theta Functions have the connection with Jacobi's functions given by

$$\mathbf{sn}(\omega) = \frac{H(\omega)}{\sqrt{k}\Theta(\omega)} = \frac{\vartheta_1(u)}{\sqrt{k}\,\vartheta_0(u)},$$

$$\mathbf{cn}(\omega) = \sqrt{\frac{k'}{k}}\frac{H_1(\omega)}{\Theta(\omega)} = \sqrt{\frac{k'}{k}}\frac{\vartheta_2(u)}{\vartheta_0(u)},$$

$$\mathbf{dn}(\omega) = \sqrt{k'}\frac{\Theta_1(\omega)}{\Theta(\omega)} = \sqrt{k'}\frac{\vartheta_3(u)}{\vartheta_0(u)},$$

$$\mathbf{tn}(\omega) = \frac{H(\omega)}{\sqrt{k'}\,H_1(\omega)} = \frac{\vartheta_1(u)}{\sqrt{k'}\,\vartheta_2(u)}.$$

Thus, the periods of the Theta Functions can be obtained by the periods of Jacobi's functions. The Weierstrass function is given by the function $\vartheta_1$ as

$$\wp(\omega) = \frac{1}{3\omega_1}\frac{\vartheta_1'''(0)}{\vartheta_1'(0)} - \frac{d^2}{d\omega^2}\Big[\log\vartheta_1(\frac{\omega}{\omega_1})\Big].$$

## 2.4 Some frequently used moduli

In this section we compile the most frequently used moduli and reduced moduli which we will apply further in extremal problems.

### 2.4.1 Moduli of doubly connected domains

Denote by $M(D)$ the modulus of a doubly connected hyperbolic domain $D$ with respect to the family of curves that separate its boundary components.

★ Let $D = \mathbb{C} \setminus \{[0,1] \cup [R, \infty)$. Then,

$$M(D) = \frac{\mathbf{K}'}{2\mathbf{K}}\left(\frac{1}{\sqrt{R}}\right).$$

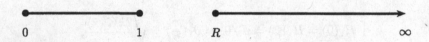

$$
\begin{array}{cccc}
0 & 1 & R & \infty
\end{array}
$$

**Fig. 2.8.** $D = \mathbb{C} \setminus \{[0,1] \cup [R, \infty)\}$

*Proof.* We slit $D$ along the segment $[1, R]$, $D' = D \setminus [1, R]$. By Example 2.1.2 and Remark thereafter in Section 2.1.3 the modulus of the quadrilateral $D'$ with respect to the family $\Gamma'$ of curves that connect the edges of the slit along $[1, R]$ is equal to $M(D) = m(D', \Gamma')$. The function $w = g_1(z) := \sqrt{z}$ maps conformally $D'$ onto the upper half-plane $H^+$ and the integral

$$\zeta = g_2(w) := \int_{w_0}^{w} \frac{dw}{\sqrt{(1 - w^2)(1 - \frac{1}{R}w^2)}}$$

maps the quadrilateral $H^+$ with the vertices $\pm 1$, $\pm 1/\sqrt{R}$ onto the rectangle with the vertices $\pm \mathbf{K}\left(1/\sqrt{R}\right)$, $\pm \mathbf{K}\left(1/\sqrt{R}\right) + i\mathbf{K}'\left(1/\sqrt{R}\right)$ (see the properties of the function $\mathbf{F}(\omega, k)$). Finally, taking into account the correspondence of the vertices under the conformal map $\zeta = g_2 \circ g_1(w)$, we calculate

$$M(D) = \frac{\mathbf{K}'}{2\mathbf{K}}\left(\frac{1}{\sqrt{R}}\right).$$

$\square$

For other moduli one can construct relevant conformal maps that we assume as exercises.

★ *The Mori domain.* Let $D = \mathbb{C} \setminus \{z : |z| = 1, \arg z \in [\pi - \varphi, \pi + \varphi],$ $0 < \varphi < \pi/2\}$. Then,

$$M(D) = \frac{\mathbf{K}'}{4\mathbf{K}} \left( \frac{\lambda}{\sqrt{4 + 2\lambda} + \sqrt{4 - 2\lambda}} \right), \quad \lambda = 2 \sin \varphi.$$

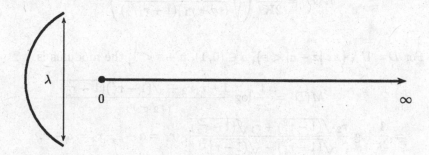

**Fig. 2.9.** The Mori domain

Let $D = \mathbb{C} \setminus \{z : |z| = 1, \arg z \in [\varphi, 2\pi - \varphi], 0 < \varphi < \pi/2.$ Then,

$$M(D) = \frac{\mathbf{K}'}{4\mathbf{K}} \left( \frac{\sqrt{2 + \sqrt{4 - \lambda^2}}}{2} \right) = \frac{\mathbf{K}'}{4\mathbf{K}} \left( \cos \frac{\varphi}{2} \right), \quad \lambda = 2 \sin \varphi.$$

★ For $D = U \setminus [0, r]$, $0 < r < 1$ the modulus is

$$M(D) = \frac{\mathbf{K}'}{4\mathbf{K}} (r).$$

★ For $D = U \setminus [r_1, r_2]$, $0 < r_1 < r_2 < 1$ the modulus is

$$M(D) = \frac{\mathbf{K}'}{4\mathbf{K}} \left( \frac{r_2 - r_1}{1 - r_1 r_2} \right).$$

★ For $D = U \setminus \{(-1, 0] \cup [r_1, r_2]\}$ the modulus is

$$M(D) = \frac{\mathbf{K}}{2\mathbf{K}'} \left( \sqrt{\frac{r_1}{r_2}} \frac{1 - r_2}{1 - r_1} \right).$$

★ For $D = U \setminus \{[0, r_1] \cup [r_2, 1)\}$ the modulus is

$$M(D) = \frac{\mathbf{K}'}{2\mathbf{K}} \left( \sqrt{\frac{r_1}{r_2} \frac{1 + r_2}{1 + r_1}} \right).$$

★ For $D = U \setminus \{(-1, -r_1] \cup [0, r_2]\}$ the modulus is

$$M(D) = \frac{\mathbf{K}}{2\mathbf{K}'} \left( \sqrt{\frac{r_1(1 - r_2)^2}{(r_2 + r_1)(1 + r_1 r_2)}} \right).$$

★ For $D = U \setminus \{z : |z - a| < \varepsilon\}$, $a \in (0, 1)$, $a + \varepsilon < 1$, the modulus is

$$M(D) = \frac{1}{2\pi} \log \frac{1 - r_1 r_2 + \sqrt{(1 - r_1^2)(1 - r_2^2)}}{r_2 - r_1} =$$

$$= \frac{1}{2\pi} \log \frac{r_2 \sqrt{(1 - r_1^2)} + r_1 \sqrt{(1 - r_2^2)}}{\sqrt{(1 - r_1^2)} - \sqrt{(1 - r_2^2)}}, \quad r_1 = a - \varepsilon, \ r_2 = a + \varepsilon.$$

### 2.4.2 Moduli of quadrilaterals

★ We consider the quadrilateral $U \setminus \{(-1, r] \cup [R, 1)\}$, $-1 < r < R < 1$ with the upper semicircle $\partial U^+$ and the lower semicircle $\partial U^-$ as the opposite sides. Then, the modulus of the family of curves $\Gamma$ that connect $\partial U^+$ and $\partial U^-$ in $D$ is

$$m(D, \Gamma) = \frac{\mathbf{K}}{\mathbf{K}'} \left( \frac{R - r}{1 - Rr} \right).$$

### 2.4.3 Reduced moduli

★ For $D = \mathbb{C} \setminus (-\infty, 0]$ the reduced modulus with respect to the point $R > 0$ is

$$m(D, R) = \frac{1}{2\pi} \log 4R.$$

★ For $D = U$ the reduced modulus with respect to the point $r \in [0, 1)$ is

$$m(D, r) = \frac{1}{2\pi} \log (1 - r^2).$$

★ For $D = U \backslash (-1, 0]$ the reduced modulus with respect to the point $r \in [0, 1)$ is

$$m(D, r) = \frac{1}{2\pi} \log \frac{4r(1-r)}{1+r}.$$

★ For $D = U \backslash [r, 1)$ the reduced modulus with respect to the origin is

$$m(D, r) = \frac{1}{2\pi} \log \frac{4r}{(1+r)^2}.$$

★ For $D = U \backslash (-1, a]$, $a \in [0, 1)$ the reduced modulus with respect to the point $r \in (a, 1)$ is

$$m(D, r) = \frac{1}{2\pi} \log \frac{4(r-a)(1-ra)(1-r)}{(1+r)(1-a)^2}.$$

### 2.4.4 Reduced moduli of digons

★ Let $D = \mathbb{C} \backslash [0, \infty)$, then $m(D, 0, \infty) = 0$.

★ Let $D = U \backslash \{(-1, 0] \cup [r, 1)\}$, then

$$m(D, 0, r) = \frac{1}{2\pi} \log \frac{r^2}{1-r^2}.$$

★ Let $D = U \backslash [0, 1)$, then

$$m(D, r, r) = \frac{2}{\pi} \log \frac{4r(1+r)}{1-r}.$$

★ Let $D = U \backslash (-1, r]$, $r > 0$, then

$$m(D, 0, 0) = \frac{2}{\pi} \log \frac{4r}{(1-r)^2}.$$

★ Let $D = \mathbb{C} \backslash [0, \infty)$, then for $R > 0$

$$m(D, R, R) = \frac{2}{\pi} \log 4R.$$

## 2.5 Symmetrization and polarization

A good collection of results about symmetrization and polarization is the contents of the survey by V. Dubinin [30] and the series of articles by A. Solynin [133]–[136]. Here we present only a part of them that we will apply further.

### 2.5.1 Circular symmetrization

**a)** Let $D$ be a simply connected hyperbolic domain in $\overline{\mathbb{C}}$. Define a point $O$ as a center of symmetrization and a ray $l^+$ starting at $O$ as the direction of symmetrization. A domain $D^*$ is said to be the result of *circular symmetrization* with the center at $O$ and the direction $l^+$ if any circle centered on $O$ lies or does not in $D$ and $D^*$ simultaneously. If this circle has an intersection with $D$ of linear Lebesgue measure $m$, then it has the intersection with $D^*$ that is a symmetric arc about $l^+$ of the same measure. Obviously, $D$ and $D^*$ have the same area.

**b)** Now let $D$ be a doubly connected hyperbolic domain and $K_1$ and $K_2$ be its complements with respect to $\overline{\mathbb{C}}$. Denote by $K_1^*$ and $K_2^*$ the results of circular symmetrization of $K_1$ and $K_2$ with the direction $l^+$ and $l^-$ respectively where the ray $l^-$ starts at $O$ in the antipodal direction to $l^+$. Then, $D^* = \overline{\mathbb{C}} \setminus \{K_1^* \cup K_2^*\}$ is said to be the result of *circular symmetrization of the doubly connected domain D*.

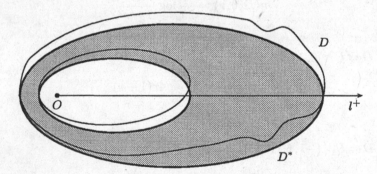

**Fig. 2.10.** Circular symmetrization of a doubly connected domain

**c)** Let $D$ be a quadrilateral with the opposite sides on the unit circle, $D'$ be its reflection under the map $1/z$. Then, the domain $D \cup D'$ with their common boundary is a doubly connected domain $G$. Denote by $G^*$ the result of circular symmetrization (b) with respect to the origin and the positive real axis. Then, $D^* = G^* \cap U$ is said to be the result of *circular symmetrization of the quadrilateral D*. One can obviously extend this definition to an arbitrary disk instead of the unit one.

There are other types of symmetrization as, e.g., the Steiner symmetrization that one can imagine as a circular symmetrization with the infinite center, the Schwarz symmetrization, the elliptic symmetrization and others. For these we refer to the excellent survey by V. Dubinin [30].

The main result on symmetrization is the monotonic change of moduli defined for the above mentioned domains.

**Theorem 2.5.1** (Polya [108], Jenkins [64]). *Let $D$ be a doubly connected hyperbolic domain in $\overline{\mathbb{C}}$ and $D^*$ be the result of circular symmetrization as in (b). Let $M(D)$ be the modulus of $D$ with respect to the family of curves that separate its boundary components. Then, $M(D) \leq M(D^*)$ with the equality only in the case when $D^*$ is a rotation of $D$ with respect to the center of symmetrization.*

The limiting form of this theorem is the following.

**Corollary 2.5.1.** *Let $D$ be a simply connected hyperbolic domain and $c \in D$. Let $D^*$ be the result of circular symmetrization as in (a). Denote by $c^*$ the circular projection of $c$ onto $l^+$. Then, the reduced modulus $m(D, c) \leq m(D^*, c^*)$ with the same condition for the equality sign.*

**Theorem 2.5.2** (Jenkins [64]). *Let $D$ be a quadrilateral in the disk $U(O, R)$ with two opposite sides on the circumference $\partial U(O, R)$ and $D^*$ be the result of circular symmetrization as in (c). Let $M(D)$ be the modulus of $D$ with respect to the family of curves that connect its circular boundary components. Then, $M(D) \leq M(D^*)$ with the equality sign only in the case when $D^*$ is a rotation of $D$ with respect to the center of symmetrization $O$.*

The proofs, however, would take us to far afield. One can find them in the famous monograph by J. Jenkins [64]. The limiting form of Theorem 2.5.2 that leads to symmetrization of digons was obtained by A. Yu. Solynin in [135].

### 2.5.2 Polarization

Another powerful tool close to symmetrization is polarization.

**d)** We begin with polarization of an arbitrary set. Let $l^+$ and $l^-$ be a straight line $l$ with two opposite orientations. Consider a set $A$ and denote by $A^*$ the set that is symmetric to $A$ with respect to the line $l$, by $A^+-$ the part of $A$ left to $l^+$, by $A^--$ the part of $A$ left to $l^+$. The set $P^+(A) := (A \cup A^*)^+ \cup (A \cap A^*)^- \cup (A \cap l)$ we call the result of polarization of the set $A$ with respect to the straight line $l^+$. One can consider the set $P^-(A) = (A \cup A^*)^- \cup (A \cap A^*)^+ \cup (A \cap p)$ which is also the result of polarization of $A$ with respect to $l^-$.

**e)** *Polarization of a doubly connected domain $D$.* We consider a doubly connected hyperbolic domain $D$ and assume $\overline{\mathbb{C}} \setminus D = E_1 \cup E_2$. Denote by $E'_j$ the

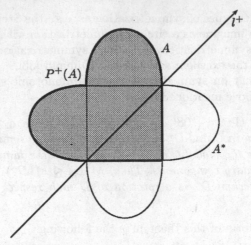

**Fig. 2.11.** Linear polarization of a simply connected domain

domain that contains a connected component of $P^+(E_j)$ left to $l^+$ (of course, $E'_j$ can lie in both half-planes). Then, the set $P^+(D) = \overline{\mathbb{C}} \setminus \{E'_1 \cup (E'_2)^*\}$ is said to be the result of polarization of the doubly connected domain $D$ with respect to the straight line $l^+$. The same definition is valid for $P^-(D)$. Observe that $P^-(D)$ is symmetric to $P^+(D)$ substituting $E_1 \to E_2$, $E_2 \to E_1$.

**f)** *Polarization of a quadrilateral $D$.* Let $D$ be a quadrilateral with the opposite sides on the unit circle, $D'$ be its reflection under the map $1/z$. The domain $D \cup D'$ with their common boundary is a doubly connected domain $G$. Denote by $P^+(G)$ the result of polarization (e) with respect to the straight line $l^+$ passing through the origin. Then, $P^+(D) = P^+(G) \cap U$ is said to be the result of polarization of the quadrilateral $D$ with respect to $l^+$. One can obviously extend this definition to an arbitrary disk instead of the unit one.

**Theorem 2.5.3** (Wolontis [176], Dubinin [30]). *Let $C = \{D_1, D_2\}$ be a condenser in $\overline{\mathbb{C}}$ and $P(C) = \{P^+(D_1), P^-(D_2)\}$ be the result of polarization as in (d). Then,* cap $C \geq$ cap $P(C)$ *with the equality sign only in the case when $P(C)$ coincides with $C$ or is symmetric to $C$ with respect to $l$.*

This result implies the following theorem.

**Theorem 2.5.4** (Solynin [137]). *Let $D$ be a doubly connected hyperbolic domain in $\overline{\mathbb{C}}$ and $P^+(D)$ be the result of polarization as in (e). Let $M(D)$ be the modulus of $D$ with respect to the family of curves that separate its boundary components. Then, $M(D) \leq M(P^+(D))$ with the equality sign only in the case when $P^+(D)$ coincides with $D$ or is symmetric to $D$ with respect to $l$.*

**Theorem 2.5.5** (Solynin [137]). *Let $D$ be a quadrilateral in the disk $U(O, R)$ with two opposite sides on the circumference $\partial U(O, R)$ and $P^+(D)$ be the*

*result of polarization as in (f). Let $M(D)$ be the modulus of $D$ with respect
to the family of curves that connect its circular boundary components. Then,
$M(D) \leq M(P^+(D))$ with the equality sign only in the case when $P^+(D)$
coincides with $D$ or is symmetric to $D$ with respect to $l$.*

The following exercises use polarization or symmetrization.

★ Let $D(\theta) = U \setminus \{[0, r_1] \cup [0, r_2 e^{i\theta}]\}$, $0 < r_k < 1$, $0 \leq \theta < 2\pi$. Prove
that $M(D(\theta))$ decreases in $0 \leq \theta \leq \pi$ as a function of $\theta$ and increases in
$\pi \leq \theta \leq 2\pi$.

★ The limiting form of Theorem 2.5.4. Let $D'(\theta) = U \setminus \{[r_1, 1) \cup [r_2 e^{i\theta}, e^{i\theta} 0]\}$,
$0 < r_k < 1$, $0 \leq \theta < 2\pi$. Prove that the reduced modulus $m(D'(\theta), 0)$
decreases in $0 \leq \theta \leq \pi$ as a function of $\theta$ and increases in $\pi \leq \theta \leq 2\pi$.
(Solynin [137])

★ Polarization+ symmetrization. Let $D = \mathbb{C} \setminus \{(-\infty, 0] \cup [1, R]\}$, $D_1 =
\mathbb{C} \setminus \{\gamma_1, \gamma_2\}$ where $\gamma_1$ is a curve that connects 0 with $\infty$ and $\gamma_2$ connects
1 with $R$ such that the curve $\gamma_2$ is not homotopic to the segment $[1, R]$ on
$\mathbb{C} \setminus \{0, 1, R\}$ and $\gamma_1 \cap \gamma_2 = \emptyset$. Prove that $M(D) \geq M(D_1)$ with the equality
sign only if $D = D_1$.

★ Let $D$ be a convex simply connected domain, i.e., any segment connecting
two points of $D$ lies in $D$. Is it true, that $D^*$ (symmetrization (a)) remains
convex?

## 2.6 Quadratic differentials on Riemann surfaces

A great collection of definitions and results about *quadratic differentials* one
can find in monographs by well known analysts J. Jenkins [64] and K. Strebel
[141]. Here we present only some basic definitions and facts. We also refer
the reader to [38], [105], [106], [131], [140] for the definitions and properties
of Riemann surfaces.

### 2.6.1 Riemann surfaces

A *Riemann surface* $S$ is a connected topological Hausdorff space $M$ with an
open covering $U_j$ and a system of homeomorphisms $g_j$ such that $g_j : S \to
\mathbb{C}$ and in a non-empty intersection $U_j \cap U_k$ the vicinity relation $g_j \circ g_k^{-1}$
is conformal. The set of pairs (*charts*) $\{U_j, g_j\}$ is said to be a conformal
structure on $S$. Two structures $\{U_j^1, g_j^1\}$ and $\{U_j^2, g_j^2\}$ are called equivalent
if their union is also a conformal structure on $S$. So, two Riemann surfaces
with equivalent conformal structures and the same underlying topological
space $M$ are thought of as the same Riemann surface. The space $S$ is a
connected component of 1-dimensional complex analytic manifold and itself
is an oriented manifold. Any finite compact Riemann surface is topologically

equivalent to a sphere with a finite number of handles that is said to be the genus of this surface. The genus is a topological invariant.

A bordered Riemann surface $S$ is also a connected Hausdorff space $M$ with charts containing an open covering $U_j$ and a system of homeomorphisms $g_j$ with the properties as above but $g_j$ acts from $M$ into the closed upper half-plane and all vicinity relations are again conformal. If there is a neighbourhood $U_j$ of a point $p \in M$ with a real interval as a part of its boundary, and $g_j : \operatorname{Im} g_j(p) = 0$, then this point is supposed to be from the border of $S$. All such points form the border of $S$.

Let $p \in S$ and let us consider a class $[c]$ of all closed curves homotopic on $S$ (continuous quotient maps from $[0,1]$ into $S$) with endpoints at $p$ and of given orientation. Suppose that $c$ represents $[c]$. We define the product operation $c_1 c_2$ as a union of $c_1$ and $c_2$ with the corresponding orientation. The inverse operation $c^{-1}$ is defined as the same curve with the reverse orientation. A point is considered to be a 0-curve. The same operations are easily understandable for homotopy classes $[c_1][c_2] = [c_1 c_2]$, $[c]^{-1} = [c^{-1}]$, $[1]$ is a class of curves homotopic to $p$. These classes with operations defined form a so called *fundamental group* $\pi(S, p)$ of the surface $S$ with respect to the point $p$ chosen. A curve $\gamma$ connecting the point $p$ with another point $p_1$ produces an isomorphism $[\gamma c \gamma^{-1}]$ of fundamental groups $\pi(S, p)$ and $\pi(S, p_1)$. So, we can speak about a single fundamental group $\pi(S)$ up to an isomorphism.

For any subgroup $F$ of the fundamental group $\pi(S)$ of the given surface $S$ one can construct a new Riemann surface $S^*$ which is non-branched, unbounded over $S$, and $\pi(S^*) = F$. This surface is a covering surface of $S$ and the deck homeomorphism $\sigma$ of $S^*$ onto itself leaves the traces of points from $S^*$ invariant. For the trivial subgroup $F = [1]$ the covering surface $\tilde{S}$ is said to be the *universal covering* of $S$. The surface $\tilde{S}$ is simply connected and, by the uniformization theorem of Koebe and Poincaré, there is a conformal homeomorphism $h$ of $\tilde{S}$ onto either $U$, $\mathbb{C}$, or else $\overline{\mathbb{C}}$. So, the surface is said to be *hyperbolic*, *parabolic*, or *elliptic* respectively. In particular, a hyperbolic simply connected domain has more than two boundary points. The homeomorphism $\sigma$ produces a conformal automorphisms $h(\sigma)$ of one of the canonical domains and the function $J \circ h^{-1}$ with the projection $J : \tilde{S} \to S$ is an automorphic analytic function with an invariant group of automorphisms of the canonical domain. Further, the universal cover is to be identified with its canonical conformal image under $h$. We denote by $G$ the group of automorphisms of the canonical domain that corresponds to the group of deck transformations for $\tilde{S}$. So, any Riemann surface $S$ is to be identified with the quotient $S = D/G$ up to conformal equivalence, where $D$ is a canonical domain. We say that $G$ *uniformizes* the surface $S$.

All elliptic Riemann surfaces are conformally equivalent to the Riemann sphere $\overline{\mathbb{C}}$. The group $G$ in this case consists of fractional linear conformal automorphisms of $\overline{\mathbb{C}}$. All parabolic Riemann surfaces are conformally equivalent either to the plane $\mathbb{C}$, the punctured plane $\mathbb{C} \setminus \{0\}$, or the torus. The group $G$

in this case is either trivial or else consists of transformations $z' = z + n\omega$, or $z' = z + n\omega_1 + m\omega_2$ where $\omega, \omega_1, \omega_2$ are complex numbers with Im $\omega_2/\omega_1 \neq 0$ and $n, m$ are some integers. All other Riemann surfaces are of hyperbolic type and $G$ is a subgroup of the Möbius group **Möb** of all conformal automorphisms of $U$. They are of the form

$$z \to e^{i\theta} \frac{z - a}{1 - z\bar{a}}, \ a \in U, \ 0 \leq \theta < 2\pi.$$

In this case the group $G$ is called the *Fuchsian group*. The fundamental domain $D$ of the Fuchsian group $G$ is a factor set of the points of $U$ that are not equivalent with respect to the actions from $G$. Geometrically this fundamental domain is a hyperbolic polygon in $U$ bounded by arcs of orthocircles or $\partial U$. If this polygon has a finite number of sides, then the group $G$ is finitely generated. A Riemann surface $S$ with a finitely generated uniformizing group $G$ is said to be a *finite Riemann surface*. If it is of genus $g$, then there are generators $A_1, B_1, \ldots, A_g, B_g$ of the group $G$ with hyperbolic fixed points on $\partial U$ corresponding to $4g$ from $4g + 2(n + r)$ sides of $D$. If it has, moreover, $n$ punctures and $r$ branch points, then there are generators $C_1, \ldots, C_{n+r}$ of the group $G$ corresponding to $n$ parabolic fixed points of $C_1, \ldots, C_n$ (cusps of $\partial D$) and $r$ elliptic points of $C_{n+1}, \ldots, C_{n+r}$ (vertices of $D$ from $U$). These generators satisfy the normalization

$$C_{n+r} \circ \cdots \circ C_1 \circ B_g^{-1} \circ A_g^{-1} \circ B_g \circ A_g \circ \cdots \circ B_1^{-1} \circ A_1^{-1} \circ B_1 \circ A_1 = id.$$

If a surface $S$ has additionally $l$ hyperbolic boundary components, then this surface is said to be of finite type $(g, n, l)$.

### 2.6.2 Quadratic differentials

Let $g, n, l, m$ be non-negative integers and $S_0$ be a Riemann surface of finite type $(g, n, l)$ or $(g, n)$ when $l = 0$, i.e., it is of genus $g$ and has $n$ punctures and $l$ hyperbolic boundary components (with the border if $l \neq 0$). $S_0$ can have a finite number of branch points. If $6g - 6 + 2n + 3l > 0$, then $S_0$ is of hyperbolic conformal type and its universal covering is conformally equivalent to the unit disk $U = \{z : |z| < 1\}$. The deck mappings that replace the sheets of the universal covering induce a corresponding Fuchsian group $G_0$ of Möbius automorphisms of $U$. The elements of $G_0$ map the fundamental polygons of $S_0$ onto themselves. One says that the Fuchsian group $G_0$ uniformizes the Riemann surface $S_0$ and this Riemann surfaces is to be idetified with the quotient $S_0 = U/G_0$.

We say that a holomorphic (meromorphic) quadratic differential $\varphi$ is defined on $S_0$ if for any local parameter $\zeta$ of $S_0$ there is a holomorphic (meromorphic) function $\varphi(\zeta)$ defined in the parametric neighbourhood. This function satisfies the condition of the invariance respectively to a change of parameters, i.e., if $\varphi^*$ and $\varphi$ are two representations of the differential in terms of the parameters $\zeta^*$ and $\zeta$, then

$$\varphi^*(\zeta^*) = \varphi(\zeta) \left( \frac{d\zeta}{d\zeta^*} \right)^2$$

for the vicinity relation $\zeta(\zeta^*)$ defined in the intersection of the parametric neighbourhoods. We denote these quadratic differentials simply by $\varphi$ or $\varphi(\zeta)d\zeta^2$ in terms of the local parameter $\zeta$. Let $J_0(z)$ be the projection of the universal covering of $S_0$ (which is realized as $U$) onto $S_0$. Define the pullback $q(z)dz^2$ of the quadratic differential $\varphi$ onto the universal cover $U$ by the formula $q(z) = \varphi(J_0(z))(J_0'(z))^2$. Since $J_0$ is an automorphic function with respect to the group $G_0$, the invariance condition for the differential $q(z)dz^2$ is satisfied as $q(\gamma(z))(\gamma'(z))^2 = q(z)$ in so far as $z \in U, \gamma \in G_0$.

Let us consider the conformal invariant metric $\sqrt{|\varphi(\zeta)|}|d\zeta|$ associated with a quadratic differential $\varphi$. The element of length in this metric is $|dw| = \sqrt{|\varphi(\zeta)|}|d\zeta|$. A maximal regular curve on $S_0$ such that the inequality $\varphi(\zeta)d\zeta^2 > 0$ holds on it we call a *trajectory* of $\varphi$; an *orthogonal trajectory* if it satisfies the reverse inequality $\varphi(\zeta)d\zeta^2 < 0$. The trajectories and orthogonal trajectories are connected with the inner differential structure and do not depend on a local parameter. We call a trajectory (or an orthogonal one) *critical* if there is a zero or a pole in its closure. Other trajectories are *regular*. The rigorous description of the local and global trajectory structure will be given in the next subsections and can be found in J. Jenkins [64] and K. Strebel [141].

A trajectory of a differential $\varphi$ on the Riemann surface $S_0$ is called *finite* (or *closed* in the case of a loop following Strebel's terminology) if its length in $\sqrt{|\varphi(\zeta)|}|d\zeta|$-metric is finite.

One says that a holomorphic (meromorphic) differential has finite trajectories if its non-finite ones run through a plane null-set.

The next propositions give some useful information about the space of quadratic differentials with finite trajectories.

**Proposition 2.6.1** (Strebel [141], p. 23). *A holomorphic quadratic differential on $S_0$ of finite $L_1$-norm can have at most simple poles on the closure $\overline{S_0}$ of the surface $S_0$.*

**Proposition 2.6.2** (Strebel [141], Theorem 21.2). *The set of holomorphic quadratic differentials on $S_0$ of fixed $L_1$-norm and with finite trajectories of given homotopy type is compact (in the local uniform topology on $S_0$).*

**Proposition 2.6.3** ([141] Theorem 25.2, see also [27] ). *On a Riemann surface of type $(g,n)$ quadratic differentials with finite trajectories form an everywhere dense set in the space of all quadratic differentials of finite $L_1$ norm.*

### 2.6.3 Local trajectory structure

Now we study the behaviour of trajectories in the neighbourhoods of singularities of a quadratic differential or the *critical points*. Many of basic results with detail proofs can be found in [141], Chapter III, [64], Chapter 2 . Here we will summarize the relevant information about the local trajectory structure.

Consider a simplest quadratic differential $dz^2$ on the Riemann sphere $\overline{\mathbb{C}}$. Changing the parameter $z \to 1/z^*$ one can see that it has the representation $(dz^*)^2/(z^*)^4$ in a neighbourhood of 0 in terms of the parameter $z^*$. So, this differential has a unique singularity at the point $\infty$ which is the pole of order 4. The trajectories of $dz^2$ are, obviously, horizontal lines.

For a quadratic differential $\varphi(\zeta)d\zeta^2$ we introduce a so called *natural parameter*

$$z = \int^{\zeta} \sqrt{\varphi(\zeta)}d\zeta$$

in a neighbourhood of a regular point of $\varphi(\zeta)d\zeta^2$. So, one can select a single valued branch of the square root and locally represent

$$\sqrt{\varphi(\zeta)} = a_0 + a_1\zeta + \ldots, \quad a_0 \neq 0.$$

Therefore, integrating term by term (this is possible due to the uniform convergence on compacts inside the neighbourhood chosen) we obtain a conformal map $\zeta \to z$. Thus, the local trajectory structure for $\varphi(\zeta)d\zeta^2$ near a regular point is the same as for $dz^2$ up to shift and rotation.

Let $p_0$ be a zero of the differential $\varphi(\zeta)d\zeta^2$ of even order $n$. We consider it in terms of a local parameter $\zeta$ near $p_0$ which represents this zero as 0. In a sufficiently small neighbourhood of $p_0$ one can select a single valued branch of the square root and locally outside $\zeta = 0$ represent

$$\sqrt{\varphi(\zeta)} = \zeta^{n/2}(a_0 + a_1\zeta + \ldots).$$

Integrating again termwise we obtain a function $w = \zeta^{(n+2)/2}(b_0 + b_1\zeta + \ldots)$ that maps the local neighbourhood of the parameter $\zeta$ onto an branched element over $w$-plane. Introducing a new parameter $z : z^{(n+2)/2} = w$ we obtain a conformal map $\zeta \to z$ which is unique up to the factor $\exp(\frac{2k\pi}{n+2})$, $k = 0, \ldots, n + 1$. This map satisfies the equation

$$\varphi(\zeta)d\zeta^2 = \left(\frac{n+2}{2}\right)^2 z^n dz^2.$$

Thus, the trajectory structure of $\varphi(\zeta)d\zeta^2$ in a neighbourhood of zero is the same as for $z^n dz^2$ up to shift, rotation, and the factor mentioned. The function $w = z^{(n+2)/2}$ maps each sector

$$\frac{2\pi}{n+2}k < \arg z < \frac{2\pi}{n+2}(k+1), \quad k = 0, \ldots, n+1$$

onto the upper or lower half-plane and all horizontal arcs are the images of the trajectories of $\varphi(\zeta)d\zeta^2$. For an odd $n$ one can construct the additional

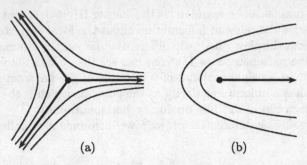

**Fig. 2.12.** The local trajectory structure near (a) simple zero, (b) simple pole

parameter $z = \omega^2$ where $z$ now is defined in a doubly covered neighbourhood of 0.

Analogously, if $p_0$ is a simple pole this procedure leads to the same form of trajectories in the sector $0 < \arg z < 2\pi$.

Let $p_0$ be a pole of the differential $\varphi(\zeta)d\zeta^2$ of order 2. This case is distinct because the function $\sqrt{\varphi(\zeta)}$ after integration gives the logarithmic singularity. Further exponentiating leads to the conformal map $\zeta \to z$ that satisfies the equation

$$\varphi(\zeta)d\zeta^2 = \frac{a_{-2}}{z^2}dz^2.$$

According to the coefficient $a_{-2}$ trajectories have the radial form ($a_{-2} > 0$), the circular form ($a_{-2} < 0$), or the spiral form (Im $a_{-2} \neq 0$).

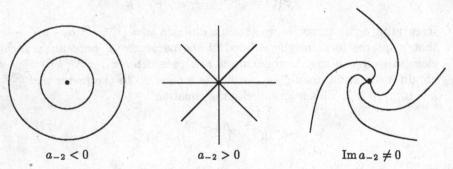

$a_{-2} < 0$ $\qquad\qquad$ $a_{-2} > 0$ $\qquad\qquad$ Im $a_{-2} \neq 0$

**Fig. 2.13.** The local trajectory structure near a double pole

★ The last possible case of the pole of order greater than 2 we leave as an exercise. All trajectories near such a pole starting and ending at this singularity have $n - 2$ tangent directions that bound congruent sectors of angles $2\pi/(n-2)$.

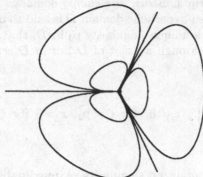

**Fig. 2.14.** The local trajectory structure near a pole of 5-th order

### 2.6.4 Trajectory structure in the large

The trajectories and orthogonal trajectories of a given differential $\varphi(\zeta)d\zeta^2$ produce a transversal foliation of the Riemann surface.

We give here a classification of domains of special types associated with a quadratic differential $\varphi(\zeta)d\zeta^2$.

First we suppose that the Riemann surface $S_0$ is of hyperbolic type $(g, n)$. We consider a holomorphic differential $\varphi(\zeta)d\zeta^2$ on $S_0$ which can be meromorphic on the closure $\overline{S_0}$ of $S_0$.

1) With a differential $\varphi(\zeta)d\zeta^2$ with **finite trajectories and of finite $L^1$-norm** we associate *ring domains*. A maximal doubly connected hyperbolic domain $D$ is said to be a ring domain for $\varphi(\zeta)d\zeta^2$ if there is no singularity in $D$, a trajectory passing through a point of $D$ lies in $D$, and there is a number $R > 1$ and a map

$$z = \exp\left(c\int^{\zeta} \sqrt{\varphi(\zeta)}\,d\zeta\right), \quad c \neq 0$$

of $D$ onto the ring $1 < |z| < R$.

2) With a differential $\varphi(\zeta)d\zeta^2$ of **finite $L^1$-norm** we associate *ring domains and spiral domains*. To describe a spiral domain we need some more definitions. Let $p_0$ be a regular point of the differential $\varphi(\zeta)d\zeta^2$ on $S_0$ and the trajectory $\gamma(p_0)$ of $\varphi$ passing through the point $p_0$ is not closed and never runs into a critical point. We consider the oriented trajectory $\gamma^+(p_0)$. Denote by $A$ the limit set of $\gamma^+(p_0)$. Then, [43], Section 2.5; [141], Section 10.2, the interior of $A$ is a simply connected domain $D$ bounded by finite critical trajectories of $\varphi(\zeta)d\zeta^2$. Moreover, $\overline{\gamma^+(p_0)} = A$ and, if $p \in A$, then $\overline{\gamma^+(p)} = A$. Hereof the domain $D$ is called a spiral domain.

3) With a differential $\varphi(\zeta)d\zeta^2$ with **finite trajectories** we associate *ring domains, circular domains, strip domains, and ending domains.*

A maximal simply connected hyperbolic domain $D$ is said to be a circular domain for $\varphi(\zeta)d\zeta^2$ if there is a unique singularity $p_0$ in $D$, that is a pole of order 2, a trajectory passing through a point of $D$ lies in $D$ separating $p_0$ from $\partial D$, and there is a map

$$z = \exp\left(c\int^\zeta \sqrt{\varphi(\zeta)}\,d\zeta\right), \quad c \neq 0, \text{ for } \zeta \neq p_0, \ z = 0 \text{ for } \zeta = p_0$$

of $D$ onto the disk $|z| < R$.

Other types of domains contain no singularity. A maximal simply connected hyperbolic domain $D$ is said to be a strip domain for $\varphi(\zeta)d\zeta^2$ if there are two singularities $p_1$ and $p_2$ in $\partial D$, that are poles of order 2, a trajectory passing through a point of $D$ lies in $D$ connecting $p_1$ and $p_2$, and there is a map

$$z = \int^\zeta \sqrt{\varphi(\zeta)}\,d\zeta$$

of $D$ onto strip $a < \operatorname{Im} z < b$.

A maximal simply connected hyperbolic domain $D$ is said to be an ending domain for $\varphi(\zeta)d\zeta^2$ if there is a pole $p_0$ of order $> 2$ in $\partial D$, a trajectory passing through a point of $D$ starting and ending in $p_0$ lies in $D$, and there is a map

$$z = \int^\zeta \sqrt{\varphi(\zeta)}\,d\zeta$$

of $D$ onto the upper or lower half-plane according to the branch of the root.

4) With a differential $\varphi(\zeta)d\zeta^2$ without the above restrictions we associate all types of the domains: *ring domains, spiral domains, circular domains, strip domains, and ending domains.*

Now we suppose a Riemann surface $S_0$ to be of hyperbolic type $(g, n, l)$ with $l > 0$, the differential $\varphi(\zeta)d\zeta^2$ has the border of $S_0$ as a trajectory or an orthogonal trajectory.

5) Suppose that the differential $\varphi(\zeta)d\zeta^2$ has no singularities on the border of $S_0$ which is a trajectory of $\varphi$. Then, the global trajectory structure is described by the same domains as for the case of compact surfaces with possible punctures.

6) Suppose that the differential $\varphi(\zeta)d\zeta^2$ has no singularities on the border of $S_0$ and a connected component of the border of $S_0$ is an orthogonal trajectory of $\varphi$. Then, the global trajectory structure is described by the same domains as for the case of compact surfaces plus *quadrangles and triangles*

as "semi-rings" and "semi-strips". Namely, one can extend symmetrically the differential onto the Riemann surface $S'$ obtained by the inversion of $S_0$ through one of its boundary components.

A maximal simply connected hyperbolic domain $D$ with two non-intersected connected boundary components lying on the border of $S_0$ is said to be a *quadrangle* for $\varphi(\zeta)d\zeta^2$ if there is no singularity in $D$, a trajectory passing through a point of $D$ lies in $D$, and there are numbers $a < b$, $c < d$, and a map

$$z = \int^\zeta \sqrt{\varphi(\zeta)}\,d\zeta,$$

of $D$ onto a rectangle $\operatorname{Re}\zeta \in (a, b)$, $\operatorname{Im}\zeta \in (c, d)$.

A maximal simply connected hyperbolic domain $D$ with a connected boundary component lying on the border of $S_0$ is said to be a *triangle* for $\varphi(\zeta)d\zeta^2$ if there is a pole $p_0$ of order 2 in $\partial D$, a trajectory passing through a point of $D$ lies in $D$, starts at $p_0$, ends in the boundary component of $D$ on the border of $S_0$ and there is a map

$$z = \exp\left(c\int^\zeta \sqrt{\varphi(\zeta)}\,d\zeta\right), \quad c \neq 0, \ \text{for} \ \zeta \neq p_0, \ z = 0 \ \text{for} \ \zeta = p_0$$

of $D$ onto $U \setminus [0, 1)$.

Poles of order 2 on the border of $S_0$ can also produce "semi-circular" domains if the border is an orthogonal trajectory of the differential locally near such a pole.

The trajectory structure of $\varphi(\zeta)d\zeta^2$ contains a finite number of non-overlapping domains of the above types. We do not consider here the case of more complicated critical points of $\varphi$ in the border of $S_0$. The information about the trajectory structure in the large is given by the Jenkins Main Structure Theorem ([64], Theorem 3.5, and its more complete version in [66] (see also [139]; [141], Chapter IV).

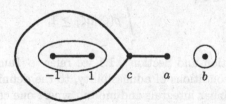

**Fig. 2.15.** The trajectory structure (in the large) of the differential $\varphi_0(\zeta)d\zeta^2$

The next example (Fig. 2.15) deals with the holomorphic differential

$$\varphi_0(\zeta)d\zeta^2 \equiv \frac{(c-\zeta)d\zeta^2}{(\zeta-b)^2(\zeta^2-1)(\zeta-a)}$$

on the Riemann surface $S_0 = \overline{\mathbb{C}} \setminus \{-1, 1, a, b\}, 1 < c < a < b$.

## 2.7 Free families of homotopy classes of curves and extremal partitions

In the development of the modulus method one may distinguish two approaches which equally connect the moduli of families of curves and quadratic differentials with finite trajectories. The first one refers to J. Jenkins [63], [67] and was developed by P. Tamrazov [146], [149], and G. Kuz'mina [78]–[85]. It comes from the concept of the modulus of a family of curves. This notion was initially defined by L. Ahlfors and A. Beurling [12] as the reciprocal of extremal length. The other one is due to K. Strebel [141] (see also [36], [137], [134], [137], [138], [139]) and comes from the extremal partition of a Riemann surface by domains of special shape.

### 2.7.1 The case of ring domains and quadrangles

Let $\Gamma$ be a family of curves on $S_0$, $\rho$ be a conformal invariant metric $\rho(\zeta)|d\zeta|$ (in terms of a local parameter $\zeta$), i.e., for any local parameter $\zeta$ on the surface $S_0$ there is a real valued non-negative measurable function $\rho(\zeta)$ such that the following invariance condition is satisfied

$$\rho^*(\zeta^*) = \rho(\zeta)\left|\frac{d\zeta}{d\zeta^*}\right|.$$

This means that we define the metric $\rho$ for all conformally equivalent complex structures on $S_0$ (see also Section 2.1.1). Let $P$ be a family of conformal invariant metrics $\rho$ on $S_0$ satisfying the following admissible condition [63], [146], for any $\gamma \in \Gamma$,

$$\int_\gamma \rho(\zeta)|d\zeta| \geq 1. \tag{2.10}$$

As in Introduction and Section 2.1.1 we refer to Tamrazov's [146], [149] approach to the conditions of admissibility, to the definitions of metrics, families of curves, linear integrals and moduli where one can omit rectifiability of curves and Borelian metrics and can consider only measurable metrics and arbitrary (not necessary rectifiable) curves. Here we consider in (2.10) the lower Darboux integrals and take the Lebesgue plane integrals in (2.11).

If $P \neq \emptyset$, then one can say that for $\Gamma$ on $S_0$ the modulus problem is defined and the modulus in this problem is given as

$$m(S_0, \Gamma) = \inf_{\rho \in P} \iint\limits_{S_0} \rho^2(\zeta) d\sigma_\zeta, \qquad (2.11)$$

where $d\sigma_\zeta$ is the area element on $S_0$.

The main properties of the modulus are the same as in Section 2.1.2, i.e., its conformal invariance: if $f$ is a conformal mapping $S_0 \to S$, then $m(S_0, \Gamma) = m(S, f(\Gamma))$; and the uniqueness of the extremal metric $\rho^*$. In other words, the modulus remains the same under a conformal change of the complex structure of $S_0$. We indicate here an important inequality $m(S_0, \Gamma_1) \leq m(S_0, \Gamma_2)$ when $\Gamma_1 \subset \Gamma_2$. However, the problem on the existence and on the form of the extremal metric is still the most difficult one. General modulus problems were considered in [63], [67]; [78], Chapter 0, and the existence has been established as well as the form of the extremal metric.

Avoiding supefluous notations we denote by $M(D)$ either the modulus of a doubly connected domain or else of a quadrilateral according to the type of $D$. $M(D) = m(D, \Gamma)$ where $\Gamma$ is the family of curves that separate the boundary components of a doubly connected domain $D$ or a family of curves that connect the opposite boundary components of a quadrilateral $D$.

Now we define rigorously the family $\Gamma$. On a Riemann surface $S_0$ we set a collection of curves $\gamma = (\gamma_1, \ldots, \gamma_m)$ of two types I, II. The first one (I) consists of simple loops on $S_0$ that are not freely homotopic pairwise. The second one (II) exists when $l > 0$ and consists of simple arcs on $S_0$ ending on the hyperbolic boundary components of $S_0$. All curves are neither homotopically trivial nor intersected. Such a collection $\gamma$ is called the *admissible system of curves*.

A set $\Gamma_j$ of curves on $S_0$ is said to be a *homotopy class* generated by $\gamma_j$ from the admissible system $(\gamma_1, \ldots, \gamma_m)$ if this set consists of all curves of type I or II that are freely homotopic to $\gamma_j$ on $S_0$.

A collection of the homotopy classes of curves $\Gamma := (\Gamma_1, \ldots, \Gamma_m)$ generated by the admissible system $(\gamma_1, \ldots, \gamma_m)$ on $S_0$ is said to be the *free family of homotopy classes of curves*.

Assume $\alpha := (\alpha_1, \ldots, \alpha_m)$ to be a non-zero vector with non-negative coordinates. Let $P$ be a family of conformal invariant metrics $\rho$ on $S_0$ satisfying the following admissible condition [63], [146]. For any $\gamma \in \Gamma_j$,

$$\int_\gamma \rho(\zeta) |d\zeta| \geq \alpha_j. \qquad (2.12)$$

If $P \neq \emptyset$, then one can say that for the family $\Gamma$ on $S_0$ and for the vector $\alpha$ the modulus problem is defined and the modulus in this problem is given by

$$m(S_0, \Gamma, \alpha) = \inf_{\rho \in P} \iint\limits_{S_0} \rho^2(\zeta) d\sigma_\zeta, \qquad (2.13)$$

In (2.12) we equally take the lower Dardoux integral as in (2.10) and (2.1). This definition of the modulus satisfies all important properties of the definition (2.10)–(2.11) or (2.1)–(2.2). Moreover, it is known [63], [67], that there exists a unique extremal metric $\rho^*$ in this modulus problem. It is defined by the unique holomorphic quadratic differential with finite trajectories $\varphi(\zeta)d\zeta^2$ on $S_0$ and $\rho^*(\zeta) = \sqrt{|\varphi(\zeta)|}$.

We call a doubly connected domain $D_j$ to be *associated* with the homotopy class $\Gamma_j$ of the first type (I) if any simple loop separating the boundary components of $D_j$ belongs to $\Gamma_j$. Then we equally call a quadrilateral $D_j$ with four distinct vertices 1,2,3,4 on the hyperbolic boundary components of $S_0$ to be *associated* with the homotopy class $\Gamma_j$ of the second type (II) if any simple arc connecting the boundary components on 1,4 and 2,3 $\in \partial S_0$ of $D_j$ belongs to $\Gamma_j$.

The critical trajectories of $\varphi(\zeta)d\zeta^2$ split $S_0$ into at most $m$ ring domains and quadrangles $\mathfrak{D}^* = (D_1^*, \ldots, D_m^*)$ associated respectively with homotopy classes of curves (some of $D_j^*$ can degenerate). Any collection of non-overlapping admissible doubly connected domains and quadrilaterals $\mathfrak{D} = (D_1, \ldots, D_m)$ of types I,II associated with the homotopy classes of the free family $\Gamma$ satisfies the following inequality

$$\sum_{j=1}^m \alpha_j^2 M(D_j) \le \sum_{j=1}^m \alpha_j^2 M(D_j^*)$$

with the equality only for $\mathfrak{D} = \mathfrak{D}^*$ where $M(D_j)$ (according to $j$) is the modulus of a doubly connected domain with respect to the family of curves which separate its boundary components or the modulus of a quadrilateral with respect to the family of curves connecting its opposite boundary components on $\partial S_0$.

Each $D_j^*$ is either a ring domain or else a quadrangle in the trajectory structure of $\varphi(\zeta)d\zeta^2$. If $D_j^*$ is a ring domain, then there is a conformal map $g_j(\zeta)$, $\zeta \in D_j^*$ that satisfies the differential equation

$$\alpha_j^2 \left( \frac{g_j'(\zeta)}{g_j(\zeta)} \right)^2 = -4\pi^2 \varphi(\zeta).$$

It maps $D_j^*$ onto the annulus $1 < |w| < \exp(2\pi M(D_j^*))$. If $D_j^*$ is a quadrangle, then there is a conformal map $g_j(\zeta)$, $\zeta \in D_j^*$ that satisfies the differential equation

$$\alpha_j^2 (g_j'(\zeta))^2 = \varphi(\zeta).$$

It maps $D_j^*$ onto the rectangle $0 < \operatorname{Re} w < 1$, $0 < \operatorname{Im} w < M(D_j^*)$. We call the system of domains $\mathfrak{D}^*$ the *characteristic system of domains* for the differential $\varphi(\zeta)d\zeta^2$.

Thus, the modulus problem is equivalent to the problem of the extremal partition of the Riemann surface $S_0$ by the domains of special types associated with certain free family of homotopy classes of curves.

Many particular cases of this modulus problem, generalization, calculation of the moduli of certain families of curves, and their applications were considered in [34], [40], [41], [78], [79], [163]. The results on extremal partition in the case of other types of associated domains such as strip-like, ending, and circular domains were obtained by G. Kuz'mina [78], [81] and E. Emel'yanov [36].

Now we formulate a slightly different problem of the extremal partition of the Riemann surface $S_0$ which is due to K. Strebel ([141], Section 20.5, Weighted sum of the reciprocals of moduli).

A holomorphic quadratic differential $\varphi$ with finite trajectories and of finite $L_1$-norm is said to to be of homotopy type $(\gamma_1, \ldots, \gamma_m)$ for a prescribed admissible system of curves on $S_0$ if its critical trajectories split $S_0$ into a system of non-overlapping ring domains and quadrangles $(D_1^*, \ldots, D_m^*)$ of homotopy type $(\gamma_1, \ldots, \gamma_m)$.

Let a quadratic differential $\varphi$ with finite trajectories be of finite norm and of homotopy type $(\gamma_1, \ldots, \gamma_m)$; $(D_1^*, \ldots, D_m^*)$ be its characteristic system of domains. If $D_j^*$ is not degenerate, then we denote by $\alpha_j$ the length of a trajectory that belongs to the domain $D_j^*$ of the differential $\varphi$ in the metric $\sqrt{|\varphi(\zeta)|}|d\zeta|$. Assume $b_j := \alpha_j \cdot M(D_j^*)$. Then, $\|\varphi\| = \sum_{j=1}^m \alpha_j \cdot b_j$. The minimum property of the differential $\varphi$ is expressed in the following propositions.

**Proposition 2.7.1** (Strebel [141], page 105). *Let $\Gamma := (\Gamma_j)_{j=1}^m$ be a free family of homotopy classes of curves generated by an admissible system $(\gamma_1, \ldots, \gamma_m)$ on $S_0$. Let $\varphi$ be a holomorphic quadratic differentials on $S_0$ of finite $L_1$-norm and with finite trajectories of given homotopy type $(\gamma_1, \ldots, \gamma_m)$. Then,*

$$m(S_0, \Gamma, \alpha) = \|\varphi\| = \sum_{j=1}^m \frac{b_j^2}{M(D_j^*)},$$

*where $(D_1^*, \ldots, D_m^*)$ is the system of characteristic domains.*

**Proposition 2.7.2** (Strebel [141], Theorem 20.5). *Let $\varphi$ be a holomorphic quadratic differential on $S_0$ of finite $L_1$-norm and with finite trajectories of given homotopy type $\gamma = (\gamma_1, \ldots, \gamma_m)$ with the characteristic domain system $\mathfrak{D}^* = (D_1^*, \ldots, D_m^*)$. If $\mathfrak{D} = (D_1, \ldots, D_m)$ is a system of non-overlapping domains on $S_0$ of homotopy type $(\gamma_1, \ldots, \gamma_m)$, then*

$$\sum_{j=1}^m \frac{b_j^2}{M(D_j)} \geq \sum_{j=1}^m \frac{b_j^2}{M(D_j^*)},$$

*where the equality sign is possible only for $\mathfrak{D} = \mathfrak{D}^*$.*

The existence theorem is given as follows.

**Proposition 2.7.3** (Strebel [141], Theorem 21.1). *Let $\gamma = (\gamma_1, \ldots, \gamma_m)$ be an admissible system of curves on $S_0$. Suppose there exists a system of doubly connected domains and quadrilaterals of this homotopy type with bounded moduli. Then, for any collection of positive numbers $(b_1, \ldots, b_m)$ there exists a unique holomorphic quadratic differential $\varphi$ on $S_0$ with finite trajectories and a system of doubly connected domains and quadrilaterals $(D_1^*, \ldots, D_m^*)$ of homotopy type $(\gamma_1, \ldots, \gamma_m)$ such that $\varphi$ is of the same homotopy type and $(D_1^*, \ldots, D_m^*)$ is the characteristic domain system for $\varphi$. Moreover, $\|\varphi\| = \sum_{j=1}^{m} \frac{b_j^2}{M(D_j^*)} < \infty$.*

These propositions give the solution to the problem of the extremal partition of a Riemann surface fixing the weight vector $\alpha$ or the height vector $(b_1, \ldots, b_m)$. The principal difference is that: if $b_k \neq 0$, then the domain $D_k^*$ never degenerates meanwhile the case $\alpha_k \neq 0$ does not imply this.

### 2.7.2 The case of circular, strip domains, and triangles

We consider the modulus problem of Section 2.7.1 as a basic one provided by J. Jenkins and K. Strebel. One can think of the problem of the extremal partition of a Riemann surface by the collection of non-overlapping domains associated with a certain free family of homotopy classes of curves as dual to the modulus problem. Recently, G. Kuz'mina has added some new types of homotopy classes and expand both the modulus problem and the problem about the extremal partition described in Section 2.7.1. It turns out to be important for the solution of various extremal problems for conformal map.

We need to add new types of curves to the definition of the admissible system. A finite number $\gamma = (\gamma_1, \ldots, \gamma_m, \gamma_{m+1}, \ldots, \gamma_s, \gamma_{s+1}, \ldots, \gamma_k)$ of simple loops and simple arcs with endpoints in the hyperbolic boundary components and punctures of $S_0$ that are not freely homotopic pairwise on $S_0$ is called the *admissible system of curves* on $S_0$ if these curves are not homotopic to a point of $S_0$ and do not intersect. We let a curve $\gamma_j$ from the admissible system be homotopic to a puncture (III type) for $j = m + 1, \ldots, s$. Another type of curves consists of those $\gamma_j$ that are not homotopically trivial, start and finish at fixed points (or the same point) which can be either punctures or else points of hyperbolic components of $S_0$ (IV type) for $j = s + 1, \ldots, k$.

Let an admissible system of curves $(\gamma_1, \ldots, \gamma_k)$ be given. A doubly connected domain $D_j$ on $S_0$ with one hyperbolic boundary component and a puncture as the other one is said to have a homotopy type III if any loop on $S_0$ separating the boundary components of $D$ is freely homotopic to a curve $\gamma_j$ of type III. A digon $D_j$ on $S_0$ with two fixed vertices on its boundary (maybe the same point) is said to be of homotopy type IV if any arc on $S_0$ connecting two vertices is homotopic (not freely) to a curve $\gamma_j$ of type IV. A triangle $D_j$ on $S_0$ with the fixed vertex at a puncture of $S_0$ is said to be of homotopy type IV if any arc on $S_0$ connecting the vertex with the opposite leg on the hyperbolic boundary component of $S_0$ is freely homotopic to a curve $\gamma_j$ of type IV.

A system of non-overlapping doubly connected domains, quadrilaterals, and simply connected domains $(D_1, \ldots, D_k)$ on $S_0$ is said to be of homotopic type $(\gamma_1, \ldots, \gamma_k)$ if $(\gamma_1, \ldots, \gamma_k)$ is an admissible system of curves on $S_0$ and for any $j \in \{1, \ldots, k\}$ the domain $D_j$ is of homotopic type $\gamma_j$ (I–IV). So, here now we let doubly connected domains be of parabolic type.

We fix a weight-height vector $\alpha = (\alpha_1, \ldots, \alpha_s, h_{s+1}, \ldots, h_k)$ and require the digons and triangles to be conformal at their vertices and to satisfy the condition of *compatibility of angles and heights*, say $\varphi_{a_j} = \pi h_j/(\sum_{k \in I_{a_j}} h_k)$, $j = s+1, \ldots, k$, $\varphi_{b_j} = \pi h_j/(\sum_{k \in I_{b_j}} h_k)$, where $I_{a_j}$ ($I_{b_j}$) is the set of indices which refer to the digons $D_j$ with the vertices at $a_j$ ($b_j$) or to the triangles $D_j$ with the vertices at $a_j$. With a given admissible system $\gamma$ and a vector $\alpha$ we associate the collection $\mathfrak{D} = (D_1, \ldots, D_k)$ of domains of the homotopy type $\gamma$, satisfying the condition of compatibility of angles and heights, which we call *associated* with $\gamma$ and $\alpha$.

With each of the domains of type III we associate the reduced modulus $m(D_k, a_k)$ of the parabolic domain $D_k$ with respect to a puncture $a_k$ and with each domain of IV type we associate the reduced modulus $m(D, a_j, b_j)$ of the digon $D_j$ with the vertices at $a_j, b_j$ or the reduced modulus $m_\Delta(D, a_j)$ of the triangle $D_j$ with the vertex at $a_j$.

Some of domains $(D_1, \ldots, D_k)$ (not all of them) can degenerate. In this case assume the modulus or the reduced one to vanish.

A general theorem by G. Kuz'mina ([78], Theorem 0.1) for circular domains, by E. Emel'yanov [36], G. Kuz'mina [81], A. Solynin [139] for the case of strip domains, A. Solynin [139] for the triangles states that any collection of non-overlapping admissible doubly connected domains, quadrilaterals, digons, and triangles $\mathfrak{D} = (D_1, \ldots, D_k)$ of types I–IV associated with the admissible system of curves $\gamma$ and the vector $\alpha$ satisfies the following inequality

$$\sum_{j=1}^{m} \alpha_j^2 M(D_j) + \sum_{j=m+1}^{s} \alpha_j^2 m(D_j, a_j) -$$
$$- \sum_{j=s+1}^{k_1} h_j^2 m(D_j, a_j, b_j) - \sum_{j=k_1+1}^{k} h_j^2 m_\Delta(D_j, a_j) \leq$$
$$\leq \sum_{j=1}^{m} \alpha_j^2 M(D_j^*) + \sum_{j=m+1}^{s} \alpha_j^2 m(D_j^*, a_j) -$$
$$- \sum_{j=s+1}^{k_1} h_j^2 m(D_j^*, a_j, b_j) - \sum_{j=k_1+1}^{k} h_j^2 m_\Delta(D_j^*, a_j),$$

with the equality sign only for $\mathfrak{D} = \mathfrak{D}^*$ where $M(D_j)$ (according to $j$) is the modulus of a doubly connected domain with respect to the family of curves that separate its boundary components or the modulus of a quadrilateral with respect to the family of curves that connect its boundary components on $\partial S_0$

for $j = 1, \ldots, m$. $m(D_j, a_j)$ is the reduced modulus of the simply connected domain $D_j \cup \{a_j\}$ with respect to the puncture $a_j$ of $S_0$ for $j = m+1, \ldots, s$. $m(D_j, a_j, b_j)$ is the reduced modulus of the digon $D_j$ with respect to the vertices $a_j$ and $b_j$ (possibly $a_j = b_j$) for $j = s+1, \ldots, k_1$. $m_\Delta(D_j, a_j)$ is the reduced modulus of the triangle $D_j$ with respect to the vertex $a_j$ for $j = k_1 + 1, \ldots, k$.

Each $D_j^*$ is either a ring domain, a quadrangle, a circular domain, a triangle, or a strip domain in the trajectory structure of a unique quadratic differential $\varphi(\zeta)d\zeta^2$. If $D_j^*$ is a circular domain, then there is a conformal mapping $g_j(\zeta)$, $\zeta \in D_j^*$ that satisfies the differential equation

$$\alpha_j^2 \left( \frac{g_j'(\zeta)}{g_j(\zeta)} \right)^2 = -4\pi^2 \varphi(\zeta), \quad j = m+1, \ldots, s$$

and maps $D_j^*$ onto the punctured disk $0 < |w| < \exp(2\pi m(D_j^*, a_j))$. If $D_j^*$ is a strip domain, then there is a conformal map $g_j(\zeta)$, $\zeta \in D_j^*$ that satisfies the differential equation

$$h_j^2 \left( \frac{g_j'(\zeta)}{g_j(\zeta)} \right)^2 = -4\pi^2 \varphi(\zeta), \quad j = s+1, \ldots, k_1$$

and maps $D_j^*$ onto the strip $\mathbb{C} \setminus [0, \infty)$. If $D_j^*$ is a triangle, then there is a conformal map $g_j(\zeta)$, $\zeta \in D_j^*$ that satisfies the differential equation

$$h_j^2 \left( \frac{g_j'(\zeta)}{g_j(\zeta)} \right)^2 = -4\pi^2 \varphi(\zeta), \quad j = k_1 + 1, \ldots, k$$

and maps $D_j^*$ onto the triangle $U \setminus [0, 1)$ with the vertex at $0$ and the unit circle as the opposite leg.

The critical trajectories of $\varphi(\zeta)d\zeta^2$ split $S_0$ into at most $m$ ring domains, quadrangles, circular domains, strip domains, and triangles $\mathfrak{D}^* = (D_1^*, \ldots, D_k^*)$ associated respectively with the admissible system of curves and the weight-height vector (some of $D_j^*$ can degenerate).

The above solves the problem of the extremal partition of $S_0$. Now we connect this problem for the general case of the admissible system of curves with the modulus problem or, in other words, we are looking for the infimum of an integral that presents the reduced area.

We suppose here that there is no curve from the admissible system of type IV. Let $P$ be the family of conformal invariant metrics $\rho$ on $S_0$ that satisfy the admissible condition (2.12) for any $\gamma \in \Gamma_j$,

$$\int\limits_\gamma \rho(\zeta)|d\zeta| \geq \alpha_j.$$

If $P \neq \emptyset$, then one can say that for the family $\Gamma$ on $S_0$ and for the vector $\alpha$ the modulus problem is defined and the modulus in this problem is given as

$$m(S_0, \Gamma, \alpha) = \inf_{\rho \in P} \iint_{S_0} \rho^2(\zeta) d\sigma_\zeta,$$

in the case when there is no curve in the admissible system homotopic to a puncture. In the case of a curve $\gamma_j$ from the admissible system that is homotopic to a puncture $a_j$, $j = m+1, \ldots, s$ we set $S(\varepsilon_{m+1}, \ldots, \varepsilon_s) = S_0 \setminus \{\{|\zeta - a_{m+1}| < \varepsilon_{m+1}\} \cup \cdots \cup \{|\zeta - a_s| < \varepsilon_s\}\}$ in terms of some parameters $\zeta$. In this case we require the limit

$$\lim_{(\varepsilon_{m+1}, \ldots, \varepsilon_s) \to (0, \ldots, 0)} \left( \iint_{S(\varepsilon_{m+1}, \ldots, \varepsilon_s)} \rho^2(\zeta) d\sigma_\zeta + \sum_{j=m+1}^{s} \frac{\alpha_j^2}{2\pi} \log \varepsilon_j \right)$$

to exist and to be finite. Then, the modulus is equal to the infimum of this limit as $\rho$ ranges over $P$. For this modulus problem the extremal metric is defined by the unique differential $\varphi(\zeta) d\zeta^2$ that has at most double poles at the punctures $a_{m+1}, \ldots a_s$ on the closure of the surface $S_0$ and at most simple poles in other punctures. This differential is the same as in the problem of the extremal partition.

The the case IV of admissible domains (strip domains and triangles) in the trajectory structure of the extremal differential is considered for the modulus problem in [81] and [139] . It makes use of more complicated definitions. In applications the results about the extremal partition are more useful so we abandon awkward cases of the modulus problems.

★ For all moduli in Section 2.4 construct relevant quadratic differentials which induce extremal metrics in the corresponding modulus problems.

★ Construct the local trajectory structure and the structure in the large for the following quadratic differentials on $\overline{\mathbb{C}}$

$$\frac{dz^2}{z(z^2 - 1)},$$

$$\frac{(z - 5)dz^2}{z^2(z - 1)(z - 2)},$$

$$\frac{(z - c)dz^2}{z^3}, \quad c \in \mathbb{C},$$

$$\frac{(z - c)dz^2}{(z - i)(z^2 - 1)}, \quad \text{for different } c : \operatorname{Re} c = 0.$$

★ For the following admissible systems of one curve $\gamma$ of type III, IV solve the problem of the extremal partition of the Riemann surface $S$. Construct a relevant quadratic differential and compare the result with Section 2.4:

$$S = U \setminus \{r\}, \ \gamma = \{z : |z - r| = \varepsilon\} \ \text{III type},$$

$$S = U \setminus \{0, r\}, \ \gamma = \{z : |z - r| = \varepsilon\} \ \text{III type},$$

$$S = U \setminus \{0, r\}, \ \gamma = \{z : |z| = \varepsilon\} \ \text{III type},$$

$$S = U \setminus \{0, r\}, \ \gamma = [0, r] \ \text{IV type},$$

$$S = U \setminus \{r\}, \ \gamma = \{z : |z| = r\} \ \text{IV type},$$

here $0 < r < 1$, $\varepsilon$ is sufficiently small.

### 2.7.3 Continuous and differentiable moduli

A study of the dependence of conformal invariants on the determining parameters started in 1966 by P. Tamrazov [145] and continued by A. Solynin [132], E. Emel'yanov [35], [37], and by the author [152], [155],[157], [170], namely, we speak about the continuous, differentiable, and harmonic properties respectively. The harmonic properties we will consider in Chapter 5; the continuous and differentiable properties we formulate in connection with the definitions of the above sections. For more general formulations see [35], [132], [139].

Here in this subsection we assume $S_0 = \overline{\mathbb{C}} \setminus \{\zeta_1^0, \ldots, \zeta_n^0, \zeta_{n+1}, \zeta_{n+2}, \zeta_{n+3}\}$, where all punctures are different. Since the modulus is a conformal invariant we can assume $\zeta_{n+1} = 0$, $\zeta_{n+2} = 1$, $\zeta_{n+3} = \infty$. Further we will consider more general surfaces and will discuss this problems from the point of view of the Teichmüller theory.

We use the notation $S = \mathbb{C} \setminus \{\zeta_1, \ldots, \zeta_n, 0, 1\}$. Let $\alpha^0 := (\alpha_1^0, \ldots, \alpha_m^0)$ be a fixed weight vector in Section 2.7.2, $G(\zeta_j^0)$ be a simply connected domain that contains the point $\zeta_j^0$, $G(\zeta_j^0) \cap G(\zeta_j^0) \neq \emptyset$, when $j \neq k$; $G := G(\zeta_1^0) \times \cdots \times G(\zeta_m^0) \times \mathbb{R}_+^m$; $\bar{x} := (\zeta_1, \ldots, \zeta_n, \alpha_1, \ldots, \alpha_m)$; $\bar{x}^0 := (\zeta_1^0, \ldots, \zeta_n^0, \alpha_1^0, \ldots, \alpha_m^0)$.

**Theorem 2.7.1** (Solynin [132]). *Let $\bar{x} \in G$ and $\Gamma^0$ be a free family of homotopy classes of curves on $S_0$ of types I-III. Then, $m(\bar{x}) := m(S, \Gamma, \alpha)$ is a continuous function with respect to $\bar{x}$. Here $\Gamma$ is a continuous deformation of $\Gamma^0$ and $S$ is a continuous deformation of $S_0$. Denote by $\varphi(\zeta, \bar{x})d\zeta^2$ the extremal differential for the modulus $m(S, \Gamma, \alpha)$. Then, $\varphi(\zeta, \bar{x})$ tends to $\varphi(\zeta, \bar{x}^0)$ locally uniformly in $\mathbb{C}$.*

Here and further on for convenience we will denote the local parameters on $S$ and $S_0$ by the same character.

**Theorem 2.7.2** (Solynin [132], Emel'yanov [35]). *Let $\bar{x} \in G$ be a vector with coordinates $(\zeta_1, \ldots, \zeta_n, \alpha_1, \ldots, \alpha_m)$ and $m(\bar{x})$ be defined as in Theorem 2.7.1. Then, for a fixed $(\alpha_1, \ldots, \alpha_m)$ the function $m(\bar{x})$ is differentiable with respect to $(\zeta_1, \ldots, \zeta_n)$ in some neighbourhood of $\bar{x}^0$ in $G$. For a fixed $(\zeta_1, \ldots, \zeta_n)$ the function $m(\bar{x})$ is differentiable with respect to $(\alpha_1, \ldots, \alpha_m)$ in some neighbourhood of $\bar{x}^0$ in $G$. Moreover, the following equalities are valid*

$$\frac{\partial m(\bar{x})}{\partial \zeta_j}\bigg|_{\bar{x}=\bar{x}^0} = \pi \text{ Res}_{\zeta=\zeta^0}\, \varphi(\zeta, \bar{x}), \quad j = 1, \ldots, n;$$

$$\frac{\partial m(\bar{x})}{\partial \alpha_j}\bigg|_{\bar{x}=\bar{x}^0} = \begin{cases} 2\alpha_j^0 M(D_j^*), & \text{for ring domains} \\ 2\alpha_j^0 m(D_j^*, a_j), & \text{for circular domains} \end{cases} \quad j = 1, \ldots, m.$$

The formulae of differentiation remain similar for the case of strip domains in the trajectory structure of the extremal differential.

Here $M(D_j^*)$ is the modulus of a ring domain, or a quadrangle $D_j^*$ in the trajectory structure of the differential $\varphi(\zeta, \bar{x}^0)d\zeta^2$. In particular, an important corollary follows from Theorem 2.7.2. Namely, the gradient of $m(\bar{x})$ by $\zeta_j$ at the point $\zeta_j^0$ has the same tangent direction with the critical trajectory of the differential $\varphi(\zeta, \bar{x}^0)d\zeta^2$ starting from $\zeta_j^0$.

This remark is very convenient for the monotonic change of the modulus when we move the point $\zeta_j$ along a prescribed curve. This is a useful addition to the method of symmetrization.

*Example 2.7.1.* Consider the free family of curves $\Gamma$ consisting of the unique homotopy class of simple loops that separate the points $\{0, 1\}$ and $\{\zeta, \infty\}$ on the surface $S_0 = \mathbb{C}\backslash\{0, 1, \zeta\}$, $|\zeta| > 1$. All curves from $\Gamma$ are homotopic on $S_0$ to the slit $[0, 1]$. As a result of circular symmetrization with respect to the center 0 and the direction $\mathbb{R}_+$ we obtain that the modulus $m(S, \Gamma)$ attains its maximum on the circle $\{z : |z| = |\zeta|\}$ at the point $\zeta = -|\zeta|$. We can learn more detail information from Theorem 2.7.2 and from the remark thereafter. Namely, this modulus increases monotonically on $a$ in the circle $\{z : z = |\zeta|e^{ia}\}$, $a \in (0, \pi)$ and decreases monotonically in $a \in (\pi, 2\pi)$. This comes from the fact that the critical trajectory of the extremal differential starting from $\zeta$ has the direction into the half-plane $\text{Im } ze^{i\arg\zeta} > 0$. So the vector of the motion has an acute angle with the gradient of the modulus.

# 3. Moduli in Extremal Problems for Conformal Mapping

Denote by $S$ the class of all holomorphic univalent functions $f(z) = z + a_2 z^2 + \ldots$ in the unit disk $U$, and by $S_R$ its subclass of functions with real coefficients $a_n$. In this chapter we start with classical theorems on conformal map like the Koebe 1/4 theorem, growth, and distortion theorems. Then we prove two-point distortion theorems, i.e., we will find the range of the system of functionals $(|f(r_1)|, |f(r_2)|)$ in the classes $S$, $S_R$ of univalent functions. We continue, furthermore, with extremal problems for other classes of univalent functions such as bounded functions, Montel functions, functions with prescribed angular derivatives.

## 3.1 Classical extremal problems for univalent functions

Here we do not present deeply the history of results and theorems because all of them are well-known nowadays. One can find details in numerous surveys and monographs, e.g., [4], [31], [48], [64], [100], [110]. We start with the famous Koebe 1/4 theorem. The Koebe function which is extremal in many important problems is $k_\theta(z) = z(1 + ze^{i\theta})^{-2}$. The transformation $f(z) \to e^{i\theta} f(e^{-i\theta} z)$ is called the *rotation* of the function $f$.

### 3.1.1 Koebe set, growth, distortion

**Theorem 3.1.1** (Koebe Theorem). *The boundary of the Koebe set*

$$\mathfrak{K}_S = \bigcap_{f \in S} f(U)$$

*in the class $S$ is the circle $|w| = 1/4$. If*

$$\inf_{w \in \partial f(U)} |w| = 1/4$$

*for some $f$, then it is the Koebe function or its rotation.*

*Proof.* Suppose, contrary to our claim, that there exists such $w_0$ with $|w_0| \leq 1/4$, that does not belong to $f(U)$ where $f$ is not the Koebe function. Then, there is a curve $\gamma \not\subset f(U)$ that connects $w_0$ and $\infty$. So, the reduced modulus is given by $0 = m(U,0) = m(f(U),0) \leq m(\mathbb{C} \setminus \gamma, 0)$. Let $D^*$ be the result of circular symmetrization of the domain $\mathbb{C} \setminus \gamma$ with respect to the origin and to the direction defined by the positive real axis. Then, $m(\mathbb{C} \setminus \gamma, 0) \leq m(D^*, 0)$. Moreover, $D^* \subset k_0(D) = \mathbb{C} \setminus [1/4, \infty)$. Then, $m(D^*, 0) \leq 0$. Therefore, we have the equality sign in all previous inequalities what is possible only if $e^{i\alpha} f(U) = D^* = k_0(D)$ for some $\alpha$ or $f \equiv k_\theta$ for some $\theta$. This contradicts our assumption and finishes the proof. $\qquad\square$

**Theorem 3.1.2** (Growth Theorem). *If $f \in S$, then*

$$\frac{r}{(1+r)^2} \leq |f(z)| \leq \frac{r}{(1-r)^2} \qquad r := |z| \tag{3.1}$$

*with the equality sign only for the Koebe function or its rotation.*

*Proof.* Starting with the lower estimation we consider the domain $U_z = U \setminus [z, e^{i \arg z})$ with the reduced modulus

$$m(U_z, 0) = \frac{1}{2\pi} \log \frac{4r}{(1+r)^2},$$

(see Section 2.4). Suppose the contrary. Let $f$ be a function from the class $S$ with $|f(z)| < r(1+r)^{-2}$. Then, denote by $D^*$ the result of circular symmetrization of the domain $f(U_z)$ with respect to the origin and to the direction defined by the positive real axis. Then, $m(f(U_z), 0) \leq m(D^*, 0)$. Moreover, $D^* \subset k_0(e^{-i \arg z} U_z) = \{\mathbb{C} \setminus [r(1+r)^{-2}, \infty)\}$. Thus,

$$m(D^*, 0) \leq \frac{1}{2\pi} \log \frac{4r}{(1+r)^2}.$$

The contradiction obtained implies the lower estimate. The uniqueness of the extremal configuration under symmetrization leads to the uniqueness of the extremal function up to rotation. Therefore, we have the equality sign in all previous inequalities only if $e^{i\alpha} f(U_z) = D^* = \mathbb{C} \setminus [r(1+r)^{-2}, \infty)$ for some $\alpha$ or $f \equiv k_\theta$.

For the upper estimate we consider the digon $U_z' = U \setminus (-e^{i \arg z}, z]$ with two vertices with the same support 0 with the equal angles $\pi$ at them. This digon is conformal at the vertices and

$$m(U_z', 0, 0) = \frac{2}{\pi} \log \frac{4r}{(1-r)^2} \quad \text{(see Section 2.4)}.$$

Again, we suppose the contrary. Let $f$ be a function from the class $S$ with $|f(z)| > r(1-r)^{-2}$. Apart from the previous consideration we need here the more powerful tool of Section 2.7.2 that consists of the extremal partition of the Riemann surface $S_0 = \mathbb{C} \setminus \{0, f(z)\}$. The digon $f(U_z')$ is conformal and admissible in the problem of minimizing the reduced modulus in the family of all digons on $S_0$ with the angles $\pi$ at 0, conformal at 0, having the homotopy type defined by a curve $\gamma$ that connects on $S_0$ the origin with itself and homotopic on $S_0$ to the slit $[0, f(z)]$. The extremal digon is $\mathbb{C}$ slit along the ray $[f(z), \infty]$ passing through the origin.

So, we have the inequality

$$\frac{2}{\pi} \log 4|f(z)| \leq \frac{2}{\pi} \log \frac{4r}{(1-r)^2}$$

(we use the fact that $f'(0) = 1$). This is equivalent to the upper estimate in the theorem. The uniqueness of the extremal configuration leads to the uniqueness of the extremal map up to rotation. □

**Theorem 3.1.3** (Distortion Theorem). *If $f \in S$, then*

$$\frac{1-r}{1+r} \leq \left| \frac{zf'(z)}{f(z)} \right| \leq \frac{1+r}{1-r}, \quad r := |z|, \tag{3.2}$$

$$\frac{1-r}{(1+r)^3} \leq |f'(z)| \leq \frac{1+r}{(1-r)^3}, \tag{3.3}$$

*with the equality sign only for the Koebe function or its rotation.*

*Proof.* We start with the upper estimate in (3.2). Consider the domain[1] $U_z = U \setminus (-1, 0]$ with the reduced modulus

$$m(U_z, z) = \frac{1}{2\pi} \log \frac{4r(1-r)}{(1+r)}.$$

The domain $f(U_z)$ has the reduced modulus $m(U_z, 0) + \frac{1}{2\pi} \log |f'(z)|$. Here we can apply either symmetrization or the results about the extremal partition. The extremal configuration for the maximal reduced modulus in both cases is the domain $D = \mathbb{C} \setminus \{w = t \exp(i \arg f(z)) : t \in (-\infty, 0]\}$ and with the puncture at the point $f(z)$. It has the reduced modulus $m(D, f(z)) = \frac{1}{2\pi} \log 4|f(z)|$. This leads to the right-hand side inequality in (3.2).

For the lower estimate in (3.2) we consider the digon $U_z' = U \setminus [0, \exp(i \arg z))$ with two vertices over the same support $z$. It is conformal at $z$, have the angles $\pi$ there and

$$m(U_z', z, z) = \frac{2}{\pi} \log \frac{4r(1+r)}{1-r} \quad \text{(see Section 2.4)}.$$

The digon $f(U_z')$ with the vertices over $f(z)$ is also conformal and has the reduced modulus

---

[1] We use the same notation for different but analogous domains from theorem to theorem

$$m(f(U'_z), f(z), f(z)) = \frac{2}{\pi} \log \frac{4r(1+r)}{1-r} + \frac{2}{\pi} \log |f'(z)|.$$

This digon is admissible in the problem of minimizing the reduced modulus over all digons on $S_0 = \mathbb{C} \setminus \{0, f(z)\}$ conformal at $f(z)$ having the angles $\pi$ there and the same homotopy type defined by a curve $\gamma$ that connects on $S_0$ the point $f(z)$ with itself and homotopic on $S$ to the slit $[0, f(z)]$. The extremal digon $D$ is $\mathbb{C}$ slit along the ray $[0, \infty]$ that passes through the point $f(z)$. Its reduced modulus is $\frac{2}{\pi} \log 4|f(z)|$. This leads to the lower estimate in (3.2). The uniqueness of the extremal configuration leads to the uniqueness of the extremal map up to some rotation.

The result of the Growth Theorem with the possibility of the equality for the Koebe function leads to the estimate (3.3).                    □

The term "distortion" comes from the geometric interpretation of $|f'(z)|$ that means the infinitesimal magnification factor of arclength under the mapping $f$.

### 3.1.2 Lower boundary curve for the range of $(|f(z)|, |f'(z)|)$

The Growth and Distortion theorems yield that the point of $\mathbb{R}^2$ with the coordinates $(x, y)$, where $x = |f(z)|$, $y = |f'(z)|$, lies within the rectangle with the vertices

$$\left( \frac{r}{(1+r)^2}, \frac{1+r}{(1-r)^3} \right), \left( \frac{r}{(1-r)^2}, \frac{1+r}{(1-r)^3} \right),$$

$$\left( \frac{r}{(1+r)^2}, \frac{1-r}{(1+r)^3} \right), \left( \frac{r}{(1-r)^2}, \frac{1-r}{(1+r)^3} \right).$$

An explicit information about the location of the point $(x, y)$ is given by theorems about the range of the system of functionals $I(f) = (|f(z)|, |f'(z)|)$. A great contribution to solution of this problem was made by J. Jenkins [65], I. Alexandrov, S. Kopanev [5], V. Popov [114], V. Gutlyanskiĭ [56].

Due to rotation it suffices to consider the point $z = r \in (0, 1)$. The range of a continuous system of functionals in a connected compact class of functions ($S$ in our case) is connected and closed (see, e.g., [58]). Since the range of the functional $|f'(r)|$ for $f \in S$ with $|f(r)|$ fixed is also connected and closed, the range $\mathfrak{M}_f$ of $I(f)$ is simply connected and $\partial \mathfrak{M}_f$ is an arc consisting of two parts: $\Gamma^+$, the arc of $y = \max |f'(r)|$ as $f \in S$ with $|f(r)| = x$ fixed, and $\Gamma^-$, the arc of $y = \min |f'(r)|$ as $f \in S$ with $|f(r)| = x$ fixed. We determine here $\mathfrak{M}_f$ by the modulus method.

**Theorem 3.1.4.** *The boundary curve $\Gamma^-$ of the range $\mathfrak{M}_f$ of the system of functionals $I(f)$ in the class $S$ consists of the points*

$$\left( \frac{r}{1 - ur + r^2}, \frac{1 - r^2}{(1 - ur + r^2)^2} \right), \quad -2 \le u \le 2.$$

*Each point is given only by the function* $g(z) = z(1-uz+z^2)^{-1}$ *or its rotation.*

*Proof.* Since

$$\frac{r}{(1+r)^2} \leq \frac{r}{1-ur+r^2} \leq \frac{1}{(1-r)^2},$$

for $u \in [-2,2]$, we choose for $|f(r)| = x$ the unique $u = u_f = r + 1/r - 1/x$ such that $|g(r)| = |f(r)|$.

Consider the digon $U'_z = U \setminus \{(-1,0] \cup [r,1)\}$ with two vertices $0, r$. It is conformal at its vertices and have the equal angles $2\pi$. The reduced modulus is given as

$$m(U'_z, 0, r) = \frac{1}{2\pi} \log \frac{r^2}{1-r^2}.$$

The digon $f(U'_z)$ with the vertices $0, f(r)$ is conformal at them and has the reduced modulus

$$m(f(U'_z), 0, f(r)) = \frac{1}{2\pi} \log \frac{r^2}{1-r^2} |f'(r)|.$$

It is admissible in the problem of minimizing the reduced modulus over all digons of homotopy type defined by $\gamma = [0, f(r)]$ which are conformal at their vertices and have the angles $2\pi$. The extremal reduced modulus $\frac{1}{2\pi} \log |f(r)|^2$ is given by the digon obtained from $\mathbb{C}$ slitting along two rays starting from $0$ and $f(r)$ lying on the straight line in the opposite directions passing through these two points. Therefore, $|f'(r)| \geq \frac{1-r^2}{r^2} x^2$. The uniqueness of the extremal configuration $e^{i \arg f(r)} g(U')$ leads to the uniqueness of the extremal map up to rotation. $\qquad\square$

The upper boundary curve is much more complicated and its calculation falls into subsections below.

### 3.1.3 Special moduli

The moduli that we are going to calculate will be used in various extremal problems. Let $S_0 = \mathbb{C} \setminus \{0, 1\}$ be the punctured Riemann sphere. On $S_0$ we consider the admissible system $(\gamma_1, \gamma_2)$ of curves of type III where $\gamma_1 = \{w : |w| = 1/\varepsilon\}$ and $\gamma_2 = \{w : |w| = \varepsilon\}$. Here $\varepsilon$ is sufficiently small such that $1$ belongs to the doubly connected domain between $\gamma_1$ and $\gamma_2$ on $\mathbb{C}$. Let $\mathfrak{D}$ be a set of all pairs $(D_1, D_2)$ of simply connected domains of homotopy type $(\gamma_1, \gamma_2)$. Then, the problem of extremal partition of $S_0$ consists of maximizing the sum $\alpha_1^2 m(D_1, \infty) + \alpha_2^2 m(D_2, 0)$ over all $(D_1, D_2) \in \mathfrak{D}$. Without loss of generality we assume $\alpha_1 = 1$, $\alpha_2 = \alpha$ and the maximum of this sum we denote by $M(\alpha)$. From Section 2.7.2 it follows that there exists a unique pair $(D_1^*, D_2^*)$ which is extremal in this problem. $D_1^*$ and $D_2^*$ are the circular domains in the trajectory structure of the differential

$$\varphi(z)dz^2 = -A\frac{(z-c)dz^2}{z^2(z-1)}, \quad A > 0,$$

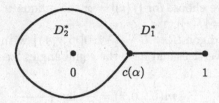

**Fig. 3.1.** The trajectory structure of the differential $\varphi(\zeta)d\zeta^2$

Here $A$ and $c$ are functions with respect to $\alpha$. If $\alpha = 1$, then the lengths of trajectories of the differential $\varphi$ from $D_1^*$ and $D_2^*$ are equal and $c(1) = 1$. In this case $M(1) = 0$ and $D_1^* = \{z : |z| > 1\}$, $D_2^* = U$. If $\alpha = 0$ or $\alpha \to \infty$, then the domains $D_2^*$ and $D_1^*$ are respectively degenerate and $M(0) = M(\infty) = (1/2\pi)\log 4$. In other cases we have the following theorem.

**Theorem 3.1.5.** *Let $0 < \alpha < \infty$, $\alpha \neq 1$. Then*

$$m(D_1^*, \infty) = \frac{1}{2\pi}\log\frac{4|1-\alpha|^{\alpha-1}}{|1+\alpha|^{\alpha+1}}, \tag{3.4}$$

$$m(D_2^*, 0) = \frac{1}{2\pi}\log\frac{4\alpha^2|1-\alpha|^{(1/\alpha)-1}}{|1+\alpha|^{(1/\alpha)+1}}. \tag{3.5}$$

*Proof.* We consider the map $u = u(z)$ whose inverse is

$$z = (c-1)\frac{1-\cos u}{(1-c) - (1+c)\cos u} + 1, \tag{3.6}$$

and obtain the representation of the differential $\varphi$ in terms of the parameter $u$ in regular points

$$\varphi(z)dz^2 = Q(u)du^2 = -A(c-1)^2c\frac{(1+\cos u)^2}{((1-c) - (1+c)\cos u)^2\cos^2 u}du^2. \tag{3.7}$$

Now we study the trajectory structure of this quadratic differential which is a complete square of a linear one. The differential $Q(u)du^2$ has the zeros of order 4 at the points $\pi + 2\pi k$ that are the images of $c$ under the map $u(z)$.

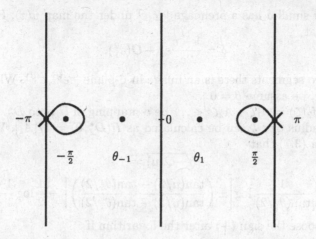

**Fig. 3.2.** The trajectory structure of the differential $Q(u)du^2$, $0 < \alpha < 1$

Further, $u(0) = \pi/2 + k\pi$ and $u(\infty) = \theta_k = \arccos(1-c)/(1+c)$ are the poles of the second order. The points $u(1) = 2\pi k$ are regular for this differential.

Let us consider a fixed branch of the function $u(z)$ that maps $\overline{\mathbb{C}} \setminus [c, 1]$ onto the strip $0 < \operatorname{Re} u < \pi$. For $0 < \alpha < 1$ the circular domain $D_2^u = u(D_2^*)$ is bounded by the critical trajectory of $Q(u)du^2$ starting and ending at $\pi$ enclosing the point $\pi/2$, $\theta_1 \in (0, \pi/2)$. The circular domain $D_2^u = u(D_2^*)$ is bounded by the same trajectory and the straight lines $\operatorname{Re} u = 0, \pi$. For $\alpha > 1$ the boundary of the domain $D_1^u$ is the critical trajectory of $Q(u)du^2$ starting and ending at $\pi$ enclosing the point $\theta_1$, $\theta_1 \in (\pi/2, \pi)$. The circular domain $D_2^u = u(D_2^*)$ is bounded by the same trajectory and the straight lines $\operatorname{Re} u = 0, \pi$.

Let $\zeta_j(u)$, $j = 0, 1$ be conformal maps of the domains $D_j^u$ onto the unit disk $U$ such that $\zeta_1(\pi/2) = \zeta_2(\theta_1) = 0$ in the case of $\alpha < 1$ or $\zeta_1(\pi/2) = \zeta_1(\theta_1) = 0$ in the case of $\alpha > 1$. Each function $\zeta_j$ satisfies in the domain $D_j^u$ the differential equation

$$\alpha_j \frac{d\zeta_j(u)}{\zeta_j(u)} = 2\pi\sqrt{-Q(u)}du, \quad \alpha_1 = 1, \ \alpha_2 = \alpha, \tag{3.8}$$

or in terms of the parameter $z$

$$\alpha_j^2 \left(\frac{d\zeta_j(u(z))}{\zeta_j(u(z))}\right)^2 = -4\pi^2 \varphi(z) dz^2. \tag{3.9}$$

Letting $z \to \infty$ in (3.9) in the case of $j = 1$, or $z \to 0$ in the case of $j = 2$ we obtain $A = 1/4\pi^2$ and $c = \alpha^2$.

To calculate the reduced moduli we first consider the case $\alpha < 1$. The part $[\pi/2 + \delta, \pi]$ of the orthogonal trajectory of the differential $Q(u)du^2$ for a sufficiently small $\delta$ has a preimage $[\varepsilon_1, c]$ under the map $u(z)$. From (3.6) we derive that

$$\delta = \frac{1-c}{2c}\varepsilon_1 + O(\varepsilon_1^2). \tag{3.10}$$

For these two segments there is an image in $\zeta$-plane $[\varepsilon e^{i\beta}, e^{i\beta}]$. Without loss of generality, we assume $\beta = 0$.

Let $z = f_2(\zeta) = A_1\zeta + A_2\zeta^2 + \ldots$ be a mapping of $U$ onto $D_2^*$. Then, the conformal radius of $D_2^*$ can be calculated as $R(D_2^*, 0) = 1/|A_1|$. We deduce directly from (3.7) that

$$\sqrt{-Q(u)} = \tag{3.11}$$

$$= \pm\frac{\alpha}{2\pi}\frac{d}{du}\left(\frac{1}{\tan(\theta_1/2)}\log\left[\pm\left(\frac{\tan(u/2) - \tan(\theta_1/2)}{\tan(u/2) + \tan(\theta_1/2)}\right)\right] + \frac{1}{2}\log\frac{1 + \sin u}{1 - \sin u}\right),$$

where we choose the sign $(+)$ after the logarithm if

$$(\tan(u/2) - \tan(\theta_1/2))/(\tan(u/2) + \tan(\theta_1/2)) > 0,$$

and $(-)$ otherwise. Moreover, $\tan(\theta_1/2) = \sqrt{c} = \alpha$. We fix the branch of the root in the left-hand side of (3.11) choosing $(-)$ after the equality sign. Integrating (3.8) along the segments described ($j = 2$), we derive that

$$\varepsilon = \left(\frac{\tan(\pi/4 + \delta/2) + \alpha}{\tan(\pi/4 + \delta/2) - \alpha}\right)^{1/\alpha} \cdot \left(\frac{1 - \cos\delta}{1 + \cos\delta}\right)^{1/2} = \left(\frac{1 + \alpha}{1 - \alpha}\right)^{1/\alpha}\frac{\delta}{2} + O(\delta^2),$$

and finally, using (3.10), we obtain

$$A_1 = \frac{4c}{1 - c}\left(\frac{1 - \alpha}{1 + \alpha}\right)^{1/\alpha}.$$

This is equivalent to (3.5) in the statement of the theorem for $\alpha < 1$.

Now we pass to the case $\alpha > 1$. For the segment $[0, \pi/2 - \delta]$ in the $u$-plane there is a preimage $[\varepsilon_1, 1]$ in the $z$-plane, and consequently, a preimage $[\varepsilon, 1]$ in the $\zeta$-plane. Integrating (3.8) along these segments, we obtain

$$\varepsilon = \left(\frac{\alpha - 1}{\alpha + 1}\right)^{1/\alpha}\frac{\delta}{2} + O(\delta^2),$$

and finally, using the relation $\delta = \frac{c-1}{2c}\varepsilon_1 + O(\varepsilon_1^2)$, we obtain

$$|A_1| = \frac{4c}{1 - c}\left(\frac{\alpha + 1}{\alpha - 1}\right)^{1/\alpha}.$$

This is equivalent to (3.5) in the statement of the theorem for $\alpha > 1$.

Now we prove (3.4). If $\alpha < 1$, then the segment $[0, \theta_1 - \delta]$ in the domain $D_1^u$ has the preimage $[0, 1/\varepsilon_1]$ in the $z$-plane, and consequently, the preimage

$[\varepsilon, 1]$ in the $\zeta$-plane. Let $z = f_1(\zeta) = B_{-1}/\zeta + B_0 + B_1\zeta + \ldots$ be a map of $U$ onto $D_1^*$. From (3.6) we derive

$$\delta = \sqrt{c}\frac{1-c}{1+c}\varepsilon_1 + O(\varepsilon_1^2). \tag{3.12}$$

Fixing the branch of the root in (3.12) with $(+)$ after the equality sign, and integrating (3.8) along the segments chosen, we have

$$\varepsilon = \left(\frac{1+\alpha}{1-\alpha}\right)^\alpha \frac{1+c}{4\sqrt{c}}\delta + O(\delta^2),$$

and finally, using (3.12) we obtain

$$\frac{1}{|B_{-1}|} = \frac{4}{1-\alpha^2}\left(\frac{1-\alpha}{1+\alpha}\right)^\alpha.$$

This is equivalent to (3.4) in the statement of the theorem for $\alpha < 1$.

For $\alpha > 1$ the segment $[\theta+\delta, \pi]$ in the $u$-plane has the preimage $[1/\varepsilon_1, c]$ in the $z$-plane, and consequently, the preimage $[\varepsilon, 1]$ in the $\zeta$-plane. Integrating (3.8) along these segments we obtain

$$\frac{1}{|B_{-1}|} = \frac{4}{\alpha^2-1}\left(\frac{\alpha-1}{\alpha+1}\right)^\alpha.$$

This is equivalent to (3.4) in the statement of the theorem for $\alpha > 1$ and completes the whole proof.    $\square$

Now we consider another modulus problem which is connected with the preceding one. Let $S_0 = \mathbb{C} \setminus \{0, c, 1\}$ be a punctured Riemann sphere, $c \in (0, 1)$. On $S_0$ we consider the admissible system $(\gamma_1, \gamma_2)$ of curves of type III where $\gamma_1 = \{w : |w| = 1/\varepsilon\}$ and $\gamma_2 = \{w : |w| = \varepsilon\}$. Here $\varepsilon$ is sufficiently small such that $c, 1$ belongs to the doubly connected domain between $\gamma_1$ and $\gamma_2$ on $\mathbb{C}$. Let $\mathfrak{B}$ be a set of all pairs $(B_1, B_2)$ of simply connected domains of homotopy type $(\gamma_1, \gamma_2)$. Then, the problem of the extremal partition of $S_0$ consists of maximizing the sum $\alpha_1^2 m(B_1, \infty) + \alpha_2^2 m(B_2, 0)$ over $(B_1, B_2) \in \mathfrak{B}$. Denote by $(B_1^*, B_2^*) \in \mathfrak{B}$ a unique extremal pair of domains in this problem. Without loss of generality, we assume $\alpha_1 = 1$, $\alpha_2 = \alpha$ and the maximum of this sum we denote by $\mathcal{M}(c, \alpha)$. For $\alpha \leq \sqrt{c}$ this problem is equivalent to the previous one and $\mathcal{M}(c, \alpha) = M(\alpha)$, $\alpha < 1$. For $\alpha \geq 1/\sqrt{c}$ this problem is also equivalent to the previous one with $\alpha > 1$ and $\mathcal{M}(c, \alpha) = M(\alpha)/\sqrt{c}$.

**Theorem 3.1.6.** *Let $\sqrt{c} < \alpha < 1/\sqrt{c}$. Then,*

$$m(B_1^*, \infty) = \frac{1}{2\pi}\log\frac{4(1-\sqrt{c})^{\alpha-1}}{(1+\sqrt{c})^{\alpha+1}}, \tag{3.13}$$

$$m(B_2^*, 0) = \frac{1}{2\pi}\log\frac{4c(1-\sqrt{c})^{(1/\alpha)-1}}{(1+\sqrt{c})^{(1/\alpha)+1}}. \tag{3.14}$$

*Proof.* The extremal pair $(B_1^*, B_2^*)$ consists of the circular domains $B_1^*$ and $B_2^*$ in the trajectory structure of the differential

$$\Phi(z)dz^2 = -A\frac{(z-b)^2dz^2}{z^2(z-1)(z-c)}.$$

By the analogy with the proof of Theorem 3.1.5 we deduce that $A = 1/4\pi^2$ and $b = \alpha\sqrt{c}$. We use the map of the form (3.6) and obtain the differential $\Phi$ in terms of the parameter $u$

$$\Phi(z)dz^2 = Q_1(u)du^2 = -A\frac{(bc-b+(b+bc-2c)\cos u)^2}{c((1-c)-(1+c)\cos u)^2\cos^2 u}du^2. \qquad (3.15)$$

Unlike the differential $Q(u)du^2$, the differential $Q_1(u)du^2$ has the zeros of the second order at the points

$$2\pi k \pm \arccos\left(\frac{b(1-c)}{b(1+c)-2c}\right).$$

Let $\zeta_j(u)$, $j = 0,1$ be conformal maps of the domains $B_j^u$ onto the unit disk $U$ such that $\zeta_1(\pi/2) = \zeta_2(\theta_1) = 0$. Let these functions satisfy the differential equations

$$\alpha_j^2\left(\frac{d\zeta_j(u(z))}{\zeta_j(u(z))}\right)^2 = -4\pi^2\Phi(z)dz^2, \quad \alpha_1 = 1, \alpha_2 = \alpha, \qquad (3.16)$$

in the corresponding domains $B_j^*$. Letting $z \to \infty$ in (3.16) in the case of $j = 1$ or $z \to 0$ in the case of $j = 2$ we obtain $A = 1/4\pi^2$ and $\alpha = b/\sqrt{c}$.

We calculate directly from (3.15)

$$\sqrt{-Q_1(u)} = \qquad (3.17)$$

$$= \pm\frac{\alpha}{2\pi}\left(\frac{1}{\alpha}\frac{d}{du}\left(\log\left[\pm\left(\frac{\tan(u/2)-\sqrt{c}}{\tan(u/2)+\sqrt{c}}\right)\right]\right)\right) + \frac{1}{2}\frac{d}{du}\log\frac{1+\sin u}{1-\sin u}\right),$$

where we choose the sign $(+)$ after the logarithm if

$$(\tan(u/2) - \sqrt{c})/(\tan(u/2) + \sqrt{c}) > 0,$$

and $(-)$ otherwise. Integrating the equation (3.16) along the segments by the analogy with Theorem 3.1.5, one easily derives (3.13) and (3.14). $\qquad\square$

### 3.1.4 Upper boundary curve for the range of $(|f(z)|, |f'(z)|)$

Now we use the moduli calculated in the preceding section to derive the upper boundary curve $\Gamma^+$ for the range of $I(f) = (|f'(z)|, |f(z)|)$ in the class $S$. Our proof is based on simultaneous consideration of two problems of the extremal partition of the punctured unit disk and of the punctured Riemann sphere.

Let $U_z = U \setminus \{0, r\}$ be a punctured unit disk. We consider on $U_z$ the admissible system of curves $(\gamma_1^z, \gamma_2^z)$ of type III where $\gamma_2^z = \{z : |z - r| = \varepsilon\}$ and $\gamma_1^z = \{z : |z| = \varepsilon\}$. Here $\varepsilon$ is sufficiently small such that $r + \varepsilon < 1$ and $\varepsilon < r/2$. Let $\mathfrak{D}^z$ be a set of all pairs $(D_1^z, D_2^z)$ of simply connected domains of homotopy type $(\gamma_1^z, \gamma_2^z)$. Then, the problem of the extremal partition of $U_z$ consists of maximizing the sum $m(D_1^z, 0) + \alpha^2 m(D_2^z, r)$ over all $(D_1^z, D_2^z) \in \mathfrak{D}^z$. The maximum of this sum we denote by $\mathcal{M}_z(\alpha, r)$. Under the map

$$Z(z) = 1 - \frac{r(1+z)^2}{z(1+r)^2}$$

two extremal domains $(D_1^{z*}, D_2^{z*})$ in the problem for $\mathcal{M}_z(\alpha, r)$ are transformed into two extremal domains $(B_1^*, B_2^*)$ in the problem for $\mathcal{M}(c, \alpha)$ with $c = (1 - r)^2/(1 + r)^2$. Taking into account the change of the reduced moduli under the conformal map $Z(z)$ we deduce that for $(1 - r)/(1 + r) \leq \alpha \leq (1 + r)/(1 - r)$ the relations

$$m(D_1^{z*}, 0) = \frac{1}{2\pi} \log r^\alpha, \quad m(D_2^{z*}, r) = \frac{1}{2\pi} \log r^{1/\alpha}(1 - r^2)$$

hold.

Let $\mathbb{C}_w = \mathbb{C} \setminus \{0, w_0\}$ be a punctured Riemann sphere. We consider on $\mathbb{C}_w$ the admissible system of curves $(\gamma_1^w, \gamma_2^w)$ of type III where $\gamma_1^w = \{w : |w| = \varepsilon\}$ and $\gamma_2^z = \{z : |w - w_0| = \varepsilon\}$. Here $\varepsilon$ is sufficiently small such that $\varepsilon < w_0/2$. Let $\mathfrak{D}^w$ be the set of all pairs $(D_1^w, D_2^w)$ of simply connected domains of homotopy type $(\gamma_1^w, \gamma_2^w)$. Then, the problem of the extremal partition of $\mathbb{C}_w$ consists of maximizing the sum $m(D_1^w, 0) + \alpha^2 m(D_2^w, w_0)$ over all $(D_1^w, D_2^w) \in \mathfrak{D}^w$. The maximum of this sum we denote by $M_w(\alpha)$. Under the map

$$W(w) = \frac{w_0 - w}{w_0 w}$$

two extremal domains $(D_1^{w*}, D_2^{w*})$ in the problem for $M_w(\alpha)$ are transformed into two extremal domains $(D_1^*, D_2^*)$ in the problem for $M(\alpha)$. Taking into account the change of the reduced moduli under the conformal map $W(w)$ we deduce that for $\alpha \neq 1$

$$m(D_1^{w*}, 0) = \frac{1}{2\pi} \log 4w_0 \frac{|1 - \alpha|^{\alpha-1}}{|1 + \alpha|^{\alpha+1}},$$

$$m(D_2^{w*}, w_0) = \frac{1}{2\pi} \log 4w_0 \alpha^2 \frac{|1 - \alpha|^{(1/\alpha)-1}}{|1 + \alpha|^{(1/\alpha)+1}}.$$

For $\alpha = 1$ we have

$$m(D_1^{w*}, 0) = m(D_2^{w*}, 0) = \frac{1}{2\pi} \log w_0.$$

We precede the next theorem by the following technical lemma omitting its simple proof.

**Lemma 3.1.1.** *For $(1-r)/(1+r) \leq \alpha \leq (1+r)/(1-r)$ the equation*

$$m(D_1^{w*}, 0) = m(D_1^{z*}, 0)$$

*defines the function $w_0(\alpha)$ which strictly increases with increasing $\alpha$ from $w_0 = r(1+r)^{-2}$ to $w_0 = r(1-r)^{-2}$.*

**Theorem 3.1.7.** *(i) The upper boundary curve $\Gamma^+$ of the range $\mathfrak{M}_f$ of the system of functionals $I(f) = (|f(z)|, |f'(z)|)$ in the class $S$ consists of the points $(x(\alpha), y(\alpha))$ for*

$$\alpha \in \left[\frac{1-r}{1+r}, \frac{1+r}{1-r}\right], \quad \alpha \neq 1,$$

*where*

$$x(\alpha) = \frac{1}{4} r^\alpha \frac{(1+\alpha)^{\alpha+1}}{|1-\alpha|^{\alpha-1}},$$

$$y(\alpha) = \alpha^2 r^{\alpha-1/\alpha} \left|\frac{1+\alpha}{1-\alpha}\right|^{\alpha-1/\alpha} \frac{1}{1-r^2}.$$

*If $\alpha = 1$, then $x(1) = r$, $y(1) = 1/(1-r^2)$.*

*(ii) Each point $(x(\alpha), y(\alpha))$ of $\Gamma^+$ is given by the unique function $F(z, \alpha)$ that satisfies the differential equation*

$$\varphi(z)dz^2 = \psi(w)dw^2, \quad \alpha \neq 1,$$

$$\varphi(z)dz^2 = \frac{(z-d)^2(z-\bar{d})^2}{z^2(z-r)^2(z-1/r)^2} dz^2,$$

$$\psi(z)dz^2 = \frac{w-C}{w^2(w-x(\alpha))^2} dw^2,$$

*where $|d| = 1$ and $d$ is one of two conjugated solutions of the equation*

$$\alpha \frac{1-r}{1+r} = 1 - \frac{r(1+d)^2}{d(1+r)^2},$$

$$C = C(\alpha) = \frac{x(\alpha)}{1-\alpha^2}.$$

*The function $F(z, \alpha)$ maps the unit disk onto the complex plane $\mathbb{C}$ minus the ray $[C(\alpha), +\infty)$ in the case $\alpha < 1$ or $(-\infty, C(\alpha)]$ in the case $\alpha > 1$, and two smooth arcs of the trajectories of the differential $\psi(w) dw^2$ emanating from $C(\alpha)$ symmetrically with respect to $\mathbb{R}$ so that the simply connected domain $F(U, \alpha)$ is of zero reduced modulus with respect to the origin.*
*If $\alpha = 1$, then*

$$\varphi(z)dz^2 = \frac{(z-d)^2(z-\bar{d})^2}{z^2(z-r)^2(z-1/r)^2}dz^2,$$

$$\psi(z)dz^2 = \frac{-1}{w^2(w-x(\alpha))^2}dw^2,$$

and the function $F(z,1)$ maps the unit disk onto the complex plane $\mathbb{C}$ minus two symmetric rays with respect to $\mathbb{R}$ along the straight line $\operatorname{Re} w = r/2$ with the analogous normalization.

Proof. Let $f \in S$ with a fixed value of $|f(r)| = x$. The previous lemma states that there exists such $\alpha$, that $w_0(\alpha) = x$. We consider the functions $f_1(z)$ and $f_2(z)$ satisfying the equations

$$\alpha^2 \left(\frac{df_1(z)}{f_1(z)}\right)^2 = 4\pi^2 \varphi(z)dz^2, \quad \alpha^2 \left(\frac{df_2(w)}{f_2(w)}\right)^2 = 4\pi^2 \psi(w)dw^2,$$

where the differentials $\varphi(z)dz^2$ and $\psi(w)dw^2$ are defined in the statement of the theorem and $\alpha$ is chosen. The superposition $f_2^{-1} \circ f_1(z)$ maps conformally the domain $D_1^{z*}$ onto the domain $D_1^{w*}$. Continuing this map analytically into $D_2^{z*}$ we obtain the function $F(z,\alpha)$ that maps the unit disk onto the domain which is admissible with respect to the differential $\psi(w)dw^2$. After the normalization $F'(0,\alpha) = 1$ the function $F(z,\alpha)$ satisfies all conditions of the theorem. Since the pair $(D_1^{w*}, D_2^{w*})$ is extremal in the family $\mathfrak{D}^w$, we have the following chain of inequalities

$$m(f(D_1^{z*}),0) + \alpha^2 m(f(D_2^{z*}),f(r)) =$$

$$= m(D_1^{z*},0) + \alpha^2 m(D_2^{z*},r) + \frac{\alpha^2}{2\pi}\log|f'(r)| \le$$

$$\le m(D_1^{w*},0) + \alpha^2 m(D_2^{w*},x) =$$

$$= m(D_1^{z*},0) + \alpha^2 m(D_2^{z*},r) + \frac{\alpha^2}{2\pi}\log|F'(r,\alpha)|.$$

Then, $|f'(r)| \le |F'(r,\alpha)| = y(\alpha)$. The uniqueness of the extremal configuration implies the uniqueness of the extremal function. Rotation finishes the proof.  □

The boundary curves $\Gamma^+$ and $\Gamma^-$ have been obtained by J. Jenkins [65] applying his General Coefficient Theorem and by I. Alexandrov, S. Kopanev [5] using the parametric method without assertion on the uniqueness of the extremal functions. Later on, V. Goryaĭnov [50] proved the uniqueness of the extremal functions developing the Löwner-Kufarev method.

Corollary 3.1.1 (Jenkins [64], Gung San [55]). Let $f \in S$ and $|z| = r < 1$. Then, the sharp estimates

$$|f'(z)| \le \frac{1+r}{(1-r)^3}\left|\frac{(1-r)^2 f(z)}{r}\right|^\mu,$$

where $\mu = 2(1 + r^2)(1 + r)^{-2}$, and

$$|f'(z)| \leq \frac{1-r}{(1+r)^3} \left| \frac{(1+r)^2 f(z)}{r} \right|^\nu,$$

where $\nu = 2(1 + r^2)(1 - r)^{-2}$, are valid. The extremal functions are rotation of the Koebe function in both cases.

## 3.2 Two-point distortion for univalent functions

In this paragraph we study the sets of values of the system of functionals $I(f) = (|f(r_1)|, |f(r_2)|), 0 < r_1 < r_2 < 1$ in the classes $S_R$ and $S$. Descriptions of various systems of functionals dependent on the values of a function from the class $S$ defined at two different non-zero points from $U$ have been studied by several analysts. J. Jenkins [61], [64] has proved that the maximum of $|f(r_2)|$ in the class $S$ for a given value of $|f(-r_1)|$ is provided by a function $f \in S_R$. He found this upper boundary curve for the system of functionals $(|f(-r_1)|, |f(r_2)|)$ in the class $S_R$. As a corollary, he derived the upper bound of the sum $|f(-r_1 e^{i\theta})| + |f(r_2 e^{i\theta})|, f \in S$. Without the assertion on the uniqueness of the extremal functions the same result had been obtained earlier by G. Goluzin [47], [48]. We note that from Jenkins' result and method it is impossible to deduce the range of $I(f)$ in our problem. J. Krzyż [76] found the range of values of the complex valued functional $f(z_1)/f(z_2), f \in S$, $z_1, z_2 \in U$ by the variational method.

### 3.2.1 Lower boundary curve for the range of $(|f(r_1)|, |f(r_2)|)$ in $S_R$

The range $\mathfrak{N}_f$ of $I(f)$ is a simply connected closed set for $f \in S$, and $\partial \mathfrak{N}_f$ is an arc consisting of two parts $\Gamma^+$ which is the arc of $y = \max |f(r_2)|$ over $f \in S$ with $|f(r_1)| = x$ fixed, and $\Gamma^-$ which is the arc of $y = \min |f(r_2)|$ over $f \in S$ with $|f(r_1)| = x$ fixed. Denote the range of $I(f)$ in the class $S_R$ by $\mathfrak{N}_f^R$ and its boundary curves by $\Gamma_R^+$ and $\Gamma_R^-$ respectively.

**Theorem 3.2.1.** The boundary curve $\Gamma_R^-$ of the range $\mathfrak{N}_f^R$ of the system of functionals $I(f)$ in the class $S_R$ consists of the points

$$\left( \frac{r_1}{1 - ur_1 + r_1^2}, \frac{r_2}{1 - ur_2 + r_2^2} \right), \quad -2 \leq u \leq 2.$$

Each point is given only by the function $g(z) = z(1 - uz + z^2)^{-1}$.

*Proof.* The Growth Theorem is the same in the classes $S$ and $S_R$. Therefore,

$$\min_{f \in S_R} |f(z)| = \frac{r}{(1+r)^2} \leq \frac{r}{1 - ur + r^2} \leq \frac{r}{(1-r)^2} = \max_{f \in S_R} |f(z)|$$

for $u \in [-2, 2]$. We choose a unique $u = u_f = r_1 + 1/r_1 - 1/x$ for a fixed $|f(r_1)| = x$ such that $g(r_1) = |f(r_1)|$.

Let us consider the digon $U'_z = U \setminus \{(-1, r_1] \cup [r_2, 1)\}$ with two vertices over 0 which is conformal at them having the equal angles $\pi$. The reduced modulus is given as

$$m(U'_z, 0, 0) = \frac{2}{\pi} \log \frac{4r_1 r_2}{(r_2 - r_1)(1 - r_1 r_2)}.$$

The digon $f(U'_z)$ with the vertices over 0 is also conformal at them and has the same reduced modulus. It is admissible in the problem of minimizing the reduced modulus over all digons conformal at 0, of the homotopy type defined by the curve $\gamma$, having the angles $\pi$. The curve $\gamma$ has two endpoints at 0 and encloses the point $f(r_1)$ having the point $f(r_2)$ in its exterior part. The extremal reduced modulus is given by the digon obtained from $\mathbb{C}$ by deleting two rays $(-\infty, f(r_1)]$ and $[f(r_2), \infty)$. Since the reduced modulus decreases as the domain extends, the inequality $f(r_2) \geq g(r_2)$ holds with the chosen value of $u$. The uniqueness of the extremal configuration $g(U')$ leads to the uniqueness of the extremal map.                    □

### 3.2.2 Special moduli

Let $A, B, C$ be different punctures of $\mathbb{C}$. Let $(\gamma_1, \gamma_2^{(k)} \big|_{k=1}^{\infty})$ be a countable set of admissible systems of curves on $S_0 = \mathbb{C} \setminus \{A, B, C\}$. Set $\gamma_1 = \{z : |z| = 1/\varepsilon\}$ for a sufficiently small $\varepsilon$ such that $\gamma_1$ is of type III and separates the puncture $\infty$ from all others, $\gamma_2^{(k)}$ is from the countable set of possible non-homotopic on $S_0$ closed curves of type I that separate the punctures $A$ and $B$ from the punctures $C, \infty$. We assume that the curve $\gamma_2^{(1)}$ is homotopic to the slit along $[A, B]$ if $C \notin [A, B]$, or else to the broken line connecting the points $A$, $C + \varepsilon e^{i(\pi/2 + \arg(C-A))}$, $B$, otherwise. Let $\mathfrak{D}^{(k)}$ be the family of all pairs $(D_1, D_2^{(k)})$ of non-overlapping domains of homotopy type $(\gamma_1, \gamma_2^{(k)})$, $D_1$ is a hyperbolic simply connected domain, $\infty \in D_1$, $D_2^{(k)}$ is a hyperbolic doubly connected domain. Degeneracy of the domain $D_2^{(k)}$ is possible in the sense that its boundary components coincide. As before, we denote by $m(D_1, \infty)$ the reduced modulus of $D_1$ with respect to $\infty$ and by $M(D_2^{(k)})$ the modulus of the domain $D_2^{(k)}$ with respect to the family of closed curves that separate its boundary components. If the domain $D_2^{(k)}$ degenerates, then $M(D_2^{(k)})$ vanishes. Consider the problem of maximizing the sum

$$m(D_1, \infty) + \alpha^2 M(D_2^{(k)}), \tag{3.18}$$

for a non-negative $\alpha$ over all $(D_1, D_2^{(k)}) \in \mathfrak{D}^{(k)}$. Denote by $M^{(k)}(\alpha, A, B, C)$ the maximum of the sum (3.18). For $\alpha = 0$ the problem of finding

$M^{(k)}(0, A, B, C)$ is the same as the problem of determination of the continuum of minimal capacity that contains the points $A, B, C$. This problem has been solved by G. Kuz'mina ([79], Theorem 1.6). For $\alpha \to \infty$ the problem is the same as finding the maximum of the modulus of doubly connected domain over all such domains on $S_0$ of homotopy type $\gamma_2^{(k)}$. The value of $M^{(k)}(\alpha, A, B, C)$ attains its maximum for the unique pair of domains $(D_1^*, D_2^{(k)^*}) \in \mathfrak{D}^{(k)}$ which are the circular and the ring domain in the trajectory structure of the quadratic differential

$$\varphi(z) \, dz^2 = A \frac{z - G}{(z - A)(z - B)(z - C)} dz^2. \qquad (3.19)$$

The domain $D_2^{(k)^*}$ is degenerate otherwise.

Let $\zeta_1(z)$ be a univalent conformal map of the domain $D_1^*$ onto the disk $|\zeta_1| < R(D_1^*, \infty)$, such that $\zeta_1(\infty) = 0$, $z \cdot \zeta_1(z)\big|_{z \to \infty} = 1$; $\zeta_2(z)$ be a univalent conformal map of the domain $D_2^{(k)^*}$ onto the annulus $1 < |\zeta_2| < M$, such that $\zeta_2(B) = 1$. These functions satisfy in the domains $D_1^*, D_2^{(k)^*}$ the differential equations

$$\alpha_j^2 \left( \frac{d\zeta_j(z)}{\zeta_j(z)} \right)^2 = -4\pi^2 \varphi(z) dz^2, \quad \alpha_1 = 1, \quad \alpha_2 = \alpha \qquad (3.20)$$

respectively.

The $\varphi$-lengths of the trajectories in the non-degenerate domains $D_1^*$ and $D_2^{(k)^*}$ are equal to 1 and $\alpha$ respectively. These conditions define a simple zero $G$. Taking the limit $z \to \infty$ in the first equation in (3.20), we obtain $A = -1/4\pi^2$.

**Lemma 3.2.1.** *In the family $\mathfrak{D}^{(k)}$ the inequality*

$$M^{(k)}(\alpha, A, B, C) < M^{(1)}(\alpha, A, B, C)$$

*holds for all $\alpha$ and $k = 2, 3, \ldots$.*

*Proof.* Without loss of generality, in the proof of this lemma we assume $\operatorname{Im} A = \operatorname{Im} B = 0$, $\operatorname{Im} C \geq 0$. We apply polarization with respect to the real axis to the extremal circular domain $D_1^*$ and the ring domain $D_2^{(k)^*}$ for $k \geq 2$. As a result, we obtain a pair of non-overlapping domains $D_1^0, D_2^0$ which are admissible with respect to the problem associated with $M^{(1)}(\alpha, A, B, C)$. Moreover, $m(D_1^*, \infty) \leq m(D_1^0, \infty)$ whenever $D_1 \neq D_1^0$, and $M(D_2^{(k)^*}) < M(D_2^0, \infty)$ whenever $D_2 \neq D_2^0$. Thus, the statement of Lemma 3.2.1 follows from the extremality of the pair of the domains $D_1^*, D_2^{(1)^*}$ for $M^{(1)}(\alpha, A, B, C)$. $\qquad \square$

**Lemma 3.2.2.** *Let* $0 < A < B$, $0 \leq \theta \leq \pi$. *The value of* $M^{(1)}(\alpha, A, Be^{i\theta}, 0)$ *strictly decreases with increasing* $\theta \in [0, \pi]$.

*Proof.* Let $0 < \alpha < 1$. The trajectories of the differential $\varphi(z)dz^2$ are piecewise analytic. By the definition of polarization with respect to the real axis, we have two domains $D_1^0$, $D_2^0$ as in the proof of Lemma 3.2.1 for $k = 1$ which are admissible for $M^{(1)}(\alpha, A, Be^{i\theta}, 0)$. But, on the other hand, $m(D_1^*, \infty) \leq m(D_1^0, \infty)$ and $M(D_2^{(1)^*}) \leq M(D_2^0, \infty)$. So, this is possible only when the polarized domains coincide with the initial ones. In this case, the trajectory of the differential $\varphi(z)dz^2$ connecting $G$ and $0$, $\theta \in (0, \pi)$ lies in the upper half-plane. The analogous application of polarization with respect to the lines $(A, Be^{i\theta})$ and $(0, Be^{i\theta})$ yields that the zero $G$ lies within the triangle with the vertices at $0, A, Be^{i\theta}$.

The same application of polarization in the case $\alpha > 1$ implies that the zero $G$ lies in the part of the angle between the lines $(0, Be^{i\theta})$ and $(0, A)$ in the lower half-plane.

In both cases the trajectory connecting the points $A, Be^{i\theta}$ lies within the part of the angle between the lines $(0, Be^{i\theta})$ and $(0, A)$ in the upper half-plane. It follows from Theorem 2.7.2 that the gradient $\nabla M^{(1)}(\alpha, A, Be^{i\theta}, 0)$, taken with respect to the third variable at the point $Be^{i\theta}$, is co-directed with the trajectory of the differential $\varphi(\zeta)d\zeta^2$ starting at this point. Thus, the scalar product of this vector and the vector of the motion of the point $Be^{i\theta}$ is negative. This finishes the proof. $\qquad\square$

We rewrite the result by G. Kuz'mina [79] and by A. Vasil'ev, S. Fedorov [163] for the case that we will use in subsequent applications. Set

$$\mu_1(a) = 1 - \frac{1}{\pi} \arccos \frac{3 - a}{a + 1}.$$

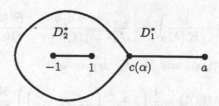

**Fig. 3.3.** The trajectory structure of the extremal differential $\varphi(\zeta)d\zeta^2$ in Theorem 3.2.2

**Theorem 3.2.2.** *Let $\alpha > 0$, $a \geq 1$. For the maximum of the sum (3.18) over the family $\mathfrak{D}^{(1)}$, $C = a$, $A = -1$, $B = 1$ the equality*

$$M^{(1)}(\alpha, -1, 1, a) = m(D_1^*, \alpha) + \alpha^2 M(D_2^*),$$

*holds for the circular domain $D_1^*$ and the ring domain $D_2^*$ in the trajectory structure of the differential*

$$\varphi(z)dz^2 = -\frac{1}{4\pi^2} \frac{z - c}{(z - a)(z^2 - 1)} dz^2.$$

*Each trajectory in $D_j^*$, $j = 1, 2$ of the differential $\varphi(z)dz^2$ in the metric $\sqrt{|\varphi(z)|}|dz|$ has the length $\alpha_j$, $\alpha_1 = 1$, $\alpha_2 = \alpha$, or else $D_1^* = \overline{\mathbb{C}} \setminus [-1, a]$, $D_2^* = \emptyset$. If $0 \leq \alpha \leq \mu_1(a)$, then we have the degenerating case. For $\mu_1(a) < \alpha < 1$ the zero $c = c(a, \alpha)$ is defined by the system*

$$\pi\left(\alpha - \frac{\omega}{\mathbf{K}(k)}\right) = -\left[\frac{\sqrt{p}(a - c)k}{\sqrt{a^2 - 1}(1 + p)} - \frac{\Theta'(\omega, k)}{\Theta(\omega, k)}\right] 2\mathbf{K}'(k),$$

$$\mathbf{dn}\,\omega = \frac{1 - p}{1 + p}, \quad p = \sqrt{\frac{c^2 - 1}{a^2 - 1}},$$

$$k = \sqrt{\frac{2p(a^2 - 1)}{p(a^2 - 1) - 1 + ac}}.$$

*In this case $m(D_1^*, \infty)$ and $M(D_2^*)$ appear as*

$$M(D_2^*) = \frac{1}{4}\left(1 - \frac{1}{\alpha}\frac{\omega}{\mathbf{K}(k)}\right)\frac{\mathbf{K}(k)}{\mathbf{K}'(k)},$$

$$m(D_1^*, \infty) = \frac{1}{2\pi} \log \frac{4k(a - c)\Theta^2(\omega, k)}{(a^2 - 1)(1 + p)^2 \Theta^2(0, k)} - \frac{1}{4}\left(\alpha - \frac{\omega}{\mathbf{K}(k)}\right)\frac{\omega}{\mathbf{K}'(k)}.$$

*Assume now $\alpha > 1$. In this case, the zero $c = c(a, \alpha)$ is defined by the equation*

$$\pi\left(\alpha - 2 + \frac{\omega}{\mathbf{K}(k)}\right) = \left[\frac{\sqrt{p}(c - a)k}{\sqrt{a^2 - 1}(1 + p)} - \frac{\Theta'(\omega, k)}{\Theta(\omega, k)}\right] 2\mathbf{K}'(k).$$

*In this case $m(D_1^*, \infty)$ and $M(D_2^*)$ appear as*

$$M(D_2^*) = \frac{1}{4}\left(1 + \frac{1}{\alpha}\left(\frac{\omega}{\mathbf{K}(k)} - 2\right)\right)\frac{\mathbf{K}(k)}{\mathbf{K}'(k)},$$

$$m(D_1^*, \infty) = \frac{1}{2\pi} \log \frac{4k(c - a)\Theta^2(\omega, k)}{(a^2 - 1)(1 + p)^2 \Theta^2(0, k)} +$$

$$+ \frac{1}{4}\left(\alpha - 2 + \frac{\omega}{\mathbf{K}(k)}\right)\frac{\omega - 2\mathbf{K}(k)}{\mathbf{K}'(k)}.$$

*For $\alpha = 1$ we have $c = a$,*

$$M(D_2^*) = \frac{1}{2\pi} \log \frac{1}{a + \sqrt{a^2 - 1}}, \quad m(D_1^*, \infty) = \frac{1}{2\pi} \log (a + \sqrt{a^2 - 1}).$$

Here we use the elliptic functions $\mathbf{dn}\, \omega$, $\mathbf{cn}\, \omega$, $\mathbf{sn}\, \omega$, Jacobi's $\theta$-functions $\Theta(\omega, k) = \vartheta_0(\omega, k)$ and the formulae for their derivatives which are defined in Section 2.3. The proof of this theorem is the same as in [79] and that for Theorem 3.2.3. So we do not attach it here. One can learn it from Theorem 3.2.3.

*Remark 3.2.1.* For the problem on $M^{(1)}(\alpha, -1, 1, a)$ with $\alpha > \mu_1(a)$ we have the following formulae of differentiation

$$\frac{\partial\, c(a, \alpha)}{\partial\, \alpha} = \frac{\pi \sqrt{p(a^2 - 1)}}{2k\mathbf{K}'(k)}, \quad \frac{\partial\, c(a, \alpha)}{\partial\, a} = -\frac{2p}{k^2(a^2 - 1)} \left( \frac{\mathbf{E}'(k)}{\mathbf{K}'(k)} - 1 \right),$$

$$\frac{\partial}{\partial\, \alpha} (\alpha M(D_2^*)) = \frac{1}{4} \frac{\mathbf{K}(k)}{\mathbf{K}'(k)}, \quad \frac{\partial}{\partial\, a} (\alpha M(D_2^*)) = \frac{\sqrt{p}}{4k\mathbf{K}'(k)\sqrt{a^2 - 1}}.$$

*Remark 3.2.2.* Let $a$ be a fixed number $1 < a < \infty$, a function $\alpha = \alpha(a, c)$ be the inverse to $c = c(a, \alpha)$. The function $\alpha(a, c)$ is defined by the conditions of Theorem 3.2.2 and continuous in $c \in (1, \infty)$. Moreover, $\alpha(a, c) \to \mu_1(a)$ as $c \to 1$. The function $\alpha(a, c)$, as a function with respect to $c$, strictly increases in $c \in (1, \infty)$.

Assume that $-1, 1, c_1, a_1$ are different punctures of $\mathbb{C}$, $1 < c_1 < a_1 < \infty$. Let $(\gamma_1', \gamma_2')$ be an admissible system of curves on $S_1 = \mathbb{C} \setminus \{-1, 1, c_1, a_1\}$; $\gamma_1' = \{z : |z| = 1/\varepsilon\}$ for a sufficiently small $\varepsilon$ such that $\gamma_1'$ is of type III and separates the puncture $\infty$ from all others; $\gamma_2'$ is a closed curve of type I and separates the punctures $-1, 1$ from the punctures $c_1, a_1, \infty$. Assume that the curve $\gamma_2'$ is homotopic to the slit along $[-1, 1]$. Let $\mathfrak{B}$ be a family of all pairs $(B_1, B_2)$ of non-overlapping domains of homotopic type $(\gamma_1', \gamma_2')$. $B_1$ is a hyperbolic simply connected domain, $\infty \in B_1$, $B_2$ is a hyperbolic doubly connected domain. The domain $B_2$ can be possibly degenerate in such sense that its boundary components coincide. As before, we denote by $m(B_1, \infty)$ the reduced modulus of $B_1$ with respect to $\infty$ and by $M(B_2)$ the modulus of the domain $B_2$ with respect to the family of simple loops that separate its boundary components. If the domain $B_2$ degenerates, then $M(B_2)$ vanishes. Consider the problem of maximizing the sum

$$m(B_1, \infty) + \alpha^2 M(B_2), \tag{3.21}$$

for a non-negative $\alpha$ over all $(B_1, B_2) \in \mathfrak{B}$. Denote by $\mathcal{M}(\alpha, -1, 1, c_1, a_1)$ the maximum of the sum (3.21). For $\alpha = 0$ the problem of finding $\mathcal{M}(0, -1, 1, c_1, a_1)$ is the same as the problem of determination of the continuum of minimal capacity containing the points $-1, 1, c_1, a_1$, which in this case is the segment $[-1, a_1]$. So, $\mathcal{M}(0, -1, 1, c_1, a_1) = \frac{1}{2\pi} \log \frac{4}{1+a_1}$ For $\alpha \to \infty$ the problem is the same as finding the maximum of the modulus of a doubly connected domain over all such domains on $S_1$ of homotopy type $\gamma_2'$.

Set

$$\mu_2(a_1, c_1) = \frac{1}{\pi} \int_{-1}^{1} \sqrt{\frac{x - c_1}{(x^2 - 1)(a_1 - x)}} dx.$$

Then, for $0 \le \alpha \le \mu_1(a_1)$ we have the degenerating case. For $\mu_1(a_1) < \alpha \le \mu_2(a_1, c_1)$ the analytic expressions for $m(B_1, \infty)$, and $M(B_2)$ are provided by Theorem 3.2.2 for the points $-1, 1, a = a_1$ as $\mu_1(a) < \alpha < 1$. For $\alpha \ge \mu_2(c_1, a_1)$ the analytic expressions for $m(B_1, \infty)$ and $M(B_2)$ are provided by Theorem 3.2.2 for the points $-1, 1, a = c_1$ as $\alpha > 1$. Otherwise, the following theorem is valid.

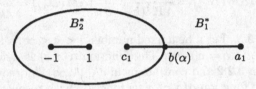

**Fig. 3.4.** The trajectory structure of the extremal differential $\varphi(\zeta)d\zeta^2$ in Theorem 3.2.3

**Theorem 3.2.3.** *Let $\mu_2(a_1, c_1) < \alpha < \mu_2(c_1, a_1)$, $1 < c_1 < a_1 < \infty$. The maximum of the sum (3.21) over the family $\mathfrak{B}$*

$$M(\alpha, -1, 1, c_1, a_1) = m(B_1^*, \alpha) + \alpha^2 M(B_2^*),$$

*is attained for the circular domain $B_1^*$ and the ring domain $B_2^*$ in the trajectory structure of the differential*

$$\varphi(z)dz^2 = -\frac{1}{4\pi^2} \frac{(z - b)^2}{(z - c_1)(z - a_1)(z^2 - 1)} dz^2.$$

*Each trajectory in $B_j^*$, $j = 1, 2$ of the differential $\varphi(z)dz^2$ in the metric $\sqrt{|\varphi(z)|}|dz|$ has the length $\alpha_j$, $\alpha_1 = 1$, $\alpha_2 = \alpha$. The zero $b = b(a_1, c_1, \alpha)$, $c_1 < b < a_1$ is defined by the system*

$$\pi\left(\alpha - \frac{\omega_1}{\mathbf{K}(k_1)}\right) = -\left[\frac{k_1(c_1 - b + p_1(a_1 - b))}{\sqrt{p_1(a_1^2 - 1)(1 + p_1)}} - \frac{\Theta'(\omega_1, k_1)}{\Theta(\omega_1, k_1)}\right] 2\mathbf{K}'(k_1),$$

$$\mathbf{dn}\,\omega_1 = \frac{1 - p_1}{1 + p_1}, \quad p_1 = \sqrt{\frac{c_1^2 - 1}{a_1^2 - 1}},$$

$$k_1 = \sqrt{\frac{2p_1(a_1^2 - 1)}{p_1(a_1^2 - 1) - 1 + a_1 c_1}}.$$

In this case $m(D_1^*, \infty)$ and $M(D_2^*)$ appear as

$$M(D_2^*) = \frac{1}{4}\left(1 - \frac{1}{\alpha}\frac{\omega_1}{\mathbf{K}(k_1)}\right)\frac{\mathbf{K}(k_1)}{\mathbf{K}'(k_1)},$$

$$m(D_1^*, \infty) = \frac{1}{2\pi}\log\frac{4k_1(a_1 - c_1)\Theta^2(\omega_1, k_1)}{(a_1^2 - 1)(1 + p_1)^2\Theta^2(0, k_1)} - \frac{1}{4}\left(\alpha - \frac{\omega_1}{\mathbf{K}(k_1)}\right)\frac{\omega_1}{\mathbf{K}'(k_1)}.$$

*Proof.* We consider the map $u = u(z)$ whose inverse is

$$z - a_1 = (c_1 - a_1)\frac{1 - \mathbf{cn}\,u}{(1 - p_1) - (1 + p_1)\mathbf{cn}\,u} + 1$$

and derive the representation of the differential $\varphi$ in terms of the parameter $u$ in regular points

$$\varphi(z)dz^2 = Q(u)du^2 = \frac{1}{p_1(1 - a_1^2)}\times$$

$$\times\left(\frac{(b - a_1)(1 - p_1) + a_1 - c_1 - ((b - a_1)(1 + p_1) + a_1 - c_1)\mathbf{cn}\,u}{(1 - p_1 - (1 + p_1)\mathbf{cn}\,u)}\right)^2 du^2.$$

Let $\alpha \in [\mu_2(a_1, c_1), \mu_2(c_1, a_1)]$. We study the trajectory structure of the

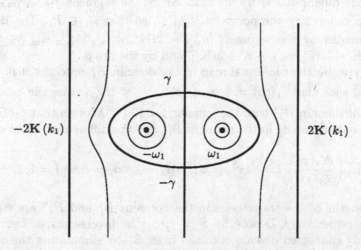

**Fig. 3.5.** The trajectory structure of the differential $Q(u)du^2$

quadratic differential $Q(u)du^2$ which is a complete square of a linear one.

Set

$$u_{m,n} = 2(2m+n)\mathbf{K}(k_1) + 2ni\mathbf{K}'(k_1), \quad \text{for integer } m,n,$$

$$\gamma = \mathbf{cn}^{-1}\frac{(b-a_1)(1-p_1)+a_1-c_1}{(b-a_1)(1+p_1)+a_1-c_1}, \quad \text{choose } \gamma \in (0, 2i\mathbf{K}'(k_1)],$$

$$\omega_1 = \mathbf{cn}^{-1}\frac{1+p_1}{1-p_1}, \quad \text{choose } \omega_1 \in (0, \mathbf{K}(k_1)].$$

The differential $Q(u)du^2$ has the poles of the second order at the points

$$\omega_{m,n}^{(1)} = \omega_1 + u_{m,n}, \quad \omega_{m,n}^{(2)} = -\omega_1 + u_{m,n},$$

and the zeros of the second order at the points

$$\gamma_{m,n}^{(1)} = \gamma + u_{m,n}, \quad \gamma_{m,n}^{(2)} = -\gamma + u_{m,n},$$

Then, the images of the points $-1, 1, c_1, a_1$ under the transformation $u(z)$ are respectively the points

$$u_{m,n}^{(0)} = -\mathbf{K}(k_1) - i\mathbf{K}'(k_1) + u_{m,n}, \quad u_{m,n}^{(1)} = -\mathbf{K}(k_1) - 3i\mathbf{K}'(k_1) + u_{m,n},$$

$$u_{m,n}^{(2)} = -2\mathbf{K}(k_1) + u_{m,n}, \quad u_{m,n}^{(1)} = -2\mathbf{K}(k_1) - 2i\mathbf{K}'(k_1) + u_{m,n}.$$

These points are regular for the differential $Q(u)du^2$. Let $\tilde{B}_2$ be obtained from the domain $B_2^*$ by slitting along the critical orthogonal trajectory $(1, c_1)$. Denote by $P_1$ the $u$-image of $B_1^*$, and by $P_2$ the $u$-image of $B_2^*$. Fix $m, n$. Then, the domain $P_1$ is the circular domain in the trajectory structure of the differential $Q(u)du^2$ bounded by the segment $[\gamma_{m,n}^{(1)}, \gamma_{m,n}^{(2)}]$ and the arc $p$ connecting the points $\gamma_{m,n}^{(1)}, \gamma_{m,n}^{(2)}$, so that $\omega_1 \in P_1$. The domain $P_2$ is bounded by the segments $[u_{m,n} - 2i\mathbf{K}', \gamma_{m,n}^{(1)}]$, $[\gamma_{m,n}^{(2)}, u_{m,n} + 2i\mathbf{K}']$, $[u_{m,n} + \mathbf{K} - 2i\mathbf{K}', u_{m,n} + \mathbf{K} + 3i\mathbf{K}']$, and by the arc $p$.

Let $\zeta_1(z)$ be the conformal map of the domain $B_1^*$ onto the disk $|\zeta_1| < R(B_1^*, \infty)$ such that $\zeta_1(\infty) = 0$, $z \cdot \zeta_1(z)\big|_{z \to \infty} = 1$; $\zeta_2(z)$ be the conformal map of the domain $B_2^*$ onto the annulus $1 < |\zeta_2| < M$ such that $\zeta_2(B) = 1$. These functions satisfy in the domains $B_1^*, B_2^*$ the differential equations

$$\alpha_j^2 \left(\frac{d\zeta_j(z)}{\zeta_j(z)}\right)^2 = -4\pi^2 \varphi(z)dz^2, \quad \alpha_1 = 1, \quad \alpha_2 = \alpha, \quad j = 1, 2.$$

The $\varphi$-lengths of the trajectories in the domains $B_1^*$ and $B_2^*$ are equal to 1 and $\alpha$ respectively. Denote by $z = f_j(\zeta)$ the inverse maps. Let $B_1^\varepsilon$ be the doubly connected domain obtained from $B_1^*$ by eliminating the set $\{z : z = f_1(\zeta), |\zeta| \geq 1/\varepsilon\}$, by $P_1^\varepsilon$ the $u$-image of $B_1^\varepsilon$. The domain $P_1^\varepsilon$ is obtained from $P_1$ eliminating the neighbourhood of $\omega_1$ which is the $u$-image of the corresponding neighbourhood of $\infty$. Thus,

$$\mathbf{cn}\,\omega_1 = \frac{1-p_1}{1+p_1}, \quad \omega_1 \in P_1,$$

$$\mathbf{sn}\,\omega_1 = \frac{2\sqrt{p_1}}{1+p_1}, \quad \mathbf{dn}\,\omega_1 = \frac{a_1 - c_1}{(1+p_1)\sqrt{a_1^2 - 1}}.$$

Consider the superpositions $u \circ f_j(\zeta)$ and the inverse ones $\zeta = \zeta_j(z(u))$. Then, we have

$$\alpha_j \frac{d\zeta_j(z(u))}{\zeta_j(z(u))} = Q_2(u)du, \quad \alpha_1 = 1,\ \alpha_2 = \alpha,$$

where

$$Q_2(u) = \frac{1}{\sqrt{p_1(1-a_1^2)}} \times$$

$$\times \frac{(a_1 - b)(1-p_1) - (a_1 - c_1) - ((a_1 - b)(1+p_1) - (a_1 - c_1))\mathbf{cn}\,u}{(1-p_1 - (1+p_1)\mathbf{cn}\,u)}.$$

Consider the domain $\tilde{P}_1^\varepsilon$ obtained from $P_1^\varepsilon$ by deleting the arc of the critical orthogonal trajectory of the differential $Q(u)du^2$ that connects the points $\omega_1$ and $u(a)$. Let $P^\varepsilon$ be the union of the domains $P_1^\varepsilon$, $P_2$, and their common boundary arc $p$. In $P^\varepsilon$ we have

$$Q_2(u) = \frac{(a_1 - b)(1+p_1) - (a_1 - c_1)}{(1+p_1)\sqrt{p_1(1-a_1^2)}} +$$

$$+ \frac{\mathbf{sn}\,\omega_1\,\mathbf{cn}\,\omega_1\,\mathbf{dn}\,\omega_1}{\mathbf{cn}^2 u - \mathbf{cn}^2 \omega_1} + \frac{\mathbf{sn}\,\omega_1\,\mathbf{cn}\,u\,\mathbf{dn}\,\omega_1}{\mathbf{cn}^2 u - \mathbf{cn}^2 \omega_1},$$

or

$$Q_2(u) = \frac{(a_1 - b)(1+p_1) - (a_1 - c_1)}{(1+p_1)\sqrt{p_1(1-a_1^2)}} - \frac{\Theta'(\omega_1, k_1)}{\Theta(\omega_1, k_1)} + \frac{dF(u)}{du},$$

where

$$F(u) = \frac{\mathbf{sn}\,\frac{\omega_1+u}{2}\,\mathbf{dn}\,\frac{\omega_1+u}{2}\,\Theta^2\left(\frac{\omega_1+u}{2}\right)}{\mathbf{sn}\,\frac{u-\omega_1}{2}\,\mathbf{dn}\,\frac{u-\omega_1}{2}\,\Theta^2\left(\frac{u-\omega_1}{2}\right)}.$$

Now we consider the problem about $\mathcal{M}(\alpha, -1, 1, c_1, a_1)$ with $\alpha \in [\mu_2(a_1, c_1), \mu_2(c_1, a_1)]$. The image of the domain $P_2$ under the transformation $\zeta = \zeta_2(z(u))$ is the domain $K = \{|\zeta| \in (1, M), \arg \zeta \in (0, 2\pi)\}$. The arc of the semicircle $\{|\zeta| = 1,\ \arg \zeta \in (0, \pi)\}$ is the image of the trajectory that connects the points $\mathbf{K} - i\mathbf{K}'$ and $\mathbf{K} + i\mathbf{K}'$ in the $u$-plane. Integrating the equation for $\zeta_2(z(u))$ we obtain

$$-i\pi\alpha = \left(\frac{(a_1 - b)(1+p_1) - (a_1 - c_1)}{(1+p_1)\sqrt{p_1(1-a_1^2)}} - \frac{\Theta'(\omega_1, k_1)}{\Theta(\omega_1, k_1)}\right) 2i\mathbf{K}'(k_1) +$$

$$+ \log \frac{F(\mathbf{K}\,(k_1) + i\mathbf{K}\,'(k_1))}{F(\mathbf{K}\,(k_1) - i\mathbf{K}\,'(k_1))}.$$

Letting $\alpha \to 0$ in this equality, we obtain that

$$\log \frac{F(\mathbf{K}\,(k_1) + i\mathbf{K}\,'(k_1))}{F(\mathbf{K}\,(k_1) - i\mathbf{K}\,'(k_1))} = i\pi \frac{\omega_1}{\mathbf{K}\,(k_1)},$$

and deduce the equation for the zero $b(\alpha)$.

Now we consider in $K$ the segment $[1, M]$. In $P_2$ it has the preimage that connects the points $\mathbf{K} + i\mathbf{K}'$ and $2i\mathbf{K}'$. Integrating the same equation along these paths we obtain the expression for $M(B_2^*)$.

Assume $\zeta_1(a_1) = R$. The segment $[R, 1/\varepsilon]$ in the $\zeta$-plain has its preimage in the $u$-plane which is the arc in $P_1^\varepsilon$ connecting the points $u(a_1) = 0$ and $u(1/\varepsilon) = \omega_1 - \delta$. Integrating $d\zeta_1(z(u))$ we obtain the expression for $m(B_1^*, \infty)$.
□

### 3.2.3 Upper boundary curve for the range of $(|f(r_1)|, |f(r_2)|)$ in $S_R$

Now we use the moduli calculated in Section 3.2.2 to define the upper boundary curve $\Gamma^+$ of the range of $(f(r_1), f(r_2))$, $0 < r_1 < r_2 < 1$ in the class $S_R$. Our proof is based on simultaneous consideration of two problems of the extremal partition of the punctured unit disk and of the punctured Riemann sphere.

Let $U_z = U \setminus \{0, r_1, r_2\}$ be the punctured unit disk. We consider on $U_z$ the admissible system $(\gamma_1^z, \gamma_2^z)$ of types III and I respectively where $\gamma_2^z$ is a simple loop that separates the points $r_1, r_2$ from $0$ and $\partial U$, $\gamma_1^z = \{z : |z| = \varepsilon\}$. Here $\varepsilon$ is sufficiently small so that at most $\varepsilon < r_1$. Let $\mathfrak{D}^z$ be the set of all pairs $(D_1^z, D_2^z)$ of a simply connected domain $D_1^z$ and a doubly connected domain $D_2^z$ of homotopy type $(\gamma_1^z, \gamma_2^z)$. Then, the problem of the extremal partition of $U_z$ consists of maximizing the sum $m(D_1^z, 0) + \alpha^2 M(D_2^z)$ over $(D_1^z, D_2^z) \in \mathfrak{D}^z$. The maximum of this sum we denote by $\mathcal{M}_z(\alpha, r_1, r_2)$. Under the transformation

$$Z(z) = \frac{2r_1 r_2}{(r_2 - r_1)(1 - r_1 r_2)} \left( \frac{(r_2 + r_1)(1 + r_1 r_2)}{2r_1 r_2} - z - \frac{1}{z} \right),$$

two extremal domains $(D_1^{z*}, D_2^{z*})$ in the problem for $\mathcal{M}_z(\alpha, r_1, r_2)$ are mapped onto two extremal domains $(B_1^*, B_2^*)$ in the problem of finding $\mathcal{M}(\alpha, -1, 1, c_1, a_1)$ where

$$c_1 = \frac{(r_2 + r_1)(1 + r_1 r_2) - 4r_1 r_2}{(r_2 - r_1)(1 - r_1 r_2)}, \quad a_1 = \frac{(r_2 + r_1)(1 + r_1 r_2) + 4r_1 r_2}{(r_2 - r_1)(1 - r_1 r_2)}.$$

Taking into account the change of the reduced modulus under the conformal map $Z(z)$ we deduce that

$$m(D_1^{z*}, 0) = \frac{1}{2\pi} \log \frac{2r_1 r_2}{(r_2 - r_1)(1 - r_1 r_2)} + m(B_1^*, \infty), \quad M(D_2^{z*}) = M(B_2^*),$$

for $\mu_2(a_1, c_1) \leq \alpha \leq \mu_2(c_1, a_1)$.

Let $\mathbb{C}_w = \mathbb{C} \setminus \{0, w_1, w_2\}$ be the punctured Riemann sphere. Let $(\gamma_1^w, \gamma_2^{w(k)}\big|_{k=1}^{\infty})$ be a countable set of admissible systems of curves on $\mathbb{C}_w$; $\gamma_1 = \{z : |z| = \varepsilon\}$ for a sufficiently small $\varepsilon$ such that $\gamma_1^w$ is of type III and separates the puncture 0 from all others; $\gamma_2^{w(k)}$ is taken from the countable set of possible non-homotopic on $\mathbb{C}_w$ simple loops of type I that separate the punctures $w_1$ and $w_2$ from the punctures $0, \infty$. Assume that the curve $\gamma_2^{w(1)}$ is homotopic to the slit along $[w_1, w_2]$. Let $\mathfrak{D}^{w(k)}$ be the family of all pairs $(D_1^w, D_2^{w(k)})$ of non-overlapping domains of homotopy type $(\gamma_1^w, \gamma_2^{w(k)})$, where $D_1^w$ is a hyperbolic simply connected domain, $0 \in D_1^w$, $D_2^{w(k)}$ is a hyperbolic doubly connected domain. The domain $D_2^{w(k)}$ can be possibly degenerate in the usual sense. We consider the problem of maximizing the sum $m(D_1^w, 0) + \alpha^2 M(D_2^{w(k)})$ for non-negative $\alpha$ over all $(D_1^w, D_2^{w(k)}) \in \mathfrak{D}^{w(k)}$. Denote by $M_w(\alpha, w_1, w_2)$ the maximum of this sum for $k = 1$. Under the transformation

$$W(w) = \frac{2w_1 w_2}{w_2 - w_1}\left(\frac{w_2 + w_1}{2w_1 w_2} - \frac{1}{w}\right)$$

two extremal domains $(D_1^{w*}, D_2^{w(k)*})$ in the problem for $M_w(\alpha, w_1, w_2)$ are mapped onto two extremal domains $(D_1^*, D_2^{(k)*})$ in the problem of finding $M^{(k)}(\alpha, -1, 1, a)$ with $a = \frac{w_2 + w_1}{w_2 - w_1}$. Taking into account the change of the reduced modulus under the conformal map $W(z)$ we deduce that

$$m(D_1^{w*}, 0) = \frac{1}{2\pi} \log \frac{2|w_1 w_2|}{|w_2 - w_1|} + m(D_1^*, \infty), \quad M(D_2^{w(k)*}) = M(D_2^{(k)*}).$$

**Lemma 3.2.3.** *For fixed numbers $a_1, c_1$ such that $1 < a_1 < c_1$ and for $a_1 \leq a \leq c_1$ the equation*

$$M(D_2^*) = M(B_2^*)$$

*defines a function $\alpha^*(a)$ which is differentiable and strictly increases in $a \in [a_1, c_1]$ from $\mu_2(a_1, c_1)$ to $\mu_2(c_1, a_1)$. Here $D_2^*$ and $B_2^*$ are the extremal ring domains for $M(\alpha, -1, 1, a)$ and $\mathcal{M}(\alpha, -1, 1, a_1, c_1)$ respectively.*

*Proof.* For convenience, we rewrite $M_2(\alpha, -1, 1, a) \equiv M(D_2^*)$ and $\mathcal{M}_2(\alpha, -1, 1, a_1, c_1) \equiv M(B_2^*)$. For a fixed $\alpha$ we set the equation

$$\alpha M_2(\alpha, -1, 1, a) = \alpha \mathcal{M}_2(\alpha, -1, 1, a_1, c_1) \tag{3.22}$$

for all possible $a$. For $\alpha = \mu_2(a_1, c_1)$ the problem on $\mathcal{M}(\alpha, -1, 1, a_1, c_1)$ is reducible to the problem on $M(\alpha, -1, 1, a)$, therefore, the equation (3.22) has the solution $a = a_1$.

Theorem 3.2.3 and the remarks to Theorem 3.2.2 imply that

$$\frac{\partial}{\partial \alpha}(\alpha \mathcal{M}_2(\alpha, -1, 1, a_1, c_1)) = \frac{\mathbf{K}(k_1)}{4\mathbf{K}'(k_1)}, \quad \frac{\partial}{\partial \alpha}(\alpha \mathcal{M}_2(\alpha, -1, 1, a)) = \frac{\mathbf{K}(k)}{4\mathbf{K}'(k)},$$

$k = k(a, \alpha)$. Thus, the equality

$$\frac{\partial}{\partial \alpha}(\alpha \mathcal{M}_2(\alpha, -1, 1, a_1, c_1)) = \frac{\partial}{\partial \alpha}(\alpha \mathcal{M}_2(\alpha, -1, 1, a)). \qquad (3.23)$$

holds for $\alpha = \mu_2(a_1, c_1)$. Explicit calculation yields

$$\frac{\partial k}{\partial \alpha} = \frac{\pi(a - c)\sqrt{p(a^2 - 1)}}{2pk^2(p(a^2 - 1) + ac - 1)^2 \mathbf{K}'(k)}.$$

Hence, $\frac{\partial k}{\partial \alpha} > 0$ as $\alpha < 1$, and $\frac{\partial k}{\partial \alpha} < 0$ as $\alpha > 1$. From (3.23) we derive that the inequality

$$\frac{\partial}{\partial \alpha}(\alpha \mathcal{M}_2(\alpha, -1, 1, a_1, c_1)) < \frac{\partial}{\partial \alpha}(\alpha \mathcal{M}_2(\alpha, -1, 1, a)), \qquad (3.24)$$

holds for $\alpha \in (\mu_2(a_1, c_1), 1]$. Taking into account (3.24) and the equality

$$\mathcal{M}_2(\mu_2(a_1, c_1), -1, 1, a_1) = \mathcal{M}_2(\mu_2(a_1, c_1), -1, 1, a_1, c_1),$$

we obtain that

$$\alpha \mathcal{M}_2(\alpha, -1, 1, a_1) < \alpha \mathcal{M}_2(\alpha, -1, 1, a_1, c_1), \quad \alpha \in (\mu_2(a_1, c_1), 1].$$

Now we examine $\alpha \mathcal{M}_2(\alpha, -1, 1, a)$ for a fixed $\alpha \in [\mu_2(a_1, c_1), 1]$ as a function with respect to $a$. Since $\mu_1(a)$ decreases from 1 to 0, there is a unique $\tilde{a} = \tilde{a}(\alpha)$ such that $\mu_1(\tilde{a}) = \alpha$. The function $\mathcal{M}_2(\alpha, -1, 1, a)$ decreases in $a \in (a_1, \tilde{a}(\alpha)]$ and $\mathcal{M}_2(\alpha, -1, 1, \tilde{a}(\alpha)) = 0$. Thus, there exists a unique solution of the equation (3.22) which we denote by $a = a^*(\alpha)$.

Since $a^*(\mu_2(a_1, c_1)) = a_1$ and $a^*(\alpha) > a_1$ for $\alpha \in (\mu_2(a_1, c_1), 1]$, the function $a^*(\alpha)$ increases for a sufficiently small $\alpha$ from the interval $(\mu_2(a_1, c_1), 1]$. Differentiating (3.22) with respect to $\alpha$ we find

$$\frac{\partial a^*(\alpha)}{\partial \alpha} = \frac{k\mathbf{K}'(k)\sqrt{a^2 - 1}}{\sqrt{p}}\left(\frac{\mathbf{K}(k_1)}{\mathbf{K}'(k_1)} - \frac{\mathbf{K}(k)}{\mathbf{K}'(k)}\right). \qquad (3.25)$$

Suppose that $a^*(\alpha)$ does not increase for some $\alpha \in (\mu_2(a_1, c_1), 1]$. Then, there exists $\tilde{\alpha}$ such that

$$\left.\frac{\partial a^*(\alpha)}{\partial \alpha}\right|_{\alpha = \tilde{\alpha}} = 0, \quad \left.\frac{\partial a^*(\alpha)}{\partial \alpha}\right|_{\alpha \in [\mu_2(a_1, c_1), \tilde{\alpha}]} \geq 0, \quad \left.\frac{\partial a^*(\alpha)}{\partial \alpha}\right|_{\alpha \in (\tilde{\alpha}, \tilde{\alpha}+\varepsilon)} < 0$$

for $\varepsilon$ sufficiently small. From (3.25) we deduce that $k(a^*(\tilde{\alpha}), \tilde{\alpha}) = k_1$. Now we apply the same argumentation that was used for the construction of the solution $a = a_1$, $\alpha = \mu_2(a_1, c_1)$. Therefore, $a^*(\alpha)$ increases with increasing $\alpha$ in the interval $(\tilde{\alpha}, \tilde{\alpha} + \varepsilon)$. This contradicts the assumption concerning the behaviour of $a^*(\alpha)$. A similar analysis for $\alpha \in [1, \mu_2(c_1, a_1)]$ completes the proof of the lemma. $\qquad \square$

**Theorem 3.2.4.** *Let* $0 < r_1 < r_2 < 1$, *and let us set*

$$c_1 = \frac{(r_2 + r_1)(1 + r_1 r_2) - 4r_1 r_2}{(r_2 - r_1)(1 - r_1 r_2)}, \quad a_1 = \frac{(r_2 + r_1)(1 + r_1 r_2) + 4r_1 r_2}{(r_2 - r_1)(1 - r_1 r_2)},$$

$\alpha \in [\mu_2(a_1, c_1), \mu_2(c_1, a_1)]$.

*(i) There exists a function* $w = f^*(z, \alpha) \in S_R$ *that satisfies in* $U$ *the differential equation*

$$\frac{(z - d)^2 (z - \bar{d})^2 dz^2}{z^2 (z - r_1)(z - r_2)(1 - r_1 z)(1 - r_2 z)} = \frac{(w - C)dw^2}{w^2 (w - w_1)(w - w_2)},$$

*where* $d = d(\alpha)$, $|d| = 1$ *is defined by the equation*

$$b(\alpha) = \frac{2r_1 r_2}{(r_2 - r_1)(1 - r_1 r_2)} \left( \frac{(r_2 + r_1)(1 + r_1 r_2)}{2r_1 r_2} - d - \frac{1}{d} \right),$$

*and* $b(\alpha)$ *is defined in Theorem 3.2.3. Set*

$$w_1 = w_1(\alpha) = \frac{(r_2 - r_1)(1 - r_1 r_2)}{2r_1 r_2} \frac{R(D_1^*, \infty)}{R(B_1^*, \infty)} \frac{1}{a^*(\alpha) + 1},$$

$$w_2 = w_2(\alpha) = \frac{(r_2 - r_1)(1 - r_1 r_2)}{2r_1 r_2} \frac{R(D_1^*, \infty)}{R(B_1^*, \infty)} \frac{1}{a^*(\alpha) - 1},$$

$$C = C(\alpha) = \frac{(r_2 - r_1)(1 - r_1 r_2)}{2r_1 r_2} \frac{R(D_1^*, \infty)}{R(B_1^*, \infty)} \frac{1}{a^*(\alpha) - c(a^*(\alpha), \alpha)},$$

*where* $R(D_1^*, \infty)$, $R(B_1^*, \infty)$ *are the conformal radii defined by the reduced moduli of Theorems 3.2.2, 3.2.3, the function* $a = a^*(\alpha)$ *is defined by Lemma 3.2.3,* $c(\alpha)$ *is defined by Theorem 3.2.2.*

*(ii) The functions* $w_1(\alpha)$, $w_2(\alpha)$ *increase monotonically with increasing* $\alpha$ *and*

$$w_1(\mu_2(a_1, c_1)) = \frac{r_1}{(1 + r_1)^2}, \quad w_1(\mu_2(c_1, a_1)) = \frac{r_1}{(1 - r_1)^2},$$

$$w_2(\mu_2(a_1, c_1)) = \frac{r_2}{(1 + r_2)^2}, \quad w_2(\mu_2(c_1, a_1)) = \frac{r_2}{(1 - r_2)^2}.$$

*(iii) The function* $w = f^*(z, \alpha)$ *maps the unit circle* $\partial U$ *onto the continuum* $E$ *consisting of the ray* $[C(\alpha), \infty)$ *and two symmetric arcs of the trajectories of the quadratic differential*

$$\frac{(w - C)dw^2}{w^2(w - w_1)(w - w_2)}$$

*so that they start from the point* $C(\alpha)$ *and* $R(\mathbb{C} \setminus E, 0) = 1$.

*(iv) The boundary curve* $\Gamma^+$ *of the range* $\mathfrak{N}_f^R$ *of the system of functionals* $I(f)$ *in the class* $S_R$ *consists of the points* $(w_1(\alpha), w_2(\alpha))$. *Each point is given only by the function* $f^*(z, \alpha)$.

*Proof.* Lemma 3.2.3 states that there exists a function $a^*(\alpha)$ defined in the segment $[\mu_2(a_1, c_1), \mu_2(c_1, a_1)]$, such that $M(D_2^*) = M(B_2^*)$ and $\alpha^*(a)$ is the inverse. Hence, there exist two functions $\zeta = \zeta_1(W)$ and $\zeta = \zeta_2(Z)$ that map the domains $D_2^*$ and $B_2^*$ respectively onto the ring $1 < |\zeta| < \exp(2\pi M(D_2^*)) = \exp(2\pi M(B_2^*))$. They satisfy in $D_2^*$ and $B_2^*$ the differential equations

$$\alpha^* \left( \frac{d\zeta_1(W)}{\zeta_1(W)} \right)^2 = -\frac{W - c(\alpha^*)}{(W - a)(W^2 - 1)} dW^2,$$

$\zeta_1(\pm 1) = \pm 1$, $\zeta_1(c) = \exp(2\pi M(D_2^*))$,

$$\alpha^* \left( \frac{d\zeta_2(Z)}{\zeta_2(Z)} \right)^2 = -\frac{(Z - b)^2}{(Z - c_1)(Z - a_1)(Z^2 - 1)} dZ^2,$$

$\zeta_2(\pm 1) = \pm 1$ $\zeta_2(c_1) = \zeta_1(c)$. The superposition

$$w = \tilde{f}(z) = \frac{1}{a - \zeta_1^{-1} \circ \zeta_2(Z(z))}$$

is a conformal homeomorphism of $D_2^{z*}$ onto $D_2^{w(1)*}$. The functions $Z(z)$ and $W(w)$ are defined in the beginning of Section 3.2.3. The equation

$$\frac{W - c(\alpha^*)}{(W - a)(W^2 - 1)} dW^2 = \frac{(Z - b)^2}{(Z - c_1)(Z - a_1)(Z^2 - 1)} dZ^2,$$

is equivalent to the equation in the statement of the theorem. It implies that the function $w = \tilde{f}(z)$ can be continued analytically onto $\partial U$, $f(-1) = \infty$, and maps $\partial U$ onto the continuum $\tilde{E}$ so that the simply connected domain $\mathbb{C} \setminus \tilde{E}$ has the conformal radius

$$\frac{2r_1 r_2}{(r_2 - r_1)(1 - r_1 r_2)} \frac{R(B_1^*, \infty)}{R(D_1^*, \infty)}.$$

Thus, the function $w = \tilde{f}(z)$ which we have constructed is holomorphic and univalent in $U$ with the expansion

$$\tilde{f}(z) = \frac{2r_1 r_2}{(r_2 - r_1)(1 - r_1 r_2)} \frac{R(B_1^*, \infty)}{R(D_1^*, \infty)} z + a_2 z^2 + \dots.$$

The function $w = f^*(z, \alpha)$ is easily defined as

$$f^*(z, \alpha) = \frac{(r_2 - r_1)(1 - r_1 r_2)}{2r_1 r_2} \frac{R(D_1^*, \infty)}{R(B_1^*, \infty)} \tilde{f}(z).$$

So, it satisfies the conditions (i) and (iii) of the statement of Theorem 3.2.4.
The functions $w_1(\alpha)$ and $w_2(\alpha)$ are continuous, hence,

$$\left[ \frac{r_1}{(1 + r_1)^2}, \frac{r_1}{(1 - r_1)^2} \right] \subset \{w_1(\alpha) : \alpha \in [\mu_2(a_1, c_1), \mu_2(c_1, a_1)]\},$$

$$\left[ \frac{r_2}{(1+r_2)^2}, \frac{r_2}{(1-r_2)^2} \right] \subset \{ w_2(\alpha) : \alpha \in [\mu_2(a_1, c_1), \mu_2(c_1, a_1)] \}.$$

Suppose that there exists

$$w_0 \in \left[ \frac{r_1}{(1+r_1)^2}, \frac{r_1}{(1-r_1)^2} \right],$$

such that $w_0 = w_1(\alpha_1)$, $w_0 = w_1(\alpha_2)$, $\alpha_1 \neq \alpha_2$, but $w_2(\alpha_1) < w_2(\alpha_2)$. Note, that if $w_2(\alpha_1) = w_2(\alpha_2)$, then we would have $a^*(\alpha_1) = a^*(\alpha_2)$, what is impossible. Let domains $D_1, D_2$ be extremal in the problem for $M^{(1)}(\alpha_1, w_0, w_2(\alpha_2))$. Then, they are admissible (but not extremal) for $M^{(1)}(\alpha_1, w_0, w_2(\alpha_1))$ and the inequality

$$M^{(1)}(\alpha_1, w_0, w_2(\alpha_2)) < M^{(1)}(\alpha_1, w_0, w_2(\alpha_1))$$

holds. On the other hand, we have $w_2(\alpha_1) < w_2(\alpha_2)$ and, because of Theorem 2.7.2, the modulus $M^{(1)}(\alpha_1, w_0, w_2(\alpha_1))$ increases when the puncture moves from from $w_2(\alpha_2)$ to $w_2(\alpha_1)$ and

$$M^{(1)}(\alpha_1, w_0, w_2(\alpha_1)) < M^{(1)}(\alpha_1, w_0, w_2(\alpha_2)).$$

This contradicts the previous inequality and proves the uniqueness of the parametric representation of the curve $\Gamma^+$ and (ii).

Now we prove the extremality of the function $f^*(z, \alpha)$ respectively to the system of functionals $(f(r_1), f(r_2))$. Let $f(z)$ be an arbitrary function from $S_R$. Choose $\alpha$ such that $f(r_1) = f^*(r_1, \alpha) = w_1(\alpha)$. Suppose the contrary, $f(r_2) > f^*(r_2, \alpha)$. Then, $(f(D_1^{z*}), f(D_2^{z*}))$ is a pair of admissible domains for $M^{(1)}(\alpha, w_1(\alpha), f(r_2))$. Further,

$$\mathcal{M}_z(\alpha, r_1, r_2) = m(f(D_1^{z*}, 0) + \alpha^2 M(f(D_1^{z*}) \leq M^{(1)}(\alpha, w_1(\alpha), f(r_2)).$$

Since $f(r_2) > w_2(\alpha)$, the inequality

$$M^{(1)}(\alpha, w_1(\alpha), f(r_2)) < M^{(1)}(\alpha, w_1(\alpha), w_2(\alpha))$$

holds because of the monotonicity of the modulus as before. But the latter modulus $M^{(1)}(\alpha, w_1(\alpha), w_2(\alpha))$ is equal to the initial one by construction. This contradiction proves the extremality of the function $f^*(z, \alpha)$ and (iv). The uniqueness, as usual, follows from the uniqueness of the extremal configurations. □

### 3.2.4 Upper boundary curve for the range of $(|f(r_1)|, |f(r_2)|)$ in $S$

Here we prove that the upper boundary curve for the range $\mathfrak{N}_f$ of the system of functionals $(|f(r_1)|, |f(r_2)|)$, $0 < r_1 < r_2 < 1$ in the class $S$ is the same as in the class $S_R$.

**Theorem 3.2.5.** *The upper boundary curve $\Gamma^+$ for the range $\mathfrak{N}_f$ of the system of functionals $(|f(r_1)|, |f(r_2)|)$ in the class $S$ consists of the points $(w_1(\alpha), w_2(\alpha))$ obtained in Theorem 3.2.4. They are given for each $\alpha \in [\mu_2(a_1, c_1), \mu_2(c_1, a_1)]$ by the unique function $f^*(z, \alpha) \in S_R$.*

*Proof.* Without loss of generality assume $\arg f(r_1) = 0$. Choose $\alpha$ so that $w_1(\alpha) = f(r_1) = |f(r_1)|$. Suppose the contrary, $|f(r_2)| > w_2(\alpha)$. Then, $(f(D_1^{z*}), f(D_2^{z*})$ is a pair of admissible domains for $M^{(k)}(\alpha, w_1(\alpha), f(r_2))$ for some $k$. So, by Lemma 3.2.1, we have the inequality

$$\mathcal{M}_z(\alpha, r_1, r_2) = m(f(D_1^{z*}, 0) + \alpha^2 M(f(D_1^{z*}) \leq$$

$$\leq M^{(k)}(\alpha, w_1(\alpha), f(r_2)) \leq M^{(1)}(\alpha, w_1(\alpha), f(r_2)).$$

Lemma 3.2.2 applied with the map $W(w)$ yields the inequality

$$M^{(1)}(\alpha, w_1(\alpha), f(r_2)) \leq M^{(1)}(\alpha, w_1(\alpha), |f(r_2)|).$$

The monotonicity of the modulus by Theorem 2.7.2 yields

$$M^{(1)}(\alpha, w_1(\alpha), |f(r_2)|) < M^{(1)}(\alpha, w_1(\alpha), w_2(\alpha)).$$

But the latter modulus is equal to the initial one in this chain of inequalities. This contradiction proves the extremality of the function $f^*(z, \alpha)$ and the statement of the theorem. The uniqueness, as usual, follows from the uniqueness of the extremal configurations. □

We note that the lower boundary curve $\Gamma^-$ for the class $S$ is different from that for the class $S_R$. Its points for the class $S$ are given by functions from $S \setminus S_R$ that map the unit disk onto the plane minus a curve with a unique finite endpoint. Since geometric structures in this case are not symmetric it seems to be not possible to use the modulus method in the form suggested . The same is for non-real initial points $r_1$, $r_2$.

## 3.3 Bounded univalent functions

We denote by $B_s$ the class of all univalent holomorphic maps $f(z) = bz + a_2 z^2 + \ldots$ from $U$ into itself with $0 < b < 1$. The compact subclass $B_s(b)$ consists of all functions from $B_s$ with the first coefficient $b$ fixed. The function that plays the same role as the Koebe function in the class $S$ is the so-called Pick function $w = p_\theta(z)$. The Pick function satisfies the equation

$$\frac{w}{(1 + e^{i\theta} w)^2} = \frac{bz}{(1 + e^{i\theta} z)^2},$$

and maps the unit disk onto the unit disk with the radial slit starting from the point

$$e^{i\theta} \left( -1 + \frac{1 - \sqrt{1 - b}}{b} \right).$$

### 3.3.1 Elementary estimates

★ By the analogy with the proof of the Koebe Theorem one can obtain the covering theorem for the class $B_s(b)$.

**Theorem 3.3.1.** *The boundary of the Koebe set*

$$\mathfrak{K}_B = \bigcap_{f \in B_s(b)} f(U)$$

*in the class $B_s(b)$ is the circle*

$$|w| = R_b = -1 + \frac{1 - \sqrt{1-b}}{b}$$

*If $\inf_{w \in \partial f(U)} |w| = R_b$ for some $f$, then $f$ is a suitable rotation of the Pick function.*

The growth theorem is given as follows.

**Theorem 3.3.2.** *If $f \in S$, then $p_0(r) \leq |f(z)| \leq p_\pi(r)$ with the equality only for a suitable rotation of the Pick function.*

★ The proof is the same as for the classical Growth Theorem. In the plane of preimages we choose the same initial extremal domain and consider the problem of the extremal partition in the unit disk instead of the complex plane.

More interesting features one can find in the distortion theorems.

**Theorem 3.3.3.** *If $f \in B_s$, then*

$$|f'(z)| \leq \frac{|f(z)|(1 - |f(z)|)}{1 + |f(z)|} \frac{1+r}{r(1-r)}, \quad r := |z|, \tag{3.26}$$

*with the Pick function $p_\theta(z)$ as the extremal one for any $p'_\theta(0) = b$. Moreover, $|f'(z)| \leq 1$ for $|z| \leq \sqrt{2} - 1$. The constant $\sqrt{2} - 1$ is sharp.*

*Proof.* Let $m(D, a)$ stand for the reduced modulus of a simply-connected hyperbolic domain $D$ with respect to a point $a \in D$. We take the domain $U \setminus (e^{i(\pi + \arg z)}, 0]$ as the initial one and calculate the reduced modulus (Section 2.4) as $m(U \setminus (e^{i(\pi + \arg z)}, 0], z) = \frac{1}{2\pi} \log 4r(1-r)/(1+r)$. The change of the reduced modulus under the map $f$ and further symmetrization immediately lead to the elementary sharp inequality (3.26) in the statement of the theorem.

Now we deduce the supremum over all such $r$, that $|f'(z)| \leq 1$. The inequality

$$\frac{|f(z)|(1 - |f(z)|)}{1 + |f(z)|} \frac{1+r}{r(1-r)} \leq 1$$

is equivalent to

$$|f(z)| \in [0, \min(\frac{1-r}{1+r}, r)] \cup [\max(\frac{1-r}{1+r}, r), 1).$$

1) $0 \le r \le \sqrt{2} - 1$. Hence, (3.26) is equivalent to

$$|f(z)| \in [0, r] \cup [\frac{1-r}{1+r}, 1),$$

because $|f(z)| < r$. Therefore, $|f'(z)| \le 1$ for all $z : |z| \le \sqrt{2} - 1$, with the obvious possibility of the equality sign.

2) $\sqrt{2} - 1 < r < 1$. Hence, (3.26) is equivalent to

$$|f(z)| \in [0, \frac{1-r}{1+r}] \cup [r, 1).$$

For $\sqrt{2} - 1 < |z| < 1$ there is a Pick function with some $b = p'_\theta(0)$ close to 1, so that $\frac{1-r}{1+r} < |f(z)| < r$ and, therefore, $|f'(z)| > 1$.  □

**Theorem 3.3.4.** *If $f \in B_s$, then*

$$|f'(z)| \ge \frac{|f(z)|(1 + |f(z)|)}{1 - |f(z)|} \frac{1-r}{r(1+r)}, \quad r := |z|.$$

*The equality occurs for a suitable rotation of the Pick function $p_\theta(z)$ with any $p'_\theta(0) = b$.*

*Proof.* For this lower estimate we consider the digon $U'_z = U \setminus [0, \exp(i \arg z))$ with two vertices over the same support $z$. It is conformal at $z$, have there the angles $\pi$ and the reduced modulus is

$$m(U'_z, z, z) = \frac{2}{\pi} \log \frac{4r(1+r)}{1-r} \quad \text{(see Section 2.4).}$$

The digon $f(U'_z)$ with the vertices over $f(z)$ has the reduced modulus

$$m(f(U'_z), f(z), f(z)) = \frac{2}{\pi} \log \frac{4r(1+r)}{1-r} + \frac{2}{\pi} \log |f'(z)|.$$

This digon is admissible in the problem of minimum of the reduced modulus over all digons on $S_0 = U \setminus \{0, f(z)\}$ conformal at $f(z)$ with corresponding angles and of homotopy type defined by a curve $\gamma$ that connects on $S_0$ the point $f(z)$ with itself and homotopic on $S_0$ to the slit $[0, f(z)]$. The extremal digon $D$ is $\mathbb{C}$ with the slit along the segment $[0, \exp(i \arg f(z))]$ passing through the point $f(z)$. Its reduced modulus is equal to $\frac{2}{\pi} \log 4|f(z)|(1 + |f(z)|)(1 - |f(z)|)^{-1}$. This leads to the lower estimate. The uniqueness of the extremal configuration leads to the uniqueness of the extremal map up to rotation. □

Theorem 3.3.4 and the inequality (3.26) are due to R. M. Robinson (see [49], page 57, Theorem 22). The statement of Theorem 3.3.3 about the constant $\sqrt{2} - 1$ is known also for non-univalent bounded analytic maps (J. Dieudonné [25]) (see e.g. [23], page 18).

## 3.3.2 Boundary curve for the range of $(|f(z)|, |f'(z)|)$ in $B_s(b)$

Now we describe the range of the system of functionals $(|f(z)|, |f'(z)|)$ in the class $B_s(b)$. By a suitable rotation we can assume $z = r \in (0, 1)$.

Let $U_z = U \setminus \{0, r\}$, $r \in (0, 1)$ be the punctured unit disk. We consider on $U_z$ the system of curves $(\gamma_1^z, \gamma_2^z)$ where $\gamma_2^z = \{z : |z - r| = \varepsilon\}$ and $\gamma_1^z = \{z : |z| = \varepsilon\}$. Here $\varepsilon$ is sufficiently small such that $r + \varepsilon < 1$ and $\varepsilon < r/2$. Let $\mathfrak{D}^z$ be a set of all pairs $(D_1^z, D_2^z)$ of simply connected domains of homotopy type $(\gamma_1^z, \gamma_2^z)$. Then, the problem of the extremal partition of $U_z$ consists of maximizing the sum $m(D_1^z, 0) + \alpha^2 m(D_2^z, r)$ over all $(D_1^z, D_2^z) \in \mathfrak{D}^z$. The maximum of this sum we denote by $M_z(\alpha, r)$. Under the transformation

$$Z(z) = 1 - \frac{r(1 + z)^2}{z(1 + r)^2}$$

two extremal domains $(D_1^{z*}, D_2^{z*})$ in the problem of $M_z(\alpha, r)$ are mapped onto the two extremal domains $(B_1^*, B_2^*)$ in the problem of $\mathcal{M}(\alpha, c)$ (we use here notations of Section 3.1.3), where $c = (1 - r)^2/(1 + r)^2$. Taking into account the change of the reduced moduli under the conformal map $Z(z)$ we deduce that for $(1 - r)/(1 + r) \le \alpha \le (1 + r)/(1 - r)$

$$m(D_1^{z*}, 0) = \frac{1}{2\pi} \log r^\alpha, \quad m(D_2^{z*}, r) = \frac{1}{2\pi} \log r^{1/\alpha}(1 - r^2).$$

Theorems 3.1.5, 3.1.6 and suitable conformal maps imply that for $0 \le \alpha \le (1 - r)/(1 + r)$

$$m(D_1^{z*}, 0) = \frac{1}{2\pi} \log \frac{r}{(1 + r)^2} \cdot \frac{4(1 - \alpha)^{\alpha - 1}}{(1 + \alpha)^{\alpha + 1}},$$

$$m(D_2^{z*}, r) = \frac{1}{2\pi} \log \frac{r(1 + r)}{1 - r} \cdot \frac{4\alpha^2(1 - \alpha)^{\frac{1}{\alpha} - 1}}{(1 + \alpha)^{\frac{1}{\alpha} + 1}}.$$

For $\alpha \ge (1 + r)/(1 - r)$

$$m(D_1^{z*}, 0) = \frac{1}{2\pi} \log \frac{r}{(1 - r)^2} \cdot \frac{4(\alpha - 1)^{\alpha - 1}}{(\alpha + 1)^{\alpha + 1}},$$

$$m(D_2^{z*}, r) = \frac{1}{2\pi} \log \frac{r(1 - r)}{1 + r} \cdot \frac{4\alpha^2(\alpha - 1)^{\frac{1}{\alpha} - 1}}{(\alpha + 1)^{\frac{1}{\alpha} + 1}}.$$

Here (*) denotes the extremality of domains.

Let $x \in (0, r)$. We consider the same modulus problem in the unit disk with $x$ in place of $r$. The extremal domains for this problem we denote by $(D_1^{w*}, D_1^{w*}) \in \mathfrak{D}^w$. The preceding formulae give the values of the corresponding reduced moduli.

We define now a problem of the extremal partition of the disk $U$ by digons. Let $0, r$ be punctures in $U$. We consider the family $\mathbb{D}$ of digons $D$ in $U$ such that the points $0, \omega$ are not in $D$ but they are two vertices of any $D \in \mathbb{D}$. All digons under consideration are supposed to be conformal with the angles $2\pi$ at the vertices. We define the problem of minimizing

$$\min_{D \in \mathbb{D}} m(D, 0, r).$$

There is a unique digon $D^* = U \setminus \{(-1, 0] \cup [r, 1)\}$ that gives this minimum. The reduced modulus is calculated in Section 2.4 as

$$m(D^*, 0, r) = \frac{1}{2\pi} \log \frac{r^2}{1 - r^2}.$$

We evaluate now the system of functionals $I(f) = (|f(r)|, |f'(r)|)$, $f \in B_s(b)$. Denote by $\Gamma^+$ the arc of $y = \max |f'(r)|$ as $f \in B_s(b)$ with $|f(r)| = x$ fixed, and by $\Gamma^-$ the arc of $y = \min |f'(r)|$ as $f \in B_s(b)$ with $|f(r)| = x$ fixed. We define here the range of $I(f)$, $f \in B_s(b)$ by the moduli calculated before. Set the functions

$$g(z) = \frac{z}{1 - uz + z^2}, \quad u \in [-2, 2],$$

and $G(z, u) = g^{-1}(b \cdot g(z))$.

**Theorem 3.3.5.** *The boundary curve $\Gamma^-$ of the range $\mathfrak{M}_f$ of the system of functionals $(|f(z)|, |f'(z)|)$ in the class $B_s(b)$ consists of the points $(x(u), y(u))$ where $x = G(r, u)$, $y = G'_z(r, u)$ for $-2 \leq u \leq 2$, $r = |z|$. Each point is given only by the function $G(z, u)$ or its rotation.*

*Proof.* Since

$$p_0(r) = G(r, -2) \leq G(r, u) \leq G(r, 2) = p_\pi(r),$$

for $u \in [-2, 2]$, we choose for a function $f \in B_s(b)$ the unique $u = u_0$ such that $x = |f(r)| = G(r, u_0)$. Consider the digon $U'_z = U \setminus \{(-1, 0] \cup [r, 1)\}$ with two vertices $0, r$. It is conformal at $0, r$ and the reduced modulus given by

$$m(U'_z, 0, r) = \frac{1}{2\pi} \log \frac{r^2}{1 - r^2}.$$

The digon $f(U'_z)$ with the vertices $0, f(r)$ has the reduced modulus

$$m(f(U'_z), 0, f(r)) = \frac{1}{2\pi} \log \frac{r^2}{1 - r^2} |f'(r)| b.$$

This digon is admissible in the problem of minimum of the reduced modulus over all digons of homotopy type defined by $\gamma = [0, x]$ which are conformal at $0, x$ and have the angles $2\pi$. The extremal reduced modulus

$$m(D^*, 0) = \frac{1}{2\pi} \log \frac{x^2}{1 - x^2}$$

is given by the digon obtained from $U$ by deleting two segments $(-1, 0]$ and $[x, 1)$. Therefore, $|f'(r)| \geq \frac{(1 - r^2)x^2}{(1 - x^2)r^2 b}$. This is equivalent to the statement of the theorem. The uniqueness of the extremal configuration $G(U'_z, u_0)$ leads to the uniqueness of the extremal map.    □

Let us define now the curve $\Gamma^+$. For this we need the following technical lemma.

**Lemma 3.3.1.** *For $(1 - r)/(1 + r) \leq \alpha \leq (1 + r)/(1 - r)$ the function $x(\alpha)$ defined by the equation*

$$m(D_1^{w*}, 0) = m(D_1^{z*}, 0) + \frac{1}{2\pi} \log b,$$

*is continuous and strictly increases from $G(r, -2)$ to $G(r, 2)$ with increasing $\alpha$.*

*For $\alpha \in [\frac{1-r}{1+r}, \alpha_1]$ the function $x(\alpha)$ is a unique solution to the equation*

$$r^\alpha b = \frac{x}{(1 + x)^2} \frac{4(1 - \alpha)^{\alpha - 1}}{(1 + \alpha)^{\alpha + 1}}.$$

*For $\alpha \in [\alpha_1, \alpha_2]$ the function $x(\alpha)$ is is defined as*

$$x(\alpha) = (r^\alpha b)^{1/\alpha}.$$

*For $\alpha \in [\alpha_2, \frac{1+r}{1-r}]$ the function $x(\alpha)$ is a unique solution to the equation*

$$r^\alpha b = \frac{x}{(1 - x)^2} \frac{4(\alpha - 1)^{\alpha - 1}}{(1 + \alpha)^{\alpha + 1}}.$$

*Here the quantities $m(D_1^{w*}, 0)$ and $m(D_1^{z*}, 0)$ are defined before, and $\alpha_1$ is a unique solution to the equation*

$$r^\alpha b = \left( \frac{1 - \alpha}{1 + \alpha} \right)^\alpha,$$

*$\alpha_2$ is a unique solution to the equation*

$$r^\alpha b = \left( \frac{\alpha + 1}{\alpha - 1} \right)^\alpha.$$

*Proof.* For

$$\alpha \in \left[ \frac{1 - x}{1 + x}, \frac{1 + x}{1 - x} \right]$$

the value of $x(\alpha) = (r^\alpha b)^{1/\alpha}$ obviously increases.

For

$$\alpha \in \left[\frac{1-r}{1+r}, \frac{1-x}{1+x}\right],$$

we have the derivative

$$\frac{x'(\alpha)(1-x)}{(1+x)x} = \log \frac{r(1+\alpha)}{(1-\alpha)}.$$

This implies $x'(\alpha) > 0$. The case

$$\alpha \in \left[\frac{1+x}{1-x}, \frac{1+r}{1-r}\right]$$

is considered analogously. The values of $\alpha_1$ and $\alpha_2$ are deduced from the equations $\alpha = (1-x(\alpha))/(1+x(\alpha))$ and $\alpha = (1+x(\alpha))/(1-x(\alpha))$ respectively. $\square$

**Theorem 3.3.6.** *(i) The upper boundary curve $\Gamma^+$ of the range $\mathfrak{M}_f$ of the system of functionals $(|f(r)|, |f'(r)|)$ in the class $B_s(b)$ consists of the points $(x(\alpha), y(\alpha))$, $\alpha \in [(1-r)/(1+r), (1+r)/(1-r)]$. Here $x(\alpha)$ is defined by Lemma 3.3.1 and*

$$y(\alpha) = \frac{x(\alpha)(1+x(\alpha))}{(1-x(\alpha))(1-r^2)r^{1/\alpha}} \cdot \frac{4\alpha^2(1-\alpha)^{\frac{1}{\alpha}-1}}{(1+\alpha)^{\frac{1}{\alpha}+1}},$$

*for $\alpha \in [(1-r)/(1+r), \alpha_1]$;*

$$y(\alpha) = \frac{x^\alpha}{r^{1/\alpha}(1-r^2)},$$

*for $\alpha \in [\alpha_1, \alpha_2]$;*

$$y(\alpha) = \frac{x(\alpha)(1-x(\alpha))}{(1+x(\alpha))(1-r^2)r^{1/\alpha}} \cdot \frac{4\alpha^2(\alpha-1)^{\frac{1}{\alpha}-1}}{(\alpha+1)^{\frac{1}{\alpha}+1}},$$

*for $\alpha \in [\alpha_2, (1+r)/(1-r)]$;*

*(ii) Each point $(x(\alpha), y(\alpha))$ of $\Gamma^+$ is given by a unique function $w = F(z, \alpha)$ satisfying the differential equation*

$$\varphi(z)dz^2 = \psi(w)dw^2,$$

where

$$\varphi(z)dz^2 = \frac{(z-d)^2(z-\bar{d})^2}{z^2(z-r)^2(z-1/r)^2}dz^2,$$

*such that $|d| = 1$ and $d$ is one of the conjugated solutions to the equation*

$$\alpha\frac{1-r}{1+r} = 1 - \frac{r(1+d)^2}{d(1+r)^2},$$

$$\psi(w)dw^2 = \frac{(w-c)(w-1/c)(w\pm 1)^2}{w^2(w-x(\alpha))^2(w-1/x(\alpha))^2}dw^2$$

where $c = c(\alpha)$ is a unique solution in $(x,1)$ to the equation

$$1-\alpha^2 = \frac{x(1+c)^2}{c(1+x)^2},$$

and we assume the sign (-) in nominator of $\psi$ for $\alpha \in [(1-r)/(1+r), \alpha_1]$; $c = c(\alpha)$ is a unique solution in $(-1,0)$ to the equation

$$1-\alpha^2 = \frac{x(1-c)^2}{c(1-x)^2},$$

and we assume the sign (+) for $\alpha \in [\alpha_2, (1+r)/(1-r)]$;

$$\psi(w)dw^2 = \frac{(w-c)^2(w-\bar{c})^2}{w^2(w-x(\alpha))^2(w-1/x(\alpha))^2}dw^2,$$

where $c = c(\alpha)$ is such that $|c| = 1$ and $c$ is one of the solutions to the equation

$$\alpha\frac{1-x}{1+x} = 1 - \frac{x(1+c)^2}{c(1+x)^2}$$

for $\alpha \in [\alpha_1, \alpha_2]$.

The function $F(z,\alpha)$ maps the unit disk $U$ onto itself with a slit along the piecewise analytic curve with two simmetric endpoints. The simply connected domain $F(U,\alpha)$ has the reduced modulus $b$ with respect to the origin.

Proof. Let $f \in B_s(b)$ with a fixed value of $|f(r)| = x$. The previous lemma asserts that there is a unique $\alpha$ such that $x(\alpha) = x$. Consider the functions $f_1(z)$ and $f_2(z)$ satisfying the equations

$$\left(\frac{df_1(z)}{f_1(z)}\right)^2 = 4\pi^2\varphi(z)dz^2, \quad \left(\frac{df_2(w)}{f_2(w)}\right)^2 = 4\pi^2\psi(w)dw^2,$$

where the differentials $\varphi(z)dz^2$ and $\psi(w)dw^2$ are defined in the statement of the theorem and $\alpha$ is chosen. Theorems 3.1.5, 3.1.6, the transformation $Z(z)$, and a suitable map from $U_w$ yield that the superposition $f_2^{-1} \circ f_1(z)$ maps conformally the domain $D_1^{z*}$ onto the domain $D_1^{w*}$ and the form of the differentials $\varphi$ and $\psi$ follows from that of the differentials $Q$ and $\Phi$ in Section 3.1.3. We continue this map analytically onto $D_2^{z*}$ and obtain the function $F(z,\alpha)$ that maps the unit disk onto the domain which is admissible with respect to the differential $\psi(w)dw^2$. The function satisfies the equality $F'(r,\alpha) = x(\alpha)$ and meets all conditions of the theorem. We have the obvious equality results from the change of the reduced moduli

$$m(f(D_1^{z*}),0) + \alpha^2 m(f(D_2^{z*}), f(r)) =$$

$$= m(D_1^{z*}, 0) + \alpha^2 m(D_2^{z*}, r) + \frac{1}{2\pi} \log b + \frac{\alpha^2}{2\pi} \log |f'(r)|.$$

We now symmetrize the pair of domains $(f(D_1^{z*}), f(D_2^{z*}))$ with respect to the origin and the positive real semiaxis. Let us denote the result of this symmetrization by $(\tilde{D}_1, \tilde{D}_2)$. This pair is admissible and the pair $(D_1^{w*}, D_2^{w*})$ is extremal in the family $\mathfrak{D}^w$, so we have the following chain of inequalities

$$m(f(D_1^{z*}), 0) + \alpha^2 m(f(D_2^{z*}), r) \leq m(\tilde{D}_1, 0) + \alpha^2 m(\tilde{D}_2, x) \leq$$

$$\leq m(D_1^{w*}, 0) + \alpha^2 m(D_2^{w*}, x) =$$

$$= m(D_1^{z*}, 0) + \alpha^2 m(D_2^{z*}, r) + \frac{1}{2\pi} \log b + \frac{\alpha^2}{2\pi} \log |F'(r, \alpha)|.$$

Then, $|f'(r)| \leq |F'(r, \alpha)| = y(\alpha)$. The uniqueness of the extremal configuration implies the uniqueness of the extremal function. $\qquad\Box$

A general form of the range of the system of functionals $(f(r), f'(r))$ in $B_s(b)$ was described by the parametric method by V. V. Goryaĭnov, V. Ya. Gutlyanskiĭ in [51]. Their result does not allow to deduce Theorem 3.3.6. One also could obtain a result on two-point distortion but huge formulations could make this useless. A collection of results in the class $B_s(b)$ by the variational, parametric, and area methods is presented in [143], [144].

## 3.4 Montel functions

Denote by $M(\omega)$ the class of all univalent holomorphic in $U$ functions $f(z) = a_1 z + a_2 z^2 + \ldots$ with the additional normalization $f(\omega) = \omega$ where $\omega \in (0, 1)$. This normalization is known as the Montel normalization. The subclass $M^M(\omega)$ means that $|f(z)| < M$ with $M > 1$ in $U$. The functions

$$K_1(z) = \frac{z(1 + \omega)^2}{(1 + z)^2}, \quad K_2(z) = \frac{z(1 - \omega)^2}{(1 - z)^2}$$

play the role of the Koebe functions for the class $M(\omega)$. The functions

$$K_1^M = M k_0^{-1} \left( M \left( \frac{1 + \omega}{M + \omega} \right)^2 k_0(z) \right), \quad K_2^M = M k_\pi^{-1} \left( M \left( \frac{1 - \omega}{M - \omega} \right)^2 k_\pi(z) \right)$$

play the role of the Pick functions for the class $M^M(\omega)$. Both classes are not invariant under rotation. This makes all problems more difficult.

### 3.4.1 Covering theorems

We start with the covering theorem for the class $M(\omega)$ that has been obtained by J. Krzyż and E. Złotkiewicz in 1971 [77]. We consider the modulus problem for the family $\Gamma$ of closed curves on $S_0 = \mathbb{C} \setminus \{0, \omega, A\}$ which are homotopic on $S_0$ to the slit along the segment $[0, \omega]$ if $A \notin [0, \omega]$ or along a broken line $[0, A + i\varepsilon] \cup [A + i\varepsilon, \omega]$ otherwise. Denote by $m(A)$ the modulus $m(S_0, \Gamma)$. In [78] the value of $m(A)$ was calculated as

$$m(A) = \frac{1}{2}\operatorname{Im} \tau \left( \frac{\omega^2}{A^2} \right),$$

where $\tau$ is the modular function given by

$$\tau(k^2) = \frac{i\mathbf{K}'(k)}{\mathbf{K}(k)}.$$

Here the elliptic integrals $\mathbf{K}(k)$ and $\mathbf{K}'(k)$ are understood to be the positive valued functions coinciding with usual elliptic integrals for $k^2 \in (0, 1)$, defined for $\operatorname{Im} k^2 \neq 0$ by the analytic continuation along any path not intersecting the real axis of the $k^2$-plane, and defined by analytic continuation along any path in the lower half-plane $\operatorname{Im} k^2 \leq 0$ for $k^2 > 1$.

We consider the set of points $A$ on $S_0$ satisfying the equation $m(A) = R = const > 0$. Then, this set is a closed smooth curve if $R > 1/4$ that encloses the points $0, \omega$. Otherwise, this set consists of two closed arcs with angles at their intersection with the real axis between $0, \omega$. The form of this curve was studied by G. Kuz'mina ([78], page 205). For $\operatorname{Im} A \geq 0$ the arc (or two arcs) of this curve is mapped by the function $\tau = \tau(k^2)$ onto one (or two) segment of the line $\operatorname{Im} \tau = R/2$.

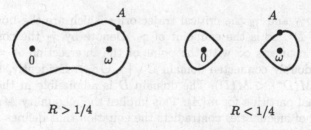

$$R > 1/4 \qquad\qquad\qquad R < 1/4$$

**Fig. 3.6.** The curve $m(A) = R$

**Theorem 3.4.1.** *The boundary of the Koebe set*

$$\mathfrak{K}_M = \bigcap_{f \in M(\omega)} f(U)$$

in the class $M(\omega)$ consists of the points that are the solutions to the equation

$$m(A) = \frac{\mathbf{K}(\omega)}{4\mathbf{K}'(\omega)}.$$

Each point $A \in \partial \mathfrak{K}_M$ is given by the unique function $f^* \in M(\omega)$ such that $A = f^*(e^{i\theta})$. It maps $U$ onto $\mathbb{C}$ with the slit along a smooth curve which is the critical trajectory of the quadratic differential

$$\varphi(w)dw^2 = e^{i\beta(A)} \frac{dw^2}{w(w - \omega)(w - A)}, \quad \beta(A) = -\arg\left(\frac{\omega^2}{A^2}\mathbf{K}^2\left(\frac{\omega}{A}\right)\right)$$

which is extremal for $m(A)$.

Proof. Denote by $U_z = U \setminus [0, \omega]$ with the modulus $M(U_z) = \mathbf{K}(\omega)/(4\mathbf{K}'(\omega))$. Suppose, there is a function $f \in M(\omega)$ such that the value $w_0 = f(e^{i\theta}) \in \mathfrak{K}_M$ for some $\theta$. The domain $f(U_z)$ is admissible in the modulus problem for one of the homotopy classes $\Gamma^{(n)}$ of curves that separate $0, \omega$ and $w_0, \infty$ on $S_0$. Lemma 3.2.1 asserts that the modulus $m(S_0, \Gamma^{(n)})$ attains its maximum over $(n)$ for $n = 1$, i.e., $m(S_0, \Gamma^{(n)}) \leq m(w_0)$. This means that

$$\frac{\mathbf{K}(\omega)}{4\mathbf{K}'(\omega)} = M(U_z) = M(f(U_z)) \leq m(w_0).$$

The extremal associated doubly connected domain $D^*$ in the problem of the extremal partition for $m(w_0)$ in $S_0 = \mathbb{C} \setminus \{0, \omega, w_0\}$ is the ring domain in the trajectory structure of the differential

$$\varphi(w)dw^2 = e^{i\beta(w_0)} \frac{dw^2}{w(w - \omega)(w - w_0)}, \quad \beta(w_0) = -\arg\left(\frac{\omega^2}{w_0^2}\mathbf{K}^2\left(\frac{\omega}{w_0}\right)\right).$$

Denote by $\gamma_1$ and $\gamma_2$ the critical trajectories which are the boundary components of $D^*$, $\infty$ is the endpoint of $\gamma_2$. Denote by $\tilde{\gamma}_2$ the connected part of $\gamma_2$ that connects $\infty$ with the point of the intersection $A = \gamma_2 \cap \partial \mathfrak{K}_M$, by $\tilde{D}$ the doubly connected domain $\mathbb{C} \setminus \{\gamma_1 \cup \tilde{\gamma}_2\}$. Evidently, $D^* \subset \tilde{D}$ and $m(w_0) = M(D^*) < M(\tilde{D})$. The domain $\tilde{D}$ is admissible in the problem of the extremal partition for $m(A)$. This implies the inequality $M(\tilde{D}) \leq m(A)$. This chain of inequalities contradicts the equation that defines $A$

$$m(A) = \frac{\mathbf{K}(\omega)}{4\mathbf{K}'(\omega)}.$$

This completes the proof.     □

An analogous theorem is valid for the bounded Montel functions. The modulus problem should be considered for the family $\Gamma$ of closed curves on $U_w = U(0, M) \setminus \{0, \omega, A\}$ which are homotopic on $U_w$ to the slit along the segment $[0, \omega]$ if $A \notin [0, \omega]$ or along a broken line $[0, A + i\varepsilon] \cup [A + i\varepsilon, \omega]$ otherwise. Denote by $m(A)$ the modulus $m(U_w, \Gamma)$.

**Theorem 3.4.2.** *The boundary of the Koebe set*

$$\mathfrak{K}_{MM} = \bigcap_{f \in M^M(\omega)} f(U)$$

*in the class $M^M(\omega)$ consists of the points which are the solutions to the equation*

$$m(A) = \frac{\mathbf{K}(\omega)}{4\mathbf{K}'(\omega)}.$$

*Each point $A \in \partial\mathfrak{K}_{MM}$ is given by the unique function $f^* \in M^M(\omega)$ such that $A = f^*(e^{i\theta})$. It maps $U$ onto $U(0, M)$ with the slit along the smooth curve which is the critical trajectory of the quadratic differential*

$$\varphi(w)dw^2 = \frac{\bar{\eta}(w - \eta)^2 dw^2}{w(w - \omega)(1 - \omega w)(w - A)(1 - \bar{A}w)},$$

*which is extremal for $m(A)$.*

Since there is no such a good description of the boundary curve of $\mathfrak{K}_{MM}$ as for the class $M(\omega)$ we give a theorem about its geometric properties.

**Theorem 3.4.3.** *The Koebe set $\mathfrak{K}_{MM}$ in the class $M^M(\omega)$ contains the union of the disk $U(0, |K_2^M(-1)|)$ and its image under the transformation*

$$\Lambda(\zeta) = \frac{\omega - \zeta}{1 - \frac{\zeta\omega}{M^2}}.$$

*Proof.* A simple application of the functions $K_1^M(z)$ and $K_2^M(z)$ yields that the points $K_2^M(-1)$ and $K_1^M(1)$ belong to $\partial\mathfrak{K}_{MM}$. Let there be a point $A \in \partial\mathfrak{K}_{MM}$, $A \in U(0, |K_2^M(-1)|)$. Let $m(A)$ be the modulus of Theorem 3.4.1 and $D^*$ be the extremal domain for $m(A)$. Denote by $D^s$ the result of circular symmetrization applied to $D^*$ with respect to the origin and the positive real axis. Evidently, $D^s \subset \{U(0, M) \setminus \{(-M, K_2^M(-1)] \cup [0, \omega]\}\}$. Hence,

$$m(A) = M(D^*) \leq M(D^s) < m(K_2^M(-1)).$$

This contradicts $K_2^M(-1) \in \partial\mathfrak{K}_{MM}$.

One can observe that $K_1^M(1) = \Lambda(K_2^M(-1))$, $K_1^M(1) \in \partial\mathfrak{K}_{MM}$. Let again $A \in \partial\mathfrak{K}_{MM}$, $A \in \Lambda(U(0, |K_2^M(-1)|))$. Then, $m(A) = M(\Lambda^{-1}(D^*))$ and $\Lambda^{-1}(A) \in U(0, |K_2^M(-1)|)$. Next we symmetrize the domain $\Lambda^{-1}(D^*)$ as in the previous case and widen the domain, which is obtained from the $\Lambda$-image of the result of symmetrization, up to the domain $U(0, M) \setminus \{[0, \omega] \cup [K_1^M(1), M)\}$. As a result, we obtain the inequality

$$M(D^*) = m(A) = M(\Lambda^{-1}(D^*)) < m(K_1^M(1)).$$

This contradicts $K_1^M(1) \in \partial \Re_{MM}$ and finishes the proof.    □

These results have been obtained by J. Krzyż [77], R. Libera, E. Złotkiewicz [92], and A. Vasil'ev, P. Pronin [165].

### 3.4.2 Distortion at the points of normalization

**Theorem 3.4.4.** *Let $f \in M(\omega)$, $f(z) = a_1 z + a_2 z^2 + \ldots$. Then,*

$$(1 - \omega)^2 \leq a_1 \leq (1 + \omega^2), \quad \frac{1 - \omega}{1 + \omega} \leq |f'(\omega)| \leq \frac{1 + \omega}{1 - \omega}.$$

*The extremal functions are only $K_1(z)$ and $K_2(z)$.*

These estimates immediately follow from the connection between the class $S$ and the class $M(\omega)$:

$$(g(z) \in S) \Longleftrightarrow \left( \frac{\omega g(z)}{g(\omega)} \in M(\omega) \right).$$

Now we learn more explicitly the mutual change of $|f'(0)|$ and $|f'(\omega)|$. Denote by $\Gamma^+$ the arc of $y = \max |f'(\omega)|$ as $f \in M(\alpha)$ with $|f'(0)| = x$ fixed, and by $\Gamma^-$ the arc of $y = \min |f'(\omega)|$ as $f \in M(\omega)$ with $|f'(0)| = x$ fixed. We determine here the range of the system of the functionals $\mathfrak{M}_f = \{(|f'(0)|, |f'(\omega)|), f \in M(\omega)\}$ by the modulus method analogously to the system of functionals $(|f(z)|, |f'(z)|)$ in the class $S$ in Section 3.1.4.

**Theorem 3.4.5.** *The boundary curve $\Gamma^-$ of the range $\mathfrak{M}_f$ of the system of functionals $(|f'(0)|, |f'(\omega)|)$ in the class $M(\omega)$ consists of the points*

$$\left( 1 - u\omega + \omega^2, \frac{1 - \omega^2}{1 - u\omega + \omega^2} \right), \quad -2 \leq u \leq 2.$$

*Each point is given only by the function*

$$g(z) = \frac{z(1 - u\omega + \omega^2)}{1 - uz + z^2}.$$

*Proof.* Since

$$(1 - \omega)^2 \leq 1 - u\omega + \omega^2 \leq (1 + \omega^2)$$

for $u \in [-2, 2]$, we choose for $|f'(0)| = x$ the unique

$$u = u_0 = \frac{1 + \omega^2 - x}{\omega},$$

so that $a_1 = x = |f'(0)| = g'(0)$. We consider the digon $U'_z = U \setminus \{(-1, 0] \cup [\omega, 1)\}$ with two vertices $0, \omega$. It is conformal at its vertices, has the angles $2\pi$, and the reduced modulus is

$$m(U'_z, 0, \omega) = \frac{1}{2\pi} \log \frac{\omega^2}{1 - \omega^2}.$$

The digon $f(U'_z)$ with the vertices $0, \omega$ has the reduced modulus

$$m(f(U'_z), 0, \omega) = \frac{1}{2\pi} \log \frac{\omega^2}{1 - \omega^2} |f'(\omega)| x.$$

It is admissible in the problem of minimum of the reduced modulus over all digons having the homotopy type defined by $\gamma = [0, \omega]$, conformal at $0, \omega$, with appropriate angles at the vertices. The extremal reduced modulus $\frac{1}{2\pi} \log \omega^2$ is given by the digon obtained from $\mathbb{C}$ by slitting along two rays starting from $0$ and $\omega$ lying on the straight line that passes through these two points. Therefore, $|f'(r)| \geq \frac{1 - \omega^2}{x}$. This is equivalent to the statement of the theorem. The uniqueness of the extremal configuration $g(U')$ leads to the uniqueness of the extremal map.    □

Now we use the moduli calculated in Section 3.1.3 to define the upper boundary curve $\Gamma^+$ for the range of $(|f'(0)|, |f'(\omega)|)$ in the class $M(\omega)$. Our proof is based on simultaneous consideration of two problems of the extremal partition in the punctured unit disk and in the punctured Riemann sphere. The proof is somewhat similar to that for the system of functionals $(|f(z)|, |f'(z)|)$ in $S$ in 3.1.4.

Let $U_z = U \setminus \{0, \omega\}$ be the punctured unit disk. We consider on $U_z$ the admissible system $(\gamma_1^z, \gamma_2^z)$ of type III where $\gamma_2^z = \{z : |z - \omega| = \varepsilon\}$ and $\gamma_1^z = \{z : |z| = \varepsilon\}$. Here $\varepsilon$ is sufficiently small so that $\omega + \varepsilon < 1$ and $\varepsilon < \omega/2$. Let $\mathfrak{D}^z$ be the set of all pairs $(D_1^z, D_2^z)$ of simply connected domains of homotopy type $(\gamma_1^z, \gamma_2^z)$. Then, the problem of the extremal partition of $U_z$ consists of maximizing the sum $m(D_1^z, 0) + \alpha^2 m(D_2^z, \omega)$ over $(D_1^z, D_2^z) \in \mathfrak{D}^z$. The maximum of this sum we denote by $\mathcal{M}_z(\alpha, \omega)$. Under the transformation

$$Z(z) = 1 - \frac{\omega(1 + z)^2}{z(1 + \omega)^2}$$

two extremal domains $(D_1^{z*}, D_2^{z*})$ in the problem for $\mathcal{M}_z(\alpha, \omega)$ are mapped onto two extremal domains $(B_1^*, B_2^*)$ in the problem of finding $\mathcal{M}(\alpha, c)$, where $c = (1 - \omega)^2 / (1 + \omega)^2$. Taking into account the change of the reduced modulus under the conformal map $Z(z)$, we deduce that for $(1 - \omega)/(1 + \omega) \leq \alpha \leq (1 + \omega)/(1 - \omega)$ the reduced moduli are given as

$$m(D_1^{z*}, 0) = \frac{1}{2\pi} \log \omega^\alpha, \quad m(D_2^{z*}, \omega) = \frac{1}{2\pi} \log \omega^{1/\alpha}(1 - \omega^2).$$

Let $\mathbb{C}_w = \mathbb{C} \setminus \{0, \omega\}$ be the punctured Riemann sphere. We consider on $\mathbb{C}_w$ the admissible system $(\gamma_1^w, \gamma_2^w)$ of type III where $\gamma_1^w = \{z : |w| = \varepsilon\}$ and $\gamma_2^z = \{z : |w - \omega| = \varepsilon\}$. Here $\varepsilon$ is sufficiently small so that $\varepsilon < \omega/2$. Let $\mathfrak{D}^w$ be the set of all pairs $(D_1^w, D_2^w)$ of simply connected domains of homotopy type $(\gamma_1^w, \gamma_2^w)$. Then, the problem of extremal partition of $\mathbb{C}_w$ consists of maximizing the sum $m(D_1^w, 0) + \alpha^2 m(D_2^w, \omega)$ over all $(D_1^w, D_2^w) \in \mathfrak{D}^w$. The maximum of this sum we denote by $M_w(\alpha)$. Under the transformation

$$W(w) = \frac{\omega - w}{\omega w}$$

two extremal domains $(D_1^{w*}, D_2^{w*})$ in the problem for $M_w(\alpha)$ are mapped onto two extremal domains $(D_1^*, D_2^*)$ in the problem of finding $M(\alpha)$. Taking into account the change of the reduced moduli under the conformal map $W(w)$, we deduce that for $\alpha \neq 1$

$$m(D_1^{w*}, 0) = \frac{1}{2\pi} \log 4\omega \frac{|1 - \alpha|^{\alpha-1}}{|1 + \alpha|^{\alpha+1}}, \quad m(D_2^{w*}, r) = \frac{1}{2\pi} \log 4\omega\alpha^2 \frac{|1 - \alpha|^{(1/\alpha)-1}}{|1 + \alpha|^{(1/\alpha)+1}}.$$

For $\alpha = 1$ we have

$$m(D_1^{w*}, 0) = m(D_2^{w*}, 0) = \frac{1}{2\pi} \log \omega.$$

Lemma 3.1.1 in our case is of the following form.

**Lemma 3.4.1.** *For $(1 - \omega)/(1 + \omega) \leq \alpha \leq (1 + \omega)/(1 - \omega)$ the function*

$$x(\alpha) = \exp(2\pi(m(D_1^{w*}, 0) - m(D_1^{z*}, 0)))$$

*is continuous and strictly decreases in $\alpha$ from $(1 + \omega)^2$ to $(1 - \omega)^{-2}$.*

*Proof.* For $(1 - \omega)/(1 + \omega) \leq \alpha < 1$ we have the derivative

$$\frac{x'(\alpha)}{x(\alpha)} = \frac{1}{2\pi} \log \frac{1 - \alpha}{(1 + \alpha)\omega}.$$

This implies $x'(\alpha) < 0$. It is easy to obtain that

$$\lim_{\alpha \to 1+0} x(\alpha) = \lim_{\alpha \to 1-0} x(\alpha) = x(1).$$

The case $\alpha > 1$ is proved similarly.    □

**Theorem 3.4.6.** *(i) The upper boundary curve $\Gamma^+$ of the range $\mathfrak{M}_f$ of the system of functionals $(|f'(0)|, |f'(\omega)|)$ in the class $M(\omega)$ consists of the points $(x(\alpha), y(\alpha))$ where*

$$\alpha \in \left[\frac{1 - \omega}{1 + \omega}, \frac{1 + \omega}{1 - \omega}\right], \quad \alpha \neq 1,$$

$$x(\alpha) = 4\omega^{1-\alpha}\frac{|1-\alpha|^{\alpha-1}}{|1+\alpha|^{\alpha+1}},$$

$$y(\alpha) = \frac{4\omega^{1-\frac{1}{\alpha}}\alpha^2|1-\alpha|^{(1/\alpha)-1}}{(1-\omega^2)|1+\alpha|^{(1/\alpha)+1}}.$$

If $\alpha = 1$, then $x(1) = 1$, $y(1) = 1/(1-\omega^2)$.

(ii) *Each point $(x(\alpha), y(\alpha))$ of $\Gamma^+$ is given by the unique function $F(z, \alpha)$ satisfying the differential equation*

$$\varphi(z)dz^2 = \psi(w)dw^2, \quad \alpha \neq 1,$$

$$\varphi(z)dz^2 = \frac{(z-d)^2(z-\bar{d})^2}{z^2(z-\omega)^2(z-1/\omega)^2}dz^2,$$

$$\psi(w)dw^2 = \frac{w-C}{w^2(w-x(\alpha))^2}dw^2,$$

*such that $|d| = 1$ and $d$ is one of the conjugated solutions to the equation*

$$\alpha\frac{1-\omega}{1+\omega} = 1 - \frac{\omega(1+d)^2}{d(1+\omega)^2},$$

$$C = C(\alpha) = \frac{x(\alpha)}{1-\alpha^2}.$$

*The function $F(z, \alpha)$ maps the unit disk onto the complex plane $\mathbb{C}$ minus the ray $[C(\alpha), +\infty)$ in case $\alpha < 1$ or $(-\infty, C(\alpha)]$ otherwise, and two smooth arcs of trajectories of the differential $\psi(w)\,dw^2$ symmetric with respect to $\mathbb{R}$ such that the simply connected domain $F(U, \alpha)$ has the reduced modulus $x(\alpha)$ with respect to the origin.*

*If $\alpha = 1$, then*

$$\varphi(z)dz^2 = \frac{(z-d)^2(z-\bar{d})^2}{z^2(z-\omega)^2(z-1/\omega)^2}dz^2,$$

$$\psi(w)dw^2 = \frac{-1}{w^2(w-\omega)^2}dw^2,$$

*and the function $F(z, 1)$ maps the unit disk onto the complex plane $\mathbb{C}$ slit along the symmetric with respect to $\mathbb{R}$ rays along the straight line $\mathrm{Re}\, w = \omega/2$ with the analogous normalization.*

*Proof.* Let $f \in M(\omega)$ with a fixed value of $|f'(0)| = x$. The preceding lemma asserts that there is such $\alpha$ that $x(\alpha) = x$. We consider the functions $f_1(z)$ and $f_2(z)$ satisfying the equations

$$\left(\frac{df_1(z)}{f_1(z)}\right)^2 = 4\pi^2\varphi(z)dz^2, \quad \left(\frac{df_2(w)}{f_2(w)}\right)^2 = 4\pi^2\psi(w)dw^2,$$

where the differentials $\varphi(z)dz^2$ and $\psi(w)dw^2$ are defined in the statement of the theorem and $\alpha$ is chosen. The superposition $f_2^{-1} \circ f_1(z)$ maps conformally the domain $D_1^{z*}$ onto the domain $D_1^{w*}$. Furthermore, we continue this map analytically onto $D_2^{z*}$ and obtain the function $F(z, \alpha)$ that maps the unit disk onto the domain which is admissible with respect to the differential $\psi(w)dw^2$. The function satisfies the equality $F'(0, \alpha) = x(\alpha)$ and meets all conditions of the theorem. Since the pair $(D_1^{w*}, D_2^{w*})$ is extremal in the family $\mathfrak{D}^w$, we deduce that the following chain of inequalities

$$m(f(D_1^{z*}), 0) + \alpha^2 m(f(D_2^{z*}), \omega) =$$

$$= m(D_1^{z*}, 0) + \alpha^2 m(D_2^{z*}, \omega) + \frac{1}{2\pi} \log x(\alpha) + \frac{\alpha^2}{2\pi} \log |f'(\omega)| \leq$$

$$\leq m(D_1^{w*}, 0) + \alpha^2 m(D_2^{w*}, \omega) =$$

$$= m(D_1^{z*}, 0) + \alpha^2 m(D_2^{z*}, \omega) + \frac{1}{2\pi} \log x(\alpha) + \frac{\alpha^2}{2\pi} \log |F'(\omega, \alpha)|$$

is valid. Then, $|f'(\omega)| \leq |F'(\omega, \alpha)| = y(\alpha)$. The uniqueness of the extremal configuration implies the uniqueness of the extremal function.      $\square$

The boundary curves $\Gamma^+$ and $\Gamma^-$ were obtained by S. Dëmin in [24] with some calculus errors.

We consider now the problem for the class of bounded Montel functions. From [92] it follows that if $f \in M^M(\omega)$, then

$$\frac{M^2(1 - \omega)^2}{(M - \omega)^2} \leq |f'(0)| \leq \frac{M^2(1 + \omega)^2}{(M + \omega)^2}.$$

The equality sign is attained by the function $K_1^M$ for the right-hand side inequality and by the function $K_2^M$ for the left-hand side inequality. It is not difficult to obtain the analogous estimates for $|f'(\omega)|$:

$$\frac{(M + \omega)(1 - \omega)}{(M - \omega)(1 + \omega)} \leq |f'(\omega)| \leq \frac{(M - \omega)(1 + \omega)}{(M + \omega)(1 - \omega)},$$

with the same extremal functions. Therefore, the range of the system of functionals $(|f'(0)|, |f'(\omega)|)$ lies inside the rectangle defined by the preceding inequalities. To obtain the sharp form of the boundary curve we consider some special problems of the extremal partition.

Let $U_z = U \setminus \{0, \omega\}$, $\omega \in (0, 1)$ be the punctured unit disk. We consider on $U_z$ the admissible system of curves $(\gamma_1^z, \gamma_2^z)$ where $\gamma_2^z = \{z : |z - \omega| = \varepsilon\}$ and $\gamma_1^z = \{z : |z| = \varepsilon\}$. Here $\varepsilon$ is sufficiently small so that $\omega + \varepsilon < 1$ and $\varepsilon < \omega/2$. Let $\mathfrak{D}^z$ be the set of all pairs $(D_1^z, D_2^z)$ of simply connected domains of homotopy type $(\gamma_1^z, \gamma_2^z)$. Then, the problem of the extremal partition of $U_z$ consists of maximizing the sum $m(D_1^z, 0) + \alpha^2 m(D_2^z, \omega)$ over $(D_1^z, D_2^z) \in \mathfrak{D}^z$. The maximum of this sum we denote by $M_z(\alpha, \omega)$. Under the transformation

$$Z(z) = 1 - \frac{\omega(1+z)^2}{z(1+\omega)^2}$$

two extremal domains $(D_1^{z*}, D_2^{z*})$ in the problem for $M_z(\alpha, \omega)$ are mapped onto two extremal domains $(B_1^*, B_2^*)$ in the problem of finding $\mathcal{M}(\alpha, c)$ where $c = (1-\omega)^2/(1+\omega)^2$. Taking into account the change of the reduced modulus under the conformal map $Z(z)$ we deduce that for $(1-\omega)/(1+\omega) \leq \alpha \leq (1+\omega)/(1-\omega)$

$$m(D_1^{z*}, 0) = \frac{1}{2\pi} \log \omega^\alpha, \quad m(D_2^{z*}, \omega) = \frac{1}{2\pi} \log \omega^{1/\alpha}(1-\omega^2).$$

Let $U_w = U(0, M) \setminus \{0, \omega\}$ be the punctured disk of the radius $M$. We consider on $U_w$ the admissible system of curves $(\gamma_1^w, \gamma_2^w)$ where $\gamma_2^w = \{w : |w - \omega| = \varepsilon\}$ and $\gamma_1^w = \{w : |w| = \varepsilon\}$. Here $\varepsilon$ is sufficiently small as before. Let $\mathfrak{D}^w$ be the set of all pairs $(D_1^w, D_2^w)$ of simply connected domains of homotopy type $(\gamma_1^w, \gamma_2^w)$. Then, the problem of the extremal partition of $U_w$ consists of maximizing the sum $m(D_1^w, 0) + \alpha^2 m(D_2^w, \omega)$ over $(D_1^w, D_2^w) \in \mathfrak{D}^z$. The maximum of this sum we denote by $M_w(\alpha, \omega)$. Theorems 3.1.5, 3.1.6 and suitable conformal maps imply that for $(M - \omega)/(M + \omega) \leq \alpha \leq (M + \omega)/(M - \omega)$

$$m(D_1^{w*}, 0) = \frac{1}{2\pi} \log M^{1-\alpha}\omega^\alpha,$$

$$m(D_2^{w*}, \omega) = \frac{1}{2\pi} \log \frac{\omega^{1/\alpha}}{M^{1+1/\alpha}}(M^2 - \omega^2).$$

For $0 \leq \alpha \leq (M - \omega)/(M + \omega)$

$$m(D_1^{w*}, 0) = \frac{1}{2\pi} \log \frac{M^2\omega}{(M+\omega)^2} \cdot \frac{4(1-\alpha)^{\alpha-1}}{(1+\alpha)^{\alpha+1}},$$

$$m(D_2^{w*}, \omega) = \frac{1}{2\pi} \log \frac{\omega(M+\omega)}{M-\omega} \cdot \frac{4\alpha^2(1-\alpha)^{\frac{1}{\alpha}-1}}{(1+\alpha)^{\frac{1}{\alpha}+1}}.$$

For $\alpha \geq (M + \omega)/(M - \omega)$

$$m(D_1^{w*}, 0) = \frac{1}{2\pi} \log \frac{M^2\omega}{(M-\omega)^2} \cdot \frac{4(\alpha-1)^{\alpha-1}}{(\alpha+1)^{\alpha+1}},$$

$$m(D_2^{w*}, \omega) = \frac{1}{2\pi} \log \frac{\omega(M-\omega)}{M+\omega} \cdot \frac{4\alpha^2(\alpha-1)^{\frac{1}{\alpha}-1}}{(\alpha+1)^{\frac{1}{\alpha}+1}}.$$

Here (*) denotes the extremality of domains.

We define now the problems of the extremal partition of the punctured disk $U_w$. Let $0, \omega$ be the punctures of $U(0, M)$. We consider the family $\mathbb{D}_1$ of digons $D$ in $U$ such that $0, \omega \notin D$ and they are two vertices of any $D \in \mathbb{D}_1$ with the angles $2\pi$. Al digons are supposed to be conformal at their vertices. We define the problem of finding the minimum

$$\min_{D \in \mathbb{D}_1} m(D, 0, \omega).$$

There is a unique digon $D_1^* = U(M) \setminus \{(-M, 0] \cup [\omega, M)\}$ that gives this minimum. One can calculate this reduced modulus by making use of suitable conformal maps of $D_1^*$ onto the digon $\mathbb{C} \setminus [0, \infty)$ of modulus zero with respect to its vertices $0, \infty$.

$$m(D_1^*, 0, \omega) = \frac{1}{2\pi} \log \frac{M^2 \omega^2}{M^2 - \omega^2}.$$

We evaluate now the system of the functionals $(|f'(0)|, |f'(\omega)|)$, $f \in M^M(\omega)$.

Denote by $\Gamma^+$ the arc of $y = \max |f'(\omega)|$ as $f \in M^M(\omega)$ with $|f'(0)| = x$ fixed, and by $\Gamma^-$ the arc of $y = \min |f'(\omega)|$ as $f \in M^M(\omega)$ with $|f'(0)| = x$ fixed. We determine here $\mathfrak{M}_f$ by the reduced moduli calculated before. Set the functions

$$g(z) = \frac{z}{1 - uz + z^2}, \quad u \in [-2, 2],$$

and

$$G(z) = M g^{-1} \left( M \left( \frac{1 - u\omega + \omega^2}{M^2 - Mu\omega + \omega^2} \right) g(z) \right).$$

**Theorem 3.4.7.** *The boundary curve $\Gamma^-$ of the range $\mathfrak{M}_f$ of the system of functionals $(|f'(0)|, |f'(\omega)|)$ in the class $M^M(\omega)$ consists of the points $(x(u), y(u))$, where*

$$x = \frac{M^2(1 - u\omega + \omega^2)}{M^2 - Mu\omega + \omega^2},$$

$$y = \frac{(1 - \omega^2)(M^2 - Mu\omega + \omega^2)}{(M^2 - \omega^2)(1 - u\omega + \omega^2)},$$

*as $-2 \leq u \leq 2$. Each point is given only by the function $G(z)$.*

*Proof.* Since

$$\frac{M^2(1 - \omega)^2}{(M - \omega)^2} \leq |G'(0)| = \frac{M^2(1 - u\omega + \omega^2)}{M^2 - Mu\omega + \omega^2} \leq \frac{M^2(1 + \omega)^2}{(M + \omega)^2}$$

for $u \in [-2, 2]$, we choose for a function $f \in M^M(\omega)$ the unique

$$u = u_0 = \frac{M^2(1 - x) + \omega^2(M^2 - x)}{M\omega(M - x)},$$

such that $x = |f'(0)| = G'(0)$. We consider the digon $U_z' = U \setminus \{(-1, 0] \cup [\omega, 1)\}$ with two vertices $0, \omega$ and the reduced modulus

$$m(U_z', 0, \omega) = \frac{1}{2\pi} \log \frac{\omega^2}{1 - \omega^2}.$$

The digon $f(U_z')$ with the vertices $0, \omega$ has the reduced modulus

$$m(f(U_z'), 0, \omega) = \frac{1}{2\pi} \log \frac{\omega^2}{1 - \omega^2} |f'(\omega)| x.$$

It is admissible in the problem of minimizing the reduced modulus over all digons of homotopy type defined by $\gamma = [0, \omega]$ with the angles $2\pi$ which are conformal at $0, \omega$. The extremal reduced modulus

$$m(D_1^*, 0, \omega) = \frac{1}{2\pi} \log \frac{M^2 \omega^2}{M^2 - \omega^2}$$

is given by the digon obtained from $U(M)$ by slitting along two segments $(-M, 0]$ and $[\omega, M)$. Therefore, $|f'(r)| \geq \frac{M^2(1 - \omega^2)}{(M^2 - \omega^2)x}$. This is equivalent to the statement of the theorem. The uniqueness of the extremal configuration $G(U_z')$ leads to the uniqueness of the extremal map.     □

We define now the curve $\Gamma^+$. For this we need the following technical lemma

**Lemma 3.4.2.** *For $(1 - \omega)/(1 + \omega) \leq \alpha \leq (1 + \omega)/(1 - \omega)$ the function*

$$x(\alpha) = \exp(2\pi(m(D_1^{w*}, 0) - m(D_1^{z*}, 0)))$$

*is continuous and strictly decreases with respect to $\alpha$*

$$\text{from} \quad \frac{M^2(1 + \omega)^2}{(M + \omega)^2} \quad \text{to} \quad \frac{M^2(1 - \omega)^2}{(M - \omega)^2}.$$

*Here the moduli $m(D_1^{w*}, 0)$ and $m(D_1^{z*}, 0)$ are defined before.*

*Proof.* For

$$\alpha \in \left[ \frac{M - \omega}{M + \omega}, \frac{M + \omega}{M - \omega} \right]$$

the value of $x(\alpha) = M^{1-\alpha}$ obviously decreases.
For

$$\alpha \in \left[ \frac{1 - \omega}{1 + \omega}, \frac{M - \omega}{M + \omega} \right]$$

we have the derivative

$$\frac{x'(\alpha)}{x(\alpha)} = \log \frac{1 - \alpha}{(1 + \alpha)\omega}.$$

This implies $x'(\alpha) < 0$. The case

$$\alpha \in \left[ \frac{M + \omega}{M - \omega}, \frac{1 + \omega}{1 - \omega} \right]$$

is considered analogously.     □

**Theorem 3.4.8.** *(i) The upper boundary curve $\Gamma^+$ for the range $\mathfrak{M}_f$ of the system of functionals $(|f'(0)|, |f'(\omega)|)$ in the class $M^M(\omega)$ consists of the points $(x(\alpha), y(\alpha))$, $\alpha \in [(1-\omega)/(1+\omega), (1+\omega)/(1-\omega)]$. Here*

$$x(\alpha) = \frac{4M^2\omega^{1-\alpha}}{(M+\omega)^2} \cdot \frac{(1-\alpha)^{\alpha-1}}{(1+\alpha)^{\alpha+1}},$$

$$y(\alpha) = \frac{4\omega^{1-1/\alpha}(M+\omega)}{(1-\omega^2)(M-\omega)} \cdot \frac{\alpha^2(1-\alpha)^{\frac{1}{\alpha}-1}}{(1+\alpha)^{\frac{1}{\alpha}+1}},$$

*for $\alpha \in [(1-\omega)/(1+\omega), (M-\omega)/(M+\omega)]$;*

$$x(\alpha) = M^{1-\alpha},$$

$$y(\alpha) = \frac{M^2 - \omega^2}{M^{1+1/\alpha}(1-\omega^2)},$$

*for $\alpha \in [(M-\omega)/(M+\omega), (M+\omega)/(M-\omega)]$;*

$$x(\alpha) = \frac{4M^2\omega^{1-\alpha}}{(M-\omega)^2} \cdot \frac{(\alpha-1)^{\alpha-1}}{(\alpha+1)^{\alpha+1}},$$

$$y(\alpha) = \frac{4\omega^{1-1/\alpha}(M-\omega)}{(1-\omega^2)(M+\omega)} \cdot \frac{\alpha^2(\alpha-1)^{\frac{1}{\alpha}-1}}{(\alpha+1)^{\frac{1}{\alpha}+1}},$$

*for $\alpha \in [(M+\omega)/(M-\omega), (1+\omega)/(1-\omega)]$;*

*(ii) Each point $(x(\alpha), y(\alpha))$ of $\Gamma^+$ is given by the unique function $F(z, \alpha)$ satisfying the differential equation*

$$\varphi(z)dz^2 = \psi(w)dw^2,$$

*where*

$$\varphi(z)dz^2 = \frac{(z-d)^2(z-\bar{d})^2}{z^2(z-\omega)^2(z-1/\omega)^2}dz^2,$$

*such that $|d| = 1$ and $d$ is one of the conjugated solutions to the equation*

$$\alpha\frac{1-\omega}{1+\omega} = 1 - \frac{\omega(1+d)^2}{d(1+\omega)^2},$$

$$\psi(w)dw^2 = \frac{(w-c)(w-M^2/c)(w-M)^2}{w^2(w-\omega)^2(w-M^2/\omega)^2}dw^2,$$

*where $c = c(\alpha)$ is a unique solution in $(\omega, M)$ to the equation*

$$1 - \alpha^2 = \frac{\omega(M+c)^2}{c(M+\omega)^2}$$

*for $\alpha \in [(1-\omega)/(1+\omega), (M-\omega)/(M+\omega)]$;*

$$\psi(w)dw^2 = \frac{(w-c)(w-M^2/c)(w+M)^2}{w^2(w-\omega)^2(w-M^2/\omega)^2}dw^2$$

and $c = c(\alpha)$ is a unique solution in $(-M, 0)$ to the equation

$$1 - \alpha^2 = \frac{\omega(M-c)^2}{c(M-\omega)^2}$$

for $\alpha \in [(M+\omega)/(M-\omega),\, (1+\omega)/(1-\omega)]$;

$$\psi(w)dw^2 = \frac{(w-c)^2(w-M^2/c)^2}{w^2(w-\omega)^2(w-M^2/\omega)^2}dw^2,$$

where $c = c(\alpha)$ is such that $|c| = M$ and $c$ is one of the solutions to the equation

$$\alpha\frac{1-\omega}{1+\omega} = 1 - \frac{\omega(M+c)^2}{c(M+\omega)^2}$$

for $\alpha \in [(M-\omega)/(M+\omega),\, (M+\omega)/(M-\omega)]$.

The function $F(z, \alpha)$ maps the unit disk onto the disk $U(M)$ slit along a piecewise analytic curve with two simmetric endpoints. The simply connected domain $F(U, \alpha)$ has the reduced modulus $x(\alpha)$ with respect to the origin.

Proof. Let $f \in M^M(\omega)$ with a fixed value of $|f'(0)| = x$. The previous lemma asserts that there is a unique $\alpha$ such that $x(\alpha) = x$. Consider the functions $f_1(z)$ and $f_2(w)$ satisfying the equations

$$\left(\frac{df_1(z)}{f_1(z)}\right)^2 = 4\pi^2\varphi(z)dz^2, \quad \left(\frac{df_2(w)}{f_2(w)}\right)^2 = 4\pi^2\psi(w)dw^2,$$

where the differentials $\varphi(z)dz^2$ and $\psi(w)dw^2$ are defined in the statement of the theorem and $\alpha$ is chosen. Theorems 3.1.5, 3.1.6, the transformation $Z(z)$, and a suitable map from the surface $U_w$ yield that the superposition $f_2^{-1} \circ f_1(z)$ maps conformally the domain $D_1^{z*}$ onto the domain $D_1^{w*}$. The form of the differentials $\varphi$ and $\psi$ follows from that of the differentials $Q$ and $\Phi$ of Section 3.1.3. Continuing this mapping analytically onto $D_2^{z*}$ we obtain the function $F(z, \alpha)$ that maps the unit disk onto the domain which is admissible with respect to the differential $\psi(w)dw^2$. The function satisfies the equality $F'(0, \alpha) = x(\alpha)$ and meets all requirements of the theorem. Since the pair $(D_1^{w*}, D_2^{w*})$ is extremal in the family $\mathfrak{D}^w$, we have the following chain of inequalities

$$m(f(D_1^{z*}), 0) + \alpha^2 m(f(D_2^{z*}), \omega) =$$

$$= m(D_1^{z*}, 0) + \alpha^2 m(D_2^{z*}, \omega) + \frac{1}{2\pi}\log x(\alpha) + \frac{\alpha^2}{2\pi}\log|f'(\omega)| \le$$

$$\le m(D_1^{w*}, 0) + \alpha^2 m(D_2^{w*}, \omega) =$$

$$= m(D_1^{z*}, 0) + \alpha^2 m(D_2^{z*}, \omega) + \frac{1}{2\pi}\log x(\alpha) + \frac{\alpha^2}{2\pi}\log|F'(\omega, \alpha)|.$$

Then, $|f'(\omega)| \le |F'(\omega, \alpha)| = y(\alpha)$. The uniqueness of the extremal configuration implies the uniqueness of the extremal function. $\qquad\square$

### 3.4.3 The range of $(|f(r)|, |f'(r)|)$ in $M_R(\omega)$

In the class of all holomorphic univalent in $U$ functions with normalization $f(0) = 0$ we have the inequality (see [100])

$$\frac{|f(r_2)|}{|f(r_1)|} \leq \frac{r_2(1 - r_1)^2}{r_1(1 - r_2)^2}$$

for $0 < r_1 < r_2 < 1$. So, the upper bound of $|f(r)|$ in $M(\omega)$ for $\omega < r < 1$ easily follows from this estimate

$$|f(r)| \leq \frac{r(1 - \omega)^2}{(1 - r)^2}$$

with the extremal function $K_2$. The analogous easy estimate holds for $|f(-r)|$. However, the extremal functions for the case $0 < r < \omega$ or for the complex value of $r$ have more complicated nature and do not map symmetric structures in $U$ onto those in $\mathbb{C}$. We do not intend to develop these features here and refer the reader to J. Krzyż [76] who obtained the range of $f(z)$ in the class $M(\omega)$ by the variational method. Since we operate with symmetric extremal structures we consider in this section the class $M_R(\omega)$ of functions from $M(\omega)$ with real coefficients in their Taylor expansion. This class plays the same role as $S_R$ for $S$.

For this class the estimates of $|f(r)|$ can be obviously derived from the connection with the class $S_R$. Namely,

$$K_1(r) \leq f(r) \leq K_2(r), \text{ for } \omega \leq r < 1,$$

$$K_2(r) \leq f(r) \leq K_1(r), \text{ for } 0 \leq r < \omega,$$

$$K_1(-r) \leq f(-r) \leq K_2(-r), \text{ for } -1 < r < 0.$$

Now we obtain the range of the system of functionals $(f(r), f'(r))$ for the class $M_R(\omega)$. For this we need to calculate some special moduli of extremal strip domains. We consider on the Riemann surface $S_0 = \mathbb{C} \setminus \{-1, a, 1\}$, $-1 < a < 1$ the pair of non-intersected curves $(\gamma_1, \gamma_2)$ of type III forming an admissible system on $S_0$. The curve $\gamma_1$ starting and ending at $\infty$ separates the puncture at $1$ from $-1, a$; the curve $\gamma_2$ starting and ending at $\infty$ separates the puncture at -1 from $1, a$. For a given vector $(1, \alpha)$ let $\mathfrak{D}$ be the family of all pairs $(D_1, D_2)$ of non-overlapping domains of homotopic type $(\gamma_1, \gamma_2)$, $D_j$ is a digon on $S_0$ with its vertices over the same support $\infty$ which is conformal at its vertices. It is of homotopy type $\gamma_j$, $j = 1, 2$. The condition of compatibility of angles and weights is given as follows: $D_1$ has two equal angles $\beta$ at its vertices over $\infty$. Consequently, the digon $D_2^*$ has two equal angles $\pi - \beta$, $\beta = \pi/(1 + \alpha)$. Degeneracy is permissible as (A) $D_1 = \emptyset$ or (B) $D_2 = \emptyset$. Consider the problem of minimizing the sum

$$m(D_1, \infty, \infty) + \alpha^2 m(D_2, \infty, \infty) \tag{3.27}$$

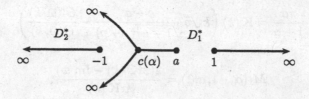

**Fig. 3.7.** Trajectory structure of the differential $\varphi(z)dz^2$

for a non-negative $\alpha$ over all $(D_1, D_2) \in \mathfrak{D}$. Denote by $M(\alpha) = M(\alpha, -1, a, 1)$ the minimum of the sum (3.27).

**Theorem 3.4.9.** *The minimum of the sum (3.27) over the family $\mathfrak{D}$*

$$M(\alpha) = M(\alpha, -1, a, 1) = m(D_1^*, \infty, \infty) + \alpha^2 m(D_2^*, \infty, \infty)$$

*is attained for the digons $D_1^*$ and $D_2^*$ which are the strip domains in the trajectory structure of the quadratic differential*

$$\varphi(z)dz^2 = A \frac{z - c}{(z - a)(z^2 - 1)} dz^2, \quad A > 0.$$

*The constant $A$ and the angle $\beta$ are defined as follows $A = 1/\beta = \alpha/(\pi - \beta)$, $\beta = \pi/(1 + \alpha)$. Each orthogonal trajectory of the differential $\varphi(z)dz^2$ lying in $D_1^*$ $(D_2^*)$ in the metric $\sqrt{|\varphi(z)|}|dz|$ has its length 1 $(\alpha)$ or the degenerating cases (A) $(\alpha \to \infty)$ or (B) $(\alpha = 0)$ appear.*
*If*

$$0 < \alpha < \frac{\pi + 2 \arcsin a}{\pi - 2 \arcsin a} =: \alpha_0,$$

*then the zero $c = c(a, \alpha)$ is defined by the system*

$$\frac{\pi \alpha}{1 + \alpha} = \mathbf{K}(k) \left( k\sqrt{p} \frac{a - c}{\sqrt{1 - a^2(1 + p)}} - \frac{\Theta'(\omega, k)}{\Theta(\omega, k)} \right),$$

$$\mathbf{dn}(\omega, k) = \frac{1 - p}{1 + p}, \quad p = \sqrt{\frac{1 - c^2}{1 - a^2}}, \quad k = \sqrt{\frac{2p(1 - a^2)}{p(1 - a^2) + 1 - ac}}.$$

*In this case*

$$m(D_1^*, \infty, \infty) = M_1(\alpha, -1, a, 1) = 2\alpha \left( \frac{\operatorname{Im} \omega}{\mathbf{K}(k)} + 1 \right) + 2,$$

$$m(D_2^*, \infty, \infty) = M_2(\alpha, -1, a, 1) = \frac{2(2\mathbf{K}'(k) - \operatorname{Im} \omega)}{\alpha \mathbf{K}(k)}.$$

*If $\alpha > \alpha_0$, then the zero $c = c(a, \alpha)$ is defined by the equation*

$$\frac{\pi\alpha}{1+\alpha} = \mathbf{K}(k)\left(k\sqrt{p}\frac{c-a}{\sqrt{1-a^2}(1+p)} - \frac{\Theta'(\omega, k)}{\Theta(\omega, k)}\right).$$

*In this case*

$$M_1(\alpha, -1, a, 1) = \frac{2(2\mathbf{K}'(k) - \operatorname{Im}\omega)}{\alpha\mathbf{K}(k)},$$

$$M_2(\alpha, -1, a, 1) = \frac{2}{\alpha}\left(\frac{\operatorname{Im}\omega}{\mathbf{K}(k)} + 1 + \alpha\right).$$

*If $\alpha = \alpha_0$, then the extremal differential is defined as*

$$\varphi(z)dz^2 = A\frac{dz^2}{z^2 - 1}.$$

*In this case*

$$M_1(\alpha, -1, a, 1) = \frac{2\pi}{\pi - 2\arcsin a},$$

$$M_2(\alpha, -1, a, 1) = \frac{2\arcsin a - \pi}{2\arcsin a + \pi}.$$

*Let $\zeta = \zeta_j(z)$ be a conformal homeomorphism of the domain $D_j^*$ onto the strip*

$$0 < \operatorname{Re}\zeta < \alpha_j, \quad \alpha_1 = 1, \quad \alpha_2 = \alpha, \quad \zeta_j(\infty) = \infty.$$

*This function satisfies in $D_j^*$ the equation*

$$(d\zeta_j(z))^2 = -\varphi(z)dz^2.$$

**Proof.** We consider the map $u = u(z)$ whose inverse is

$$z - a = (c-a)\frac{1 - \mathbf{dn}\,u}{(1-p) - (1+p)\mathbf{dn}\,u}$$

and obtain the representation of the differential $\varphi$ in terms of the parameter $u$ in regular points

$$\frac{1}{A}\varphi(z)dz^2 = Q(u)du^2 = \frac{p(c-a)^2k^2}{1-a^2}\left(\frac{1 + \mathbf{dn}\,u}{(1 - p - (1+p)\mathbf{dn}\,u)}\right)^2 du^2. \quad (3.28)$$

Let $0 < \alpha < \alpha_0$. Now we study the trajectory structure of the quadratic differential $Q$ which is a complete square of a linear one.

Set

$$u_{m,n} = 2m\mathbf{K}(k) + 4ni\mathbf{K}'(k), \quad \text{for integer } m, n,$$

$$\omega = \mathbf{dn}^{-1}\frac{1+p}{1-p}, \quad \text{choose } \omega \in (\mathbf{K}(k), \mathbf{K}(k) + 2i\mathbf{K}'(k)).$$

The differential $Q(u)du^2$ has the poles of the second order at the points

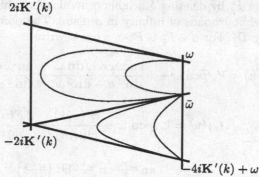

**Fig. 3.8.** Trajectory structure of the differential $Q(u)du^2$

$$\omega_{m,n}^{(1)} = \omega + u_{m,n}, \quad \omega_{m,n}^{(2)} = \overline{\omega} + u_{m,n},$$

and the zeros of the fourth order at the points

$$u_{m,n}^{(0)} = 2i\mathbf{K}'(k) + u_{m,n}.$$

Then, the images of the points $-1, 1, a$ under the transformation $u(z)$ are respectively the points

$$u_{m,n}^{(1)} = \mathbf{K}(k) + u_{m,n}, \quad u_{m,n}^{(2)} = \mathbf{K}(k) + 2i\mathbf{K}'(k) + u_{m,n}, \quad u_{m,n}^{(3)} = u_{m,n}.$$

Denote by $P_1$ the $u$-image of $D_1^*$, and by $P_2$ the $u$-image of $D_2^*$. Fix $m, n$. Then, the domain $P_1$ is a strip domain in the trajectory structure of the differential $Q(u)du^2$ bounded by the segments $[-2i\mathbf{K}', 2i\mathbf{K}']$, $[\overline{\omega}, \omega]$ and the arcs $p_1$ and $p_2$ connecting the points $2i\mathbf{K}'(k)$, $\omega$, and $-2i\mathbf{K}'(k)$, $\overline{\omega}$ respectively. The domain $P_2$ is bounded by the segment $[-4i\mathbf{K}' + \omega, \overline{\omega}]$ and by the arcs $p_2$ and $p_3$ that connect the points $-2i\mathbf{K}'(k)$, $-4i\mathbf{K}' + \omega$. The same structure one can obtain for $\alpha > \alpha_0$.

The $\varphi$-lengths of the orthogonal trajectories in the domains $D_1^*$ and $D_2^*$ are equal to 1 and $\alpha$ respectively. We denote by $z = f_j(\zeta)$ the inverse maps to $\zeta_j(z)$ defined in the theorem. Considering the maps $u \circ f_j(\zeta)$ and their inverses we obtain the following equations

$$\frac{1}{A}d\zeta_j(z(u)) = \frac{\beta_j}{\alpha_j}d\zeta_j(z(u)) = Q_2(u)du, \quad \alpha_1 = 1, \ \alpha_2 = \alpha,$$

where

$$Q_2(u) = k\sqrt{p}\,\mathbf{cn}\,\omega\,\frac{1 + \mathbf{dn}\,\omega}{\mathbf{dn}\,u - \mathbf{dn}\,\omega},$$

and

$$\mathbf{cn}\,\omega = \pm\frac{a - c}{\sqrt{1 - a^2(1 + p)}}.$$

Here the sign is $(+)$ in the case $\alpha < \alpha_0$ and $(-)$ for $\alpha > \alpha_0$. Let $P_j^\varepsilon$ be obtained from $P_j$ by deleting $\delta$-neighbourhoods of $\omega$ which are the $u$-images of the $\frac{1}{\varepsilon}$-neighbourhoods of infinity in $z$-plane. The domains in the $z$-plane we denote by $D_j^\varepsilon$. For $u \in P_1^\varepsilon \cup P_2^\varepsilon$ one can rewrite

$$Q_2(u) = k\sqrt{p}\,\mathrm{cn}\,\omega + + \frac{\mathrm{sn}\,\omega\,\mathrm{cn}\,\omega\,\mathrm{dn}\,\omega}{\mathrm{dn}^2 u - \mathrm{dn}^2 \omega} + \frac{\mathrm{sn}\,\omega\,\mathrm{cn}\,\omega\,\mathrm{dn}\,u}{\mathrm{dn}^2 u - \mathrm{dn}^2 \omega},$$

or

$$Q_2(u) = k\sqrt{p}\,\mathrm{cn}\,\omega - \frac{\Theta'(\omega,k)}{\Theta(\omega,k)} + \frac{d\,F(u)}{du},$$

where

$$F(u) = \frac{\mathrm{sn}\,\frac{\omega+u}{2}\,\mathrm{cn}\,\frac{\omega+u}{2}\,\Theta^2\left(\frac{\omega+u}{2}\right)}{\mathrm{sn}\,\frac{u-\omega}{2}\,\mathrm{cn}\,\frac{u-\omega}{2}\,\Theta^2\left(\frac{u-\omega}{2}\right)}.$$

Now we consider the problem on $M(\alpha,-1,a,1)$ with $\alpha < \alpha_0$. The image of the domain $P_2^{(\varepsilon)}$ under the transformation $\zeta = \zeta_2(z(u))$ is the rectangle

$$0 < \mathrm{Re}\,\zeta < \alpha, \quad -\frac{1}{M(D_2^{(\varepsilon)})} \leq \mathrm{Im}\,\zeta \leq \frac{1}{M(D_2^{(\varepsilon)})},$$

where $M(D_2^{(\varepsilon)})$ is the modulus of the quadrilateral $D_2^{(\varepsilon)}$. The segment $[0,i\alpha]$ in the $\zeta$-plane has the preimage $[\mathbf{K}+2i\mathbf{K}', 2i\mathbf{K}']$ in the $u$-plane. Observe that

$$\log \frac{F(2i\mathbf{K}')}{F(\mathbf{K}+2i\mathbf{K}')} = 2i\pi m$$

for an integer $m$. Then we have, by integrating the equation for $\zeta_2(z(u))$ along the above segment,

$$\frac{\alpha}{A} = \left(k\sqrt{p}\,\mathrm{cn}\,\omega - \frac{\Theta'(\omega,k_1)}{\Theta(\omega,k_1)}\right)\mathbf{K}(k) + 2i\pi m.$$

Analogously, we integrate the equation for $\zeta_1(z(u))$ along the segment $[0,\mathbf{K}(k)]$ and deduce that $A = (1+\alpha)/\pi$. Tending $\alpha$ to $0$, we obtain $m = 0$. Therefore, the equation for the zero $c$ is of the form

$$\frac{\pi\alpha}{1+\alpha} = \left(k\sqrt{p}\,\mathrm{cn}\,\omega - \frac{\Theta'(\omega,k_1)}{\Theta(\omega,k_1)}\right)\mathbf{K}(k).$$

Now we find the expressions for $M_1(\alpha,-1,a,1)$ and $M_2(\alpha,-1,a,1)$. Let $\delta > 0$ and $\omega + i\delta$ be the $u$-image of $1/\varepsilon$. The segment $[0,i/M(D_1^{(\varepsilon)})]$ in the $\zeta$-plane has the preimage $[1,1/\varepsilon]$ in the $z$-plane and the segment $[\mathbf{K}(k),\omega-i\delta]$ in the $u$-plane. Integrating the equation for $\zeta_1$, we obtain

$$\frac{i\pi}{2(1+\alpha)}\left[\frac{1}{M(D_1^{(\varepsilon)})} + 2\frac{1+\alpha}{\pi}\log\varepsilon\right] - \log\varepsilon =$$

$$= \left[k\sqrt{p}\,\mathbf{cn}\,\omega - \frac{\Theta'(\omega,k_1)}{\Theta(\omega,k_1)}\right](\omega - i\delta - \mathbf{K}(k)) + \log\frac{F(\omega-i\delta)}{F(\mathbf{K}(k))}.$$

Then, we use the relations

$$i\delta = \frac{\sqrt{p(1-a^2)}}{k}\varepsilon + o(\varepsilon),$$

$$\log\frac{F(\omega-i\delta)}{F(\mathbf{K}(k))} = \log\left(2k\mathbf{sn}\,\omega\mathbf{cn}\,\omega\frac{\Theta^2(\omega,k_1)}{\Theta^2(0,k_1)}\right) + \log\frac{1}{i\delta} + i\pi + 2i\pi m,$$

and derive that

$$M_1(\alpha,-1,a,1) = 2\alpha\left(\frac{\mathrm{Im}\,\omega}{\mathbf{K}(k)} + 1\right) + 2.$$

Similarly,

$$M_2(\alpha,-1,a,1) = \frac{2(2\mathbf{K}'(k) - \mathrm{Im}\,\omega)}{\alpha\mathbf{K}(k)}.$$

The same observations for $\alpha \geq \alpha_0$ lead to the other statements of Theorem 3.4.9. $\qquad\qquad\qquad\qquad\qquad\qquad\qquad\qquad\qquad\qquad\qquad\qquad\qquad\square$

*Remark 3.4.1.* Let $0 < x < y < \infty$. Set

$$a = \frac{2x-y}{y}, \quad W(w) = \frac{2x-y}{y} - \frac{2x(y-x)}{y(w-x)}.$$

Then, the inverse map $w(W)$ satisfies the equalities $w(a) = \infty$, $w(-1) = y$, $w(1) = 0$, $w(\infty) = x$. The extremal configuration $(D_1^*, D_2^*)$ in the problem of $M(\alpha,-1,a,1)$ is transformed under the mapping $w(W)$ onto the extremal pair $(D_1^{w*}, D_2^{w*})$ for the corresponding problem which we denote by $M(\alpha,0,x,y)$. Theorem 2.2.2 and the formulae thereafter yield that

$$M_j(\alpha,-1,a,1) = M_j(\alpha,0,x,y) + \frac{2}{\beta_j}\log\frac{2x(y-x)}{y}.$$

Consequently, we consider on the Riemann surface $S_1 = \mathbb{C}\backslash\{-1, c_1, a_1, 1\}$, $-1 < c_1 < a_1 < 1$ the pair of non-intersected curves $(\gamma_1, \gamma_2)$ of type III forming an admissible system on $S_1$. The curve $\gamma_1$ starting and ending at $\infty$ separates the puncture at 1 from $-1, c_1, a_1$; the curve $\gamma_2$ starting and ending at $\infty$ separates the puncture at -1 from $1, c_1, a_1$. Now let $\mathfrak{B}$ be a family of all pairs $(B_1, B_2)$ of non-overlapping domains of homotopic type $(\gamma_1, \gamma_2)$, $B_j$ be a digon on $S_1$ with its vertices over the same support $\infty$. It is of homotopy type $\gamma_j$, $j = 1, 2$. The condition of compatibility of angles and weights we suppose the same as in Theorem 3.4.9. Degeneracy is permissible as (A) $B_1 = \emptyset$ or (B) $B_2 = \emptyset$. We consider the problem of minimizing the sum

$$m(B_1, \infty, \infty) + \alpha^2 m(B_2, \infty, \infty) \qquad\qquad\qquad (3.29)$$

for a non-negative $\alpha$ over all $(B_1, B_2) \in \mathfrak{B}$. Denote by $\mathcal{M}(\alpha) \equiv \mathcal{M}(\alpha, -1, c_1, a_1, 1)$ the minimum of the sum (3.29).

If $0 < \alpha \le \mu_1$, $\mu_1 = J_1/(\pi - J_1)$, where

$$J_1 = \int_{-1}^{c_1} \sqrt{\frac{c_1 - x}{(1 - x^2)(a_1 - x)}} dx,$$

then this problem is reduced to Theorem 3.4.9 with $\alpha < \alpha_0$. If $\alpha \ge \mu_2$, $\mu_2 = J_2/(\pi - J_2)$, where

$$J_2 = \int_{-1}^{c_1} \sqrt{\frac{a_1 - x}{(1 - x^2)(c_1 - x)}} dx,$$

then this problem is reduced to Theorem 3.4.9 with $\alpha > \alpha_0$. Otherwise, the following theorem holds.

**Theorem 3.4.10.** *Let $\mu_1 < \alpha < \mu_2$. The minimum of the sum (3.29) over the family $\mathfrak{B}$,*

$$\mathcal{M}(\alpha) = \mathcal{M}(\alpha, -1, c_1, a_1, 1) = m(B_1^*, \infty, \infty) + \alpha^2 m(B_2^*, \infty, \infty),$$

*is attained for the digons $B_1^*$ and $B_2^*$ which are the strip domains in the trajectory structure of the quadratic differential*

$$\varphi(z)dz^2 = A\frac{(z - b)^2}{(z - a_1)(z - c_1)(z^2 - 1)}dz^2, \quad A > 0.$$

*The extremal digon $B_1^*$ has two equal angles $\beta$ at its vertices over $\infty$. Consequently, the digon $B_2^*$ has two equal angles $\pi - \beta$. The constant $A$ and the angle $\beta$ are defined as $A = 1/\beta = \alpha/(\pi - \beta)$, $\beta = \pi/(1 + \alpha)$. Each orthogonal trajectory of the differential $\varphi(z)dz^2$ lying in $B_1^*$ $(B_2^*)$ has in the metric $\sqrt{|\varphi(z)|}|dz|$ length $1$ $(\alpha)$.*

*The zero $b = b(\alpha)$ is defined by the system*

$$\frac{\pi\alpha}{1 + \alpha} = \mathbf{K}(k_1)\left(\frac{k_1(c_1 - b + p_1(a_1 - b))}{\sqrt{p_1(1 - a_1^2)(1 + p_1)}} - \frac{\Theta'(\omega_1, k_1)}{\Theta(\omega_1, k_1)}\right),$$

$$\mathbf{dn}(\omega_1, k_1) = \frac{1 - p_1}{1 + p_1}, \quad p_1 = \sqrt{\frac{1 - c_1^2}{1 - a_1^2}}, \quad k = \sqrt{\frac{2p_1(1 - a_1^2)}{p_1(1 - a_1^2) + 1 - a_1 c_1}}.$$

*In this case*

$$M(B_1^*, \infty, \infty) = \mathcal{M}_1(\alpha, -c_1, a_1, 1) = 2\alpha\left(\frac{\operatorname{Im}\omega_1}{\mathbf{K}(k_1)} + 1\right) + 2,$$

$$M(B_2^*, \infty, \infty) = \mathcal{M}_2(\alpha, -1, c_1, a_1, 1) = \frac{2(2\mathbf{K}'(k_1) - \text{Im } \omega_1)}{\alpha \mathbf{K}(k_1)}.$$

Let $\zeta = \zeta_j(z)$ be a conformal homeomorphism from the domain $B_j^*$ onto the strip

$$0 < \text{Re } \zeta < \alpha_j, \quad \alpha_1 = 1, \quad \alpha_2 = \alpha, \quad \zeta_j(\infty) = \infty.$$

This function satisfies in $B_j^*$ the equation

$$(d\zeta_j(z))^2 = -\varphi(z)dz^2.$$

The *proof* of this theorem is analogous to that for Theorem 3.4.9, so we omit it.

**Remark 3.4.2.** Let $0 < r_1 < r_2 < 1$, $R_j = r_j + 1/r_j$, $j = 1, 2$. Set

$$a_1 = 1 + \frac{2(R_1 - R_2)}{2 - R_1}, \quad c_1 = 1 - \frac{2(R_1 - R_2)}{2 + R_1},$$

$$Z(z) = 1 + \frac{2z(R_1 - R_2)}{1 + z^2 - zR_1}, \quad z \in U.$$

Then, the inverse map $z(Z)$ satisfies the equalities $z(-1) = r_2$, $z(1) = r_1$, $z(a_1) = 1$, $z(c_1) = -1$. The extremal configuration $(B_1^*, B_2^*)$ in the problem of $\mathcal{M}(\alpha, -1, c_1, a_1, 1)$ is transformed under the mapping $z(Z)$ onto the extremal pair $(D_1^{z*}, D_2^{z*})$ for the corresponding modulus problem which we denote by $\mathcal{M}(\alpha, 0, r_1, r_2)$. The change of the reduced modulus of digon yields that

$$\mathcal{M}_j(\alpha, -1, c_1, a_1, 1) = \mathcal{M}_j(\alpha, 0, r_1, r_2) + \frac{2}{\beta_j} \log \frac{2r_1(R_1 - R_2)}{1 - r_1^2}.$$

Now we need the following technical lemma.

**Lemma 3.4.3.** *Let* $0 < r_1 < r_2 < 1$ *and the values of* $a_1$, $c_1$ *be defined as in the remark to Theorem 3.4.10. Then, there is a unique function* $a(\alpha)$ *defined in* $\alpha \in [\mu_1, \mu_2]$ *by the equation*

$$M_1(\alpha, -1, a, 1) = \mathcal{M}_1(\alpha, -1, c_1, a_1, 1). \tag{3.30}$$

*Proof.* First, assume $\alpha = \alpha_0$. Then, the equation (3.30) implies

$$\alpha_0 = \frac{\pi - 2}{\pi + 2 + 2\pi \frac{\text{Im } \omega_1}{\mathbf{K}(k_1)}},$$

$$a(\alpha_0) = -\cos\left(\frac{\frac{\pi}{2} - 1}{1 + \frac{\text{Im } \omega_1}{\mathbf{K}(k_1)}}\right).$$

Now let $\alpha \in [\mu_1, \alpha_0]$. Then, the equation (3.30) is equivalent to the equation

$$\frac{\text{Im } \omega}{\mathbf{K}(k)} = \frac{\text{Im } \omega_1}{\mathbf{K}(k_1)}. \tag{3.31}$$

Obviously, at $\alpha = \mu_1$ the problem about $\mathcal{M}(\alpha, -1, c_1, a_1, 1)$ is equivalent to that about $M(\alpha, -1, a, 1)$. Hence,

$$\left. \frac{\text{Im } \omega}{\mathbf{K}(k)} \right|_{\alpha=\mu_1, \, a=a_1} = \frac{\text{Im } \omega_1}{\mathbf{K}(k_1)}.$$

Under these conditions the equation (3.30) has the solution $a(\mu_1) = a_1$. Since

$$\alpha = \int\limits_{-1}^{c} \sqrt{\frac{A(c-x)}{(1-x^2)(a-x)}} \, dx,$$

the function $c(a, \alpha)$ increases with respect to $a$ for any fixed $\alpha \in [\mu_1, \alpha_0]$, and increases with respect to $\alpha$ for any fixed $a \in [c_1, a_1]$. Therefore, for $a = a_1$ the functions $k$ and $\mathbf{K}(k)$ increase with respect to $\alpha$, the function $p$ decreases with respect to $\alpha$, and $\mathbf{dn}(\omega, k)$ increases with respect to $\alpha$. The function $\mathbf{dn}(\omega, k)$ decreases with respect to $\text{Im } \omega \in [\mathbf{K}(k), \mathbf{K}(k) + 2i\mathbf{K}'(k)]$ from $\sqrt{1-k^2}$ to $-\sqrt{1-k^2}$ for a fixed $k$. Since $\mathbf{dn}(\omega, k)$ decreases with respect to $k$ for fixed $\text{Im } \omega$, the hypothesis that $\text{Im } \omega$ increases on $\alpha$ contradicts increasing of $\mathbf{dn}(\omega, k)$ with increasing $\alpha$. Therefore,

$$\left. \frac{\text{Im } \omega}{\mathbf{K}(k)} \right|_{\alpha \in (\mu_1, \alpha_0), \, a=a_1} < \frac{\text{Im } \omega_1}{\mathbf{K}(k_1)}.$$

Analogously,

$$\left. \frac{\text{Im } \omega}{\mathbf{K}(k)} \right|_{\alpha \in (\mu_1, \alpha_0), \, a=c_1} > \frac{\text{Im } \omega_1}{\mathbf{K}(k_1)}.$$

The continuity of the left-hand side of (3.31) implies the existence of the solution to the equation (3.30) for any fixed $\alpha \in (\mu_1, \alpha_0)$. The existence of the solution for $\alpha \in (\alpha_0, \mu_2]$ is provided in the same manner.

Now we prove the uniqueness. For this we use the monotonicity of $c(a, \alpha)$ with respect to $a$ for a fixed $\alpha$. Analogously the preceding observations we obtain that $\text{Im } \omega / \mathbf{K}(k)$ decreases in $a \in [c_1, a_1]$. This yields the uniqueness. $\square$

*Remark 3.4.3.* Assume $f \in M_R(\omega)$, $r_1 = r$, $r_2 = \omega$, $0 < r < \omega$ and denote by $x(\alpha) = \frac{\omega}{2}(a(\alpha) + 1)$,

$$y(\alpha) = \omega \exp \frac{\pi(M(\alpha) - \mathcal{M}(\alpha))}{2(1+\alpha)^2} x(\alpha) \left(1 - \frac{x(\alpha)}{\omega}\right) \frac{\omega(1-r^2)}{r(\omega-r)(1-\omega r)}.$$

Then the equation (3.30) is equivalent to the equation

$$M_2(\alpha, -1, a, 1) + \frac{2(1+\alpha)}{\pi\alpha}\log\frac{1-a^2}{2} =$$

$$= M_2(\alpha, -1, c_1, a_1, 1) - \frac{2(1+\alpha)}{\pi\alpha}\log\frac{1-r^2}{2(R_1-R_2)r^2y(\alpha)},$$

where $a_1$, $c_1$, $R_1$, $R_2$ are defined in Remarks to Theorems 3.4.9, 3.4.10.

**Theorem 3.4.11.** *Let* $0 < r < \omega < 1$, $f \in M_R(\omega)$, $\alpha \in [\mu_1, \mu_2]$. *(i) There exists a function* $w = f^*(z, \alpha) \in M_R(\omega)$ *that satisfies in* $U$ *the differential equation*

$$\frac{(1-b(\alpha))(4-R_1^2)r^2\omega}{4(R_1-R_2)}\frac{(z-d)^2(z-\bar{d})^2 dz^2}{z(z-r)^2(z-\omega)(1-rz)^2(1-\omega z)} =$$

$$= 2(a(\alpha) - c(a(\alpha), \alpha))\frac{(w-C)dw^2}{w(w-x(\alpha))^2(w-\omega)}, \tag{3.32}$$

*where* $d = d(\alpha)$, $|d| = 1$, *and* $C$ *are defined by the equations*

$$d = \exp\left(i\arccos\left(\frac{R_2-R_1}{1-b(\alpha)}\right) + \frac{R_1}{2}\right),$$

$$C = \frac{x(\alpha)}{\omega}(1 + 2(1 - \frac{x(\alpha)}{\omega}))\frac{1}{a(\alpha) - c(a(\alpha), \alpha)}.$$

*Here* $c(a, \alpha)$ *and* $b(\alpha)$ *are defined in Theorems 3.4.9, 3.4.10,* $f^*(r, \alpha) = x(\alpha)$, $f_z^{*\prime}(r, \alpha) = y(\alpha)$.

*(ii) The functions* $x(\alpha)$, $y(\alpha)$ *satisfy the equalities*

$$x(\mu_1) = K_2(r), \quad x(\mu_2) = K_1(r),$$

$$y(\mu_1) = K_2'(r), \quad y(\mu_2) = K_1'(r).$$

*(iii) The function* $w = f^*(z, \alpha)$ *maps the unit circumference* $\partial U$ *onto the continuum* $E$ *consisting of the ray* $[C(\alpha), \infty]$ *and two symmetric arcs of the quadratic differential*

$$\frac{(w-C)dw^2}{w(w-x(\alpha))(w-\omega)}$$

*that originate at the point* $C(\alpha)$, *cap* $E = 1$.

*(iv) The boundary curve* $\Gamma^-$ *of the range of the system of functionals* $(f(r), f'(r))$ *in the class* $M_R(\omega)$ *consists of the points* $(x(\alpha), y(\alpha))$ *and each point is given by the only function* $f^*(z, \alpha)$. □

*Proof.* By Lemma 3.4.3 there exists such an $a(\alpha)$ that

$$M_2(\alpha, -1, a(\alpha), 1) = \mathcal{M}_2(\alpha, -1, c_1, a_1, 1).$$

Let the pairs $(D_1^*, D_2^*)$ and $(B_1^*, B_2^*)$ be extremal for $M(\alpha, -1, a, 1)$ and $\mathcal{M}(\alpha, -1, c_1, a_1, 1)$ respectively. The above equality implies that there exists a conformal map $W = F(Z)$ from the domain $B_2^*$ onto $D_2^*$ with the following expansion near $\infty$:

$$F(Z) = Z + a_0 + a_1 \frac{1}{Z} + \dots$$

We define two functions $g_1(W)$ and $g_2(Z)$ satisfying in $D_2^*$ and $B_2^*$ the equations

$$(dg_1(W))^2 = -\left(\frac{1+\alpha}{\pi}\right)^2 \frac{W - c(a(\alpha), \alpha)}{(W - a(\alpha))(W^2 - 1)} dW^2, \quad g_1(\infty) = \infty,$$

$$(dg_2(Z))^2 = -\left(\frac{1+\alpha}{\pi}\right)^2 \frac{(Z - b(\alpha))^2}{(Z - a_1)(Z - c_1)(Z^2 - 1)} dZ^2, \quad g_2(\infty) = \infty.$$

These functions map their domains onto the strip $0 < \operatorname{Re} \zeta < \alpha$. Then, the map $W = F(Z)$ has the form $F = g_1^{-1} \circ g_2$. From the equation (3.32) it follows that the function $F(Z)$ has an analytical continuation onto the segment $[c_1, a_1]$ and, furthermore, into the domain $D_1^*$. It maps the segment $[c_1, a_1]$ onto the continuum $w(E)$ where the function $w(W)$ is defined in Remark to Theorem 3.4.9. Thus, we can define the function $f^*(z, \alpha)$ as a superposition $f^*(z, \alpha) = K \cdot w(F(Z(z)))$. The constant $K$ is defined by the condition cap $E = 1$. The properties of this function obviously follow from Theorems 3.4.9, 3.4.10 and Remarks.

Now we suppose that the extremal pairs $(D_1^{z*}, D_2^{z*})$ and $(D_1^{w*}, D_2^{w*})$ are defined as in Remarks to Theorems 3.4.9, 3.4.10 with $r_1 = r$, $r_2 = \omega$. The functions $x(\alpha)$ and $y(\alpha)$ are continuous, hence,

$$\left[\frac{(1-\omega)^2 r}{(1-r)^2}, \frac{(1+\omega)^2 r}{(1+r)^2}\right] \subset \{x(\alpha) : \alpha \in [\mu_1, \mu_2]\},$$

$$\left[\frac{(1-\omega)^2(1+r)}{(1-r)^3}, \frac{(1+\omega)^2(1-r)}{(1+r)^3}\right] \subset \{y(\alpha) : \alpha \in [\mu_1, \mu_2]\}.$$

Let $f(z) \in M_R(\omega)$. Then, we can choose $\alpha$ such that $x(\alpha) = f(r)$. Thus,

$$m(f(D_1^{z*}), f(r), f(r)) + \alpha^2 m(f(D_2^{z*}), f(r), f(r)) = \qquad (3.33)$$

$$= m(D_1^{z*}, r, r) + \alpha^2 M(D_2^{z*}, r, r) + \frac{2}{\pi} \log f'(r).$$

Then, the pair $(f(D_1^{z*}), f(D_2^{z*}))$ is admissible in the problem for $M(\alpha, 0, x(\alpha), \omega)$ and the inequality

$$m(f(D_1^{z*}), f(r), f(r)) + \alpha^2 m(f(D_2^{z*}), f(r), f(r)) \geq M(\alpha, 0, x(\alpha), \omega)$$

holds. The equality (3.33) yields the inequality $f'(r) \geq y(\alpha)$. The uniqueness, as usual, follows from the uniqueness of the extremal configurations. $\quad\square$

Analogously one can obtain the theorem with another location of $r$.

**Theorem 3.4.12.** *Let $f \in M_R(\omega)$, $0 < \omega < r < 1$. Then, the lower boundary curve of the range of the system of functionals $(f(r), f'(r))$ in the class $M_R(\omega)$ consists of the points*

$$\left(x, \, x\left(\frac{x}{\omega} - 1\right)\frac{(1-r^2)\omega}{r(r-\omega)(1-\omega r)}\right), \quad x \in \left[\frac{(1+\omega)^2 r}{(1+r)^2}, \frac{(1-\omega)^2 r}{(1-r)^2}\right].$$

*Each point is given by the unique function*

$$g(z) = \frac{z(1 - u\omega + \omega^2)}{1 - uz + z^2}$$

*with the parameter $u$ such that $g(r) = x$.*

For the upper boundary curve we use the reduced moduli of circular domains.

**Theorem 3.4.13.** *Let $f \in M_R(\omega)$, $0 < r < \omega < 1$. Then, the upper boundary curve of the range of the system of functionals $(f(r), f'(r))$ in the class $M_R(\omega)$ consists of the points*

$$\left(x, \, x\left(1 - \frac{x}{\omega}\right)\frac{(1-r^2)\omega}{r(r-\omega)(1-\omega r)}\right), \quad x \in \left[\frac{(1-\omega)^2 r}{(1-r)^2}, \frac{(1+\omega)^2 r}{(1+r)^2}\right].$$

*Each point is given by the unique function*

$$g(z) = \frac{z(1 - u\omega + \omega^2)}{1 - uz + z^2}$$

*with the parameter $u$ such that $g(r) = x$.*

*Proof.* We consider the domain $U_z = U \setminus \{(-1, 0] \cup [\omega, 1)\}$ with the reduced modulus

$$m(U_z, r) = \frac{1}{2\pi} \log \frac{4r(\omega - r)(1 - \omega r)}{\omega(1 - r^2)}.$$

The domain $f(U_z)$ has the reduced modulus $m(f(U_z), f(r)) = m(U_z, r) + \frac{1}{2\pi} \log f'(r)$. This domain is admissible in the problem of the extremal partition of $\mathbb{C}$ and the extremal configuration is given by the domain $D = \mathbb{C} \setminus \{(-\infty, 0] \cup [\omega, \infty)\}$ with the reduced modulus

$$m(D, x) = \frac{1}{2\pi} \log \frac{4x(\omega - x)}{\omega}, \quad x \in (0, \omega).$$

Choose $u$ such that $g(r) = f(r) = x$. Then, we have the inequality of Theorem 3.4.13 with the equality sign only for the function $g$. $\qquad\square$

The above results have been obtained in see [157], [164]. The upper boundary curve for the system of the functionals $(f(r), f'(r))$, $0 < \omega < r < 1$ can be obtained (see [166]) by use the results of Section 3.2.2. The extremal functions map the unit disk onto the complex plane minus a symmetric slit with two finite endpoints. Principally, this result does not use more, than the results of Theorem 3.4.11, 3.4.12, and we will not develop this point here.

## 3.5 Univalent functions with the angular derivatives

We consider conformal maps $f$ of $U$ into $U$. We have already discuss some properties of the angular limits and the angular derivatives in Section 2.2.3. We briefly remind here that the angular limit

$$f(\zeta) = \lim_{z \to \zeta, z \in \Delta} f(z), \quad \Delta \text{ is any Stolz angle at } \zeta,$$

exists for almost all $\zeta \in \partial U$, the exceptional set even has zero capacity. In general, very little can be said about the existence of the angular derivative

$$f'(\zeta) = \lim_{z \to \zeta, z \in \Delta} f'(z),$$

see, e.g., ([112], Chapter 6) for the discussion.

The situation becomes much better when we restrict ourselves to the set

$$A = \{\zeta \in \partial U : f(\zeta) \text{ exists and } |f(\zeta)| = 1\},$$

because the angular derivative exists for every $\zeta \in A$ by the Julia-Wolff Lemma ([112], Proposition 4.13), even without the assumption that $f$ is injective in $U$. It may, however, be infinite. In our case of univalent functions, it follows from the McMillan Twist Theorem [98] that $f'(\zeta) \neq \infty$ for almost all $\zeta \in A$.

Here we denote by $S^1(\beta)$ the class of all holomorphic, univalent in the unit disk $U$ functions $f(z) = az + a_2 z^2 + \dots$, $f(1) = 1$ that map $U \to U$ and have the finite angular derivative $|f'(1)| = \beta$ fixed. The class $S^1(\beta)$ is different comparing with other classes of univalent functions and possess some new features. Set the canonical Pick function (only one) for this class

$$K_a(z) = k_\pi^{-1}(ak_\pi(z)), \quad a \in (0,1), \quad a = \frac{4h}{(1+h)^2}, \quad k_\pi(z) = \frac{z}{(1-z)^2}.$$

It maps $U$ onto the unit disk slit along $(-1, -h]$. For $a = 1/\beta^2$ the function $K_a(z)$ belongs to $S^1(\beta)$, and $|K_a'(0)| = a = 1/\beta^2$. The class $S^1(\beta)$ is not compact. Indeed, consider the sequence of functions $f_n$ from $S^1(\beta)$ that map $U$ onto $U$ slit along two segments starting form the points $\exp(\pm i\pi/n)$ which are parallel to the real axis and of length $1/n$. Obviously, $\lim_{n \to \infty} f_n = id$, which does not belong to the class $S^1(\beta)$.

### 3.5.1 Estimates of the angular derivatives

For the class $S^1(\beta)$ the obvious estimate $0 < |f'(0)| < 1$ holds. The right-hand side inequality is never reachable, but sharp due to the previous example. In order to provide the sharp form of the left-hand side inequality we prove the following theorem.

**Theorem 3.5.1.** *For all functions from $S^1(\beta)$ the sharp estimate $a := |f'(0)| \geq 1/\beta^2$ holds with the extremal function $K_a(z)$.*

*Proof.* Let $f \in S^1(\beta)$. We consider the digon $U_z = U \setminus (-1,0]$ with two vertices $0,1$. It is conformal at $0,1$, with the angles $2\pi$, $\pi$ at the vertices $0,1$ respectively, and its reduced modulus is $m(U_z, 0, 1) = 0$. The digon $f(U_z)$ with the vertices $0,1$ is also conformal at $0,1$, has the same angles and the reduced modulus

$$m(f(U_z), 0, 1) = \frac{1}{2\pi} \log |f'(0)| + \frac{1}{\pi} \log \beta.$$

It is admissible in the problem of minimum of the reduced modulus over all digons having the homotopy type defined by $\gamma = [0,1]$ (with the angles $2\pi$ and $\pi$ at the vertices $0$, $1$ respectively where they are conformal). The extremal reduced modulus $m(D_1^*, 0, 1) = 0$ is given by the same digon $U_w = U_z$. Therefore, $|f'(0)| \geq 1/\beta^2$. The uniqueness of the extremal configuration $K_a(U_z)$, $a = 1/\beta^2$. leads to the uniqueness of the extremal function.    □

This theorem was earlier proved by A. Solynin in [135].

Considering the digon $U \setminus \{(-1,0] \cup [r,1)\}$ with two vertices over 1 and applying analogous observations we arrive at the following result.

**Theorem 3.5.2.** *For all functions from $S^1(\beta)$ the sharp estimate $K_{1/\beta^2}(r) \leq |f(r)| < r$ holds with the extremal function $K_{1/\beta^2}(z)$ for the left-hand side inequality. The right-hand side inequality is sharp but never reachable.*

The next theorem generalizes Theorem 3.5.1 for the case of the angular derivatives at two points of $\partial U$.

**Theorem 3.5.3.** *Let $\alpha_1, \alpha_2$ be positive numbers such that $\alpha_1 + \alpha_2 = 1$. Let $w = f(z)$ be a conformal map of the unit disk into itself, $f(0) = 0$, such that for two points $z_1, z_2 \in \partial U$ the angular limits $w_1 = f(z_1)$ and $w_2 = f(z_2)$ also lie in $\partial U$. Suppose also that the angular derivatives $f'(z_1), f'(z_2)$ are finite. For every such function the inequality*

$$\left| \frac{f(z_1) - f(z_2)}{z_1 - z_2} \right|^{2\alpha_1\alpha_2} \geq \frac{1}{\sqrt{|f'(0)|}|f'(z_1)|^{\alpha_1^2}|f'(z_2)|^{\alpha_2^2}} \tag{3.34}$$

*holds. For all $a \in (0, a_0]$,*

$$a_0 = \left| \frac{z_1}{w_1} \right|^{\alpha_1} \left| \frac{z_2}{w_2} \right|^{\alpha_2},$$

*the extremal map $w = f^*(z)$ is the unique solution to the equation*

$$\frac{w\, w_1^{\alpha_1} w_2^{\alpha_2}}{(w - w_1)^{2\alpha_1}(w - w_2)^{2\alpha_2}} = \frac{a\, z\, z_1^{\alpha_1} z_2^{\alpha_2}}{(z - z_1)^{2\alpha_1}(z - z_2)^{2\alpha_2}}. \tag{3.35}$$

*In terms of quadratic differentials the extremal function maps the unit disk*
*U onto itself minus one or two slits along the trajectories of the differential*

$$\psi(w)dw^2 = \frac{1}{4\pi^2} \frac{(w - e^{i\eta})^2 (w - e^{i\xi})^2}{w^2 (w - w_1)^2 (w - w_2)^2} dw^2,$$

*where $\xi, \eta$ are calculated by the equations $\xi + \eta = \pi + \arg w_1 + \arg w_2$ and*

$$\alpha_2 - \alpha_1 = \frac{e^{i\xi} + e^{i\eta}}{w_1 - w_2}.$$

**Corollary 3.5.1.** *Let $w = f(z)$ be a conformal map of the unit disk into itself, $f(0) = 0$, such that $z_1, z_2 \in \partial U$ and $z_1 = f(z_1)$, $z_2 = f(z_2)$. Suppose also that the angular derivatives $f'(z_1), f'(z_2)$ are finite. The sharp inequality*

$$|f'(0)| \geq \frac{1}{\sqrt{|f'(z_1)f'(z_2)|}}$$

*holds with the extremal function given by Theorem 3.5.3 with $\alpha_1 = \alpha_2 = 1/2$, $a = |f'(0)|$.*

First we give some preliminary observations. Let $U' = U \setminus \{0\}$. On the boundary $\partial U$ of $U$ we fix two points $\zeta_1 = e^{i\theta_1}$, $\zeta_2 = e^{i\theta_2}$, $0 \leq \theta_1 < \theta_2 < 2\pi$.

The next problem about the extremal partition we formulate as follows. Set a non-zero weight vector $(\alpha_1, \alpha_2)$ with non-negative coordinates. We construct two intervals $\gamma_1 = (0, e^{i\theta_1})$ and $\gamma_2 = (0, e^{i\theta_2})$ in $U'$. This is an admissible system $(\gamma_1, \gamma_2)$ of curves of type IV. Now we consider the pairs of non-overlapping digons $(D_1, D_2)$ on $U'$ associated with the system of curves $(\gamma_1, \gamma_2)$ and the vector $(\alpha_1, \alpha_2)$, where the digon $D_j$ has its vertices at $0, e^{i\theta_j}$, $j = 1, 2$. We require the digons $D_1$ and $D_2$ to be conformal at their vertices and to satisfy the condition of compatibility of angles and weights, i.e., $\varphi_0^{(j)} = \frac{2\pi\alpha_j}{\alpha_1 + \alpha_2}$ is the inner angle of the digon $D_j$ at the origin and $\varphi_{\zeta_j}^{(j)} = \pi$ is the inner angle of the digon $D_j$ at the boundary point $\zeta_j$, $j = 1, 2$. We also require that the reduced moduli $m(D_j, 0, \zeta_j)$ exist.

Any collection of non-overlapping admissible digons associated with the system of curves $(\gamma_1, \gamma_2)$ and the vector $(\alpha_1, \alpha_2)$ satisfies the following inequality

$$\alpha_1^2 m(D_1, 0, e^{i\theta_1}) + \alpha_2^2 m(D_2, 0, e^{i\theta_2}) \geq \quad (3.36)$$
$$\geq \alpha_1^2 m(D_1^*, 0, e^{i\theta_1}) + \alpha_2^2 m(D_2^*, 0, e^{i\theta_2})$$

with the equality sign only for $D_j = D_j^*$. We denote this minimum by $\mathcal{M}_1(\alpha_1, \alpha_2, 0, e^{i\theta_1}, e^{i\theta_2})$ showing its dependence on parameters.

Each $D_j^*$ is a strip domain in the trajectory structure of a unique quadratic differential

$$\varphi(z)dz^2 = A \frac{(z - e^{i\gamma})^2 (z - e^{i\beta})^2}{z^2 (z - e^{i\theta_1})^2 (z - e^{i\theta_2})^2} dz^2, \quad A > 0, \quad (3.37)$$

associated with the problem about the extremal partition. The factor $A$ is positive because of strip domains and the local trajectory structure close to the origin ($\varphi(z) = \frac{A}{z^2}(1 + \dots)$). Here $A$, $\gamma$, and $\beta$ are functions of $\alpha_1, \alpha_2$. For $D_j^*$ there is a conformal map $g_j(z)$, $z \in D_j^*$ satisfying the differential equation

$$\alpha_j^2 \left( \frac{g_j'(z)}{g_j(z)} \right)^2 = \pi^2 \varphi(z), \tag{3.38}$$

that maps $D_j^*$ onto the digon $\mathbb{H}^+ = \{z : \operatorname{Im} z > 0\}$ with the vertices at 0 and $\infty$.

The critical trajectories of $\varphi(z)dz^2$ split $U'$ into at most two strip domains $\{D_j^*\}$ associated with the admissible system (one of $D_j^*$ can degenerate).

**Lemma 3.5.1.** *Let $\alpha_1, \alpha_1 \in (0, \infty)$, $0 \le \theta_1 < \theta_2 < 2\pi$, $\theta_2 - \theta_1 \le \pi$. Then,*

$$m(D_1^*, 0, e^{i\theta_1}) = \frac{\alpha_2}{\pi\alpha_1} \log \frac{1}{2|\sin \frac{\theta_1 - \theta_2}{2}|},$$

$$m(D_2^*, 0, e^{i\theta_2}) = \frac{\alpha_1}{\pi\alpha_2} \log \frac{1}{2|\sin \frac{\theta_1 - \theta_2}{2}|}.$$

*In the differential (3.37) we have*

$$A = \frac{(\alpha_1 + \alpha_2)^2}{4\pi^2}, \quad \beta = \arcsin\left( \frac{\alpha_2 - \alpha_1}{\alpha_2 + \alpha_1} \sin \frac{\theta_1 - \theta_2}{2} \right) + \frac{\theta_1 + \theta_2}{2},$$

*$\gamma = \pi - \beta + \theta_1 + \theta_2$, and two critical trajectories of the differential $\varphi(z)dz^2$ starting from the origin under the angle $\varphi_0^{(1)} = \frac{2\pi\alpha_1}{\alpha_1 + \alpha_2}$, ending at the points $e^{i\beta}$ and $e^{i\gamma}$.*

*Proof.* First we observe that the unit circle $\partial U$ should be the trajectory of the differential (3.37). Therefore, $-\varphi(e^{i\theta})e^{2i\theta}d\theta^2 > 0$ in regular points of $\partial U$. This implies that $\beta + \gamma = \pi + \theta_1 + \theta_2$.

Since we can rotate the configuration $z \to ze^{i\frac{\theta_1 + \theta_2}{2}}$, without loss of generality, assume that $\theta_2 \in (0, \pi/2]$ and $\theta_1 = -\theta_2$.

The map $g_1(z)$ has the expansion $g_1 = z^{\pi/\varphi_0^{(1)}}(c_1 + \dots)$ about the origin. Tending in (3.38) $z \to 0$ for $j = 1$, we obtain $\frac{\alpha_1^2}{(\varphi_0^{(1)})^2} = A$. The map $g_2(z)$ has the expansion $g_2 = z^{\pi/\varphi_0^{(2)}}(c_2 + \dots)$ about the origin. Tending in (3.38) $z \to 0$ for $j = 2$, we obtain $\frac{\alpha_2^2}{(\varphi_0^{(2)})^2} = A$. Thus, we calculate $A = \frac{(\alpha_1 + \alpha_2)^2}{4\pi^2}$ and the extremal domains $D_1^*, D_2^*$ satisfy the condition of compatibility of weights and angles. Close to the point $e^{i\theta_1}$ the map $g_1(z)$ has the expansion $g_1 = (z - e^{i\theta_1})(d_1 + \dots)$. Tending in (3.38) $z \to e^{i\theta_1}$ for $j = 1$, we obtain $\frac{2\alpha_1}{\alpha_1 + \alpha_2} = 1 - \frac{\sin \beta}{\sin \theta_1}$ or $\beta = \arcsin\left( \frac{\alpha_2 - \alpha_1}{\alpha_2 + \alpha_1} \sin \theta_1 \right)$.

Suppose $j = 1$. We take the square root in (3.38) and integrate it putting the branch of the root so that $g(0) = \infty$ and $g(e^{i\theta_2}) = 0$. We normalize the function $g(z)$ so that the part of the the boundary of $D_1^*$ lying on the unit circle is mapped into the real axis. Thus, after the normalization of the imaginary constant in the logarithm and exponentiating we obtain

$$g_1(z) = \frac{(z - e^{i\theta_1})(z - e^{-i\theta_1})^{\alpha_2/\alpha_1}}{z^{\frac{\alpha_1+\alpha_2}{2\alpha_1}}} e^{i\left(\theta_1 \frac{\alpha_2-\alpha_1}{2\alpha_1} - \frac{\pi}{2}\frac{\alpha_1+\alpha_2}{\alpha_1}\right)}. \tag{3.39}$$

Suppose $j = 2$. By the analogy with the previous case, we obtain

$$g_2(z) = \frac{(z - e^{i\theta_1})^{\alpha_1/\alpha_2}(z - e^{-i\theta_1})}{z^{\frac{\alpha_1+\alpha_2}{2\alpha_2}}} e^{i\left(\theta_1 \frac{\alpha_2-\alpha_1}{2\alpha_2} - \frac{\pi}{2}\frac{\alpha_1+\alpha_2}{\alpha_2}\right)}. \tag{3.40}$$

The reduced modulus of the digon $\mathbb{H}^+$ with respect to its vertices at 0 and $\infty$ is equal to zero. Therefore, calculating the derivatives of the maps $g_1$ and $g_2$ and making use the formulae on the change of the reduced moduli under a conformal map, we obtain the values of the reduced moduli $m(D_1^*, 0, e^{i\theta_1})$, $m(D_2^*, 0, e^{i\theta_2})$ asserted in Lemma 3.5.1. This finishes the proof.     $\square$

*Remark 3.5.1.* The case $\theta_2 - \theta_1 > \pi$ is analogous and the moduli in Lemma 3.5.1 are the same, interchanging $\gamma \leftrightarrow \beta$.

Now we give the *proof* of Theorem 3.5.3 using the extremal partitions and Lemma 3.5.1. We use the inequality (3.37) to derive the inequality (3.34). Let $f$ be an arbitrary map with the properties asserted in Theorem 3.5.3. Without less of generality we again assume here that $\zeta_1 = \overline{\zeta_2}$. Fix the positive values of $\alpha_1, \alpha_2$. Let the pair of domains $(D_1^*, D_2^*)$ be the extremal pair for the minimum $\mathcal{M}(\alpha_1, \alpha_2, \zeta_1, \zeta_2)$ of the sum (3.37). Then, the pair of domains $(f(D_1^*), f(D_2^*))$ is an admissible pair for the minimum $\mathcal{M}(\alpha_1, \alpha_2, w_1, w_2)$. Therefore, the inequality

$$\alpha_1^2 m(f(D_1^*), 0, w_1) + \alpha_2^2 m(f(D_2^*), 0, w_2) \geq \tag{3.41}$$
$$\geq \alpha_1^2 m(B_1^*, 0, w_1) + \alpha_2^2 m(B_2^*, 0, w_2),$$

holds where $(B_1^*, B_2^*)$ is the extremal pair for the minimum $\mathcal{M}(\alpha_1, \alpha_2, w_1, w_2)$. We deduce that

$$\alpha_1^2 m(f(D_1^*), 0, w_1) + \alpha_2^2 m(f(D_2^*), 0, w_2) = \tag{3.42}$$
$$= \alpha_1^2 m(D_1^*, 0, w_1) + \alpha_2^2 m(D_2^*, 0, w_2) +$$
$$+ \frac{\alpha_1^2}{\pi} \log |f'(\zeta_1)| + \frac{\alpha_2^2}{\pi} \log |f'(\zeta_2)| + \frac{(\alpha_1 + \alpha_2)^2}{2\pi} \log |f'(0)|.$$

Now we use Lemma 3.5.1 and calculate the moduli in (3.42) and (3.43). Thus, normalizing $\alpha_1 + \alpha_2 = 1$, the inequalities (3.42) and (3.43) imply the inequality (3.34) in the assertion of Theorem 3.5.3.

Now we derive the extremal function $w = f^*(\zeta)$ and the equation (3.35). We examine the function $g_1(\zeta)$ given by (3.39) that maps the domain $D_1^*$ onto the upper halfplane $\mathbb{H}^+$, and the function $G_1$ that maps the domain $B_1^*$ onto $\mathbb{H}^+$. The function $G_1$ satisfies the differential equation

$$\alpha_1^2 \left( \frac{G_1'(w)}{G_1(w)} \right)^2 = \pi^2 \psi(w).$$

By the analogy with (3.39), we obtain

$$G_1(w) = \frac{(w - w_1)(w - w_2)^{\alpha_2/\alpha_1}}{w^{\frac{\alpha_1 + \alpha_2}{2\alpha_1}}} w_1^{-\alpha_1/2} w_2^{-\alpha_2/2} e^{i\left(-\frac{\pi}{2} \frac{\alpha_1 + \alpha_2}{\alpha_1}\right)}.$$

The superposition $G_1^{-1} \circ g_1(\zeta)$ exists for $a \in (0, a_0]$ and leads to the equation (3.35) defined in the domain $D_1^*$. Now we repeat the same observations for $G_2(\zeta)$ that maps the domain $B_2^*$ onto $\mathbb{H}^+$, and satisfies in $D_2^*$ the differential equation

$$\alpha_2^2 \left( \frac{G_2'(w)}{G_2(w)} \right)^2 = \pi^2 \psi(w).$$

The equation which we deduce for the superposition $G_2^{-1} \circ g_2(\zeta)$ is the same as for $f^*$. Therefore, the function $f^*$ is defined in the whole disk $U$ as a subordinating superposition, and maps the extremal configuration $(D_1^*, D_2^*)$ onto the extremal configuration $(B_1^*, B_2^*)$. Thus, it gives the equality sign to (3.42–3.43), and, therefore, also to (3.34). $\qquad\square$

## 3.5.2 The range of $(|f(r)|, |f'(0)|)$

Now we consider the problem of finding the range of the system of functionals $(|f(r)|, |f'(0)|)$ in the class $S^1(\beta)$ corresponding to the problem on $\max |f'(0)|$ as $|f(r)|$ fixed.

Let $S_0 = \mathbb{C} \setminus \{0, a\}$, $a > 0$ be the twice-punctured Riemann sphere. We consider on $S_0$ the admissible system of curves $(\gamma_1, \gamma_2)$ of types I and IV respectively where $\gamma_1 = \{w : |w| = 1/\varepsilon\}$ and $\gamma_2 = \{w : |w - a| = a\}$, so that $\varepsilon$ is sufficiently small. Let $\mathfrak{D}$ be the set of all pairs $(D_1, D_2)$ of doubly connected parabolic domains and digons associated with the admissible system $(\gamma_1, \gamma_2)$ and a weight-height vector $(\alpha_1, \alpha_2)$ given. The digons are supposed to be conformal at their vertices and the condition of the compatibility of weights and angles is satisfied with the equal angles $\pi$ at the vertices over $0$ of all digons $D_2 \in \mathfrak{D}$. Then, the problem of the extremal partition of $S_0$ consists of maximizing the sum $\alpha_1^2 m(D_1, \infty) - \alpha_2^2 m(D_2, 0, 0)$ over all $(D_1, D_2) \in \mathfrak{D}$. Without loss of generality, we assume $\alpha_1 = \alpha$, $\alpha_2 = 1$, $\alpha \in [0, \infty)$, and the maximum of this sum we denote by $M(\alpha, a)$. There is a unique pair $(D_1^*, D_2^*)$ which is extremal in this problem. $D_1^*$ is the circular domain and $D_2^*$ is the strip domain in the trajectory structure of the differential

$$\varphi(z)dz^2 = -A\frac{(z-c)dz^2}{z^2(z-a)}, \quad A > 0, \ c \le 0. \tag{3.43}$$

Here $A$ and $c$ are functions with respect to $\alpha$.

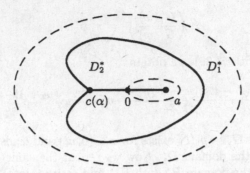

**Fig. 3.9.** Trajectory structure of the differential $\varphi(z)dz^2$ (dashed lines are sample trajectories)

If $\alpha = 0$, then $D_1^* = \emptyset$ and $D_2^* = \mathbb{C} \setminus (-\infty, a]$ is the digon with two vertices with the same support 0. In this case $M(0, a) = \frac{2}{\pi} \log 4a$. If $\alpha \to \infty$, then $D_1^* = \mathbb{C} \setminus [0, a]$. In this case $M(\infty, a) = \frac{1}{2\pi} \log 4/a$.

**Theorem 3.5.4.** *Let* $0 < \alpha < \infty$. *Then,*

$$m(D_1^*, \infty) = \frac{1}{2\pi} \log \frac{4\alpha^2}{a(1+\alpha^2)} - \frac{1}{\pi\alpha}\left(\frac{\pi}{2} - \arctan\frac{1}{\alpha}\right),$$

$$m(D_2^*, 0, 0) = \frac{2}{\pi} \log \frac{4a}{1+\alpha^2} + \frac{4\alpha}{\pi}\left(\frac{\pi}{2} - \arctan\frac{1}{\alpha}\right).$$

*Proof.* We consider the map $u = u(z)$ whose inverse is

$$z = c\frac{a + 1 + (a-1)\cos u}{(c+1) + (c-1)\cos u}, \tag{3.44}$$

and obtain the representation of the differential $\varphi$ in terms of the parameter $u$ in regular points

$$\varphi(z)dz^2 = Q(u)du^2 = \tag{3.45}$$

$$= \frac{4Ac(a-c)^2(1+\cos u)^2}{((c+1) + (c-1)\cos u)^2((a+1) + (a-1)\cos u)^2}du^2.$$

Here

$$\left|\frac{a+1}{a-1}\right| > 1 \quad \text{and} \quad \left|\frac{c+1}{c-1}\right| \le 1.$$

Now we study the trajectory structure of this quadratic differential which is a complete square of a linear one. The differential $Q(u)du^2$ has zeros of order 4 at the points $\pi + 2\pi k$ which are the images of $c$ under the map $u(z)$. Furthermore, $u(0) = \pm\eta_k$, so that $\text{Re}\,\eta_0 = 0$ in the case $a < 1$ or $\text{Re}\,\eta_0 = \pi$ in the case $a > 1$, and

$$\eta_k = \arccos\frac{1+a}{1-a}, \quad k = 1, 2, \ldots, n, \ldots.$$

For definiteness, assume now $a < 1$. Then, $u(\infty) = \theta_k = \arccos(1+c)/(1-c)$, $\theta_0 \in (0, \pi)$, and $\theta_k$ and $\eta_k$ are the poles of the second order. The points $u(a) = 2\pi k$ are regular for this differential.

Consider a fixed branch of the function $u(z)$ that maps $\overline{\mathbb{C}} \setminus [c, a]$ onto the strip $0 < \text{Re}\,u < \pi$. The circular domain $D_1^u = u(D_1^*)$ is bounded by the critical trajectory of $Q(u)du^2$ starting and ending at $\pi$ enclosing the real point $\theta_0$. The strip domain $D_2^u = u(D_2^*)$ is bounded by the same trajectory, the imaginary axis, and the straight line $\text{Re}\,u = \pi$.

Let $\zeta_j(u)$, $j = 1, 2$ be conformal maps of the domains $D_j^u$ onto the unit disk $U$ and the digon $\mathbb{C}\setminus[0, \infty)$ respectively, such that $\zeta_1(\theta_0) = 0$ and $\zeta_2(\eta_0) = 0$, $\zeta_2(-\eta_0) = \infty$. These functions satisfy in the domain $D_1^u$ the differential equation

$$\alpha\frac{d\zeta_1(u)}{\zeta_1(u)} = 2\pi\sqrt{-Q(u)}du, \tag{3.46}$$

and in the domain $D_2^u$ the differential equation

$$\frac{d\zeta_2(u)}{\zeta_2(u)} = 2\pi\sqrt{Q(u)}du, \tag{3.47}$$

or in terms of the parameter $z$,

$$\alpha\left(\frac{d\zeta_1(u(z))}{\zeta_1(u(z))}\right)^2 = -4\pi^2\varphi(z)dz^2. \tag{3.48}$$

$$\left(\frac{d\zeta_2(u(z))}{\zeta_2(u(z))}\right)^2 = 4\pi^2\varphi(z)dz^2. \tag{3.49}$$

Letting $z \to \infty$ in (3.48) in the case of $j = 1$ or $z \to 0$ in (3.49) in the case of $j = 2$, we obtain $A = \alpha/4\pi^2$ and $c = -a/\alpha^2$.

Now we calculate the reduced modulus of the circular domain. The part $[\theta_0 + \delta, \pi]$ of the orthogonal trajectory of the differential $Q(u)du^2$ for a sufficient small $\delta$ has the preimage $[-1/\varepsilon_1, c]$ under the map $u(z)$. From (3.44) we derive

$$\delta = \frac{\sqrt{-c}(a - c)}{1 - c}\varepsilon_1 + O(\varepsilon_1^2). \tag{3.50}$$

These two segments have the image $[\varepsilon e^{i\beta}, e^{i\beta}]$ in the $\zeta$-plane. Without loss of generality, assume $\beta = 0$.

Let $z = f_1(\zeta) = A_{-1}/\zeta + A_0 + A_1\zeta + \ldots$ be a function from $U$ onto $D_1^*$. Then, the reduced modulus of $D_1^*$ can be calculated as $m(D_1^*, \infty) = \frac{1}{2\pi}\log 1/|A_1|$. We derive directly from (3.45)

$$\sqrt{-Q(u)} = 2\sqrt{-cA}\left(\frac{1}{c+1+(c-1)\cos u} - \frac{1}{a+1+(a-1)\cos u}\right) \quad (3.51)$$

$$= \pm\frac{1}{2\pi}\cdot\frac{d}{du}\left(\alpha\log\frac{\alpha\tan\frac{u}{2} - \sqrt{a}}{\alpha\tan\frac{u}{2} + \sqrt{a}} - 2\arctan\frac{\tan\frac{u}{2}}{\sqrt{a}}\right).$$

We choose the branch of the root so that the sign $(+)$ occurs in front of the previous expression. Moreover, $\tan(\theta_0/2) = \sqrt{-c} = \sqrt{a}/\alpha$. Integrating (3.46) along the segments described we derive

$$\varepsilon = \frac{\alpha\tan\frac{\theta_0+\delta}{2} - \sqrt{a}}{\alpha\tan\frac{\theta_0+\delta}{2} + \sqrt{a}}\cdot\exp\left(\frac{2}{\alpha}\left(\frac{\pi}{2} - \arctan\frac{\tan\frac{\theta_0+\delta}{2}}{\sqrt{a}}\right)\right) =$$

$$= \frac{(a+\alpha^2)}{4\alpha\sqrt{a}}\exp\left(\frac{2}{\alpha}\left(\frac{\pi}{2} - \arctan\frac{1}{\alpha}\right)\right)\cdot\delta + O(\delta^2)$$

and finally, using (3.50), we obtain

$$A_1 = \frac{a(1+\alpha^2)}{4\alpha^2}\exp\left(\frac{2}{\alpha}\left(\frac{\pi}{2} - \arctan\frac{1}{\alpha}\right)\right).$$

Then the modulus $m(D_1^*, \infty)$ is of the form stated in Theorem 3.5.4.

Next we calculate the reduced modulus of the digon $D_2^*$ with respect to its two vertices with the same support 0. For this we consider the strip domain $D_2^u$ in the $u$-plane and the segment of the imaginary axis $[0, \eta_0 - i\delta]$ that belongs to the critical trajectory of the differential $Q(u)du^2$. It has the preimage $[\varepsilon_1, a]$ in the $z$-plane under the mapping $u(z)$. We calculate from (3.44) that

$$\delta = \frac{c-a}{c\sqrt{a}(a-1)}\varepsilon_1 + O(\varepsilon_1^2). \quad (3.52)$$

For these two segments there is an image in the $\zeta$-plane $[1, 1/\varepsilon]$ from the boundary of the digon $\mathbb{C}\setminus[0,\infty)$ which is the image of the domain $D_2^*$ under the map $\zeta_2(u(z))$. We find that the lengths of the segments $[\varepsilon, 1]$ and $[1, 1/\varepsilon]$ are equal in the metric $|d\zeta|/|\zeta|$ and, therefore, $1 = \zeta_2(0)$. Let $z = f_2(\zeta) = B_{-1}/\zeta + B_0 + B_1\zeta + \ldots$ be a conformal map from $\mathbb{C}\setminus[0,\infty)$ onto $D_2^*$. Then, the reduced modulus of $D_2^*$ turns out to be $m(D_2^*, 0, 0) = \frac{2}{\pi}\log|B_{-1}|$. Here this derivative is thought of as one of the angular derivatives in $D_2^*$.

By (3.51) we have

$$\sqrt{Q(u)} = \pm\frac{i}{2\pi}\cdot\frac{d}{du}\left(\alpha\log\frac{\alpha\tan\frac{u}{2} - \sqrt{a}}{\alpha\tan\frac{u}{2} + \sqrt{a}} - 2\arctan\frac{\tan\frac{u}{2}}{\sqrt{a}}\right). \quad (3.53)$$

Again, we choose the branch of the root such that the sign $(+)$ occurs in front of the right-hand side of (3.53). We have $\tan(\eta_0/2) = i\sqrt{a}$. Rewrite the equation (3.47) as

$$\frac{d\zeta}{\zeta} = i\frac{d}{du}\left(\alpha\log\frac{\alpha\tan\frac{u}{2}-\sqrt{a}}{\alpha\tan\frac{u}{2}+\sqrt{a}} - 2\arctan\frac{\tan\frac{u}{2}}{\sqrt{a}}\right)du.$$

Since we use the complex tangent, we transform the right-hand side applying the identity

$$\arctan w = \frac{1}{2i}\log\frac{1+iw}{1-iw}.$$

Then,

$$\frac{d\zeta}{\zeta} = \frac{d}{du}\left(-2\alpha\arctan\frac{i\sqrt{a}}{\alpha\tan\frac{u}{2}} - \log\frac{1+i\frac{\tan\frac{u}{2}}{\sqrt{a}}}{1-i\frac{\tan\frac{u}{2}}{\sqrt{a}}}\right). \qquad (3.54)$$

Integrating (3.54) along the segment $[1, 1/\varepsilon]$ in the left-hand side, and along the vertical segment $[0, \eta_0 - i\delta]$, $\mathrm{Im}\,\eta_0 > 0$, in the right-hand side, we deduce that

$$\varepsilon = \frac{1+i\frac{\tan\frac{\eta_0-i\delta}{2}}{\sqrt{a}}}{1-i\frac{\tan\frac{\eta_0-i\delta}{2}}{\sqrt{a}}}\cdot\exp\left(2\alpha\left(\arctan\frac{i\sqrt{a}}{\alpha\tan\frac{\eta_0-i\delta}{2}}-\frac{\pi}{2}\right)\right) =$$

$$= \frac{1-a}{4\sqrt{a}}\exp\left(2\alpha\left(\arctan\frac{1}{\alpha}-\frac{\pi}{2}\right)\right)\delta + O(\delta^2).$$

Finally, using (3.50) and substituting $c = -a/\alpha^2$, we obtain

$$|B_1| = \frac{4a}{1+\alpha^2}\exp\left(2\alpha\left(\frac{\pi}{2}-\arctan\frac{1}{\alpha}\right)\right).$$

This leads to the expression in Theorem 3.5.4. The case $a > 1$ can be obtained by applying the map $w = kz$, where $k > 1/a$. This leads to the same expressions. □

We define another problem of the extremal partition connected with the previous one. Let $S_1 = \mathbb{C}\setminus\{c_1, 0, a_1\}$, $a_1 > 0$, $c_1 < 0$, be the thrice-punctured Riemann sphere. We consider on $S_1$ the admissible system of curves $(\gamma_1, \gamma_2)$ of type I, IV respectively, where $\gamma_1 = \{w : |w| = 1/\varepsilon\}$ and $\gamma_2 = \{w : |w - a_1| = a_1\}$, so that and $\varepsilon$ is sufficiently small. Let $\mathfrak{B}$ be the set of all pairs $(B_1, B_2)$ of doubly connected parabolic domains and digons associated with the admissible system $(\gamma_1, \gamma_2)$ and a weight-height vector $(\alpha_1, \alpha_2)$ given. All digons are supposed to be conformal at their vertices and the condition of the compatibility of weights and angles is satisfied with the equal angles $\pi$ at the vertices over 0 of all digons $B_2 \in \mathfrak{B}$. Then, the problem of the extremal partition of $S_1$ consists of maximizing the sum $\alpha_1^2 m(B_1, \infty) - \alpha_2^2 m(B_2, 0, 0)$ over all $(B_1, B_2) \in \mathfrak{D}$. Without loss of generality, we assume $\alpha_1 = \alpha$, $\alpha_2 = 1$, $\alpha \in [0, \infty)$, and the maximum of this sum we denote by $\mathcal{M}(\alpha, c_1, a_1)$. There is a unique pair $(B_1^*, B_2^*)$ which is extremal in this problem. $B_1^*$ is the circular domain and $B_2^*$ is the strip domain in the trajectory structure of the differential

$$\psi(z)dz^2 = -A\frac{(z-b)^2 dz^2}{z^2(z-a_1)(z-c_1)}, \quad A > 0, \ b \le 0. \tag{3.55}$$

Here $A$ and $b$ are functions on $\alpha$. For $\alpha \in [0, \sqrt{\frac{a_1}{-c_1}}]$ the problem can be reduced to the previous case (Theorem 3.5.4) with $a = a_1$. If $\alpha \to \infty$, then $B_1^* = \mathbb{C} \setminus [c_1, a_1]$. In this case $\mathcal{M}(\infty, c_1, a_1) = \frac{1}{2\pi} \log 4/(a_1 - c_1)$.

**Theorem 3.5.5.** *Let* $\sqrt{\frac{a_1}{-c_1}} \le \alpha < \infty$. *Then,*

$$m(B_1^*, \infty) = \frac{1}{2\pi} \log \frac{4}{a_1 - c_1} - \frac{1}{\pi\alpha}\left(\frac{\pi}{2} - \arctan\sqrt{\frac{-c_1}{a_1}}\right),$$

$$m(B_2^*, 0, 0) = \frac{2}{\pi} \log \frac{4a_1 c_1}{a_1 - c_1} + \frac{4\alpha}{\pi}\left(\frac{\pi}{2} - \arctan\sqrt{\frac{-c_1}{a_1}}\right).$$

*Proof.* As in the previous theorem, we consider the map $u = u(z)$ whose inverse is

$$z = c_1 \frac{a_1 + 1 + (a_1 - 1)\cos u}{(c_1 + 1) + (c_1 - 1)\cos u},$$

and obtain the representation of the differential $\psi$ in terms of the parameter $u$ in its regular points

$$\psi(z)dz^2 = \Phi(u)du^2 =$$

$$= \frac{4A}{c_1} \frac{(c_1(a_1 + 1) - b(c_1 + 1) + (c_1(a_1 - 1) - b(c_1 - 1))\cos u)^2}{((c + 1) + (c - 1)\cos u)^2((a + 1) + (a - 1)\cos u)^2}du^2. \tag{3.56}$$

Here

$$\left|\frac{a_1 + 1}{a_1 - 1}\right| > 1, \quad \left|\frac{c_1 + 1}{c_1 - 1}\right| \le 1,$$

and

$$\left|\frac{c_1(a_1 + 1) - b(c_1 + 1)}{c_1(a_1 - 1) - b(c_1 - 1)}\right| > 1.$$

Now we study the trajectory structure of this quadratic differential which is a complete square of a linear one. The differential $Q(u)du^2$ has zeros of order 2 at the points $\pm\gamma_k + 2\pi k$ which are the images of

$$\frac{c_1(a_1 + 1) - b(c_1 + 1)}{c_1(a_1 - 1) - b(c_1 - 1)}$$

under the map $u(z)$. Furthermore, $u(0) = \pm\eta_k$, so that $\operatorname{Re} \eta_0 = 0$ in the case $a_1 < 1$, or $\operatorname{Re} \eta_0 = \pi$ in the case $a_1 > 1$, and

$$\eta_k = \arccos\frac{1+a}{1-a}, \quad k = 1, 2, \ldots, n, \ldots.$$

For definiteness, assume now $a_1 < 1$. The case $a_1 > 1$ can be considered as in Theorem 3.5.4. Then, $u(\infty) = \theta_k = \arccos(1 + c)/(1 - c)$, $\theta_0 \in (0, \pi)$, and $\theta_k$ and $\eta_k$ are the poles of the second order. The points $u(a_1) = 2\pi k$ are regular for this differential.

We consider a fixed branch of the function $u(z)$ that maps $\overline{\mathbb{C}} \setminus [c, a]$ onto the strip $0 < \operatorname{Re} u < \pi$. The circular domain $D_1^u = u(D_1^*)$ is bounded by the critical trajectory of $Q(u)du^2$ starting and ending at $\pi$ enclosing the real point $\theta_0$. The strip domain $D_2^u = u(D_2^*)$ is bounded by the same trajectory, the imaginary axis, and the straight line $\operatorname{Re} u = \pi$.

Let $\zeta_j(u)$, $j = 1, 2$ be univalent conformal maps of the domains $B_j^u$ onto the unit disk $U$ and the digon $\mathbb{C} \setminus [0, \infty)$ respectively, such that $\zeta_1(\theta_0) = 0$ and $\zeta_2(\eta_0) = 0$, $\zeta_2(-\eta_0) = \infty$. These functions satisfy in the domain $B_1^u$ the differential equation

$$\alpha \frac{d\zeta_1(u)}{\zeta_1(u)} = 2\pi \sqrt{-\Phi(u)}du, \tag{3.57}$$

and in the domain $B_2^u$ the differential equation

$$\frac{d\zeta_2(u)}{\zeta_2(u)} = 2\pi \sqrt{\Phi(u)}du, \tag{3.58}$$

or in terms of the parameter $z$

$$\alpha \left( \frac{d\zeta_1(u(z))}{\zeta_1(u(z))} \right)^2 = -4\pi^2 \psi(z)dz^2. \tag{3.59}$$

$$\left( \frac{d\zeta_2(u(z))}{\zeta_2(u(z))} \right)^2 = 4\pi^2 \psi(z)dz^2. \tag{3.60}$$

Letting $z \to \infty$ in (3.59) in the case of $j = 1$ or $z \to 0$ in (3.60) in the case of $j = 2$, we obtain $A = \alpha/4\pi^2$ and $b = -\sqrt{-a_1 c_1/\alpha}$.

As in Theorem 3.5.4 the calculation gives us

$$\sqrt{-\Phi(u)} = \frac{2\sqrt{A}}{\sqrt{-c_1}} \left( \frac{c_1}{c_1 + 1 + (c_1 - 1)\cos u} - \frac{b}{a_1 + 1 + (a_1 - 1)\cos u} \right) =$$

$$= \frac{1}{2\pi} \cdot \frac{d}{du} \left( -\alpha \log \frac{\tan \frac{u}{2} - \sqrt{-c_1}}{\tan \frac{u}{2} + \sqrt{-c_1}} + 2 \arctan \frac{\tan \frac{u}{2}}{\sqrt{a_1}} \right).$$

Taking into account

$$\delta = \frac{(a_1 - c_1)\sqrt{-c_1}}{1 - c_1} \varepsilon_1 + O(\varepsilon_1^2),$$

and integrating (3.57, 3.58) as before, we obtain the expressions in Theorem 3.5.5. $\qquad\square$

Now let $U_z = U \setminus \{0, r\}$, $r \in (0, 1)$ be the twice-punctured unit disk. We consider on $U_z$ an admissible system of curves $(\gamma_1^z, \gamma_2^z)$, where $\gamma_2^z = \{z : |z| = \varepsilon\}$ and $\gamma_2^z$ is an arc with starting and ending points at 1, enclosing $r$ such that $\gamma_1^z \cap \gamma_2^z = \emptyset$. Here $\varepsilon < r$ and sufficiently small. Let $\mathfrak{D}^z$ be a set of all pairs $(D_1^z, D_2^z)$ of parabolic doubly connected domains and digons in $U_z$ associated with $(\gamma_1^z, \gamma_2^z)$ and the vector $(\alpha, 1)$ and satisfying the condition of compatibility of weights and angles and conformality. Then, the problem of the extremal partition of $U_z$ consists of maximizing the sum $\alpha^2 m(D_1^z, 0) - m(D_2^z, 1, 1)$ over all $(D_1^z, D_2^z) \in \mathfrak{D}^z$. The maximum of this sum we denote by $M_z(\alpha, r)$. Under the transformation $Z(z) = (1 - z)^2 / z$ two extremal domains $(D_1^{z*}, D_2^{z*})$ in the problem of $M_z(\alpha, r)$ are mapped onto two extremal domains $(B_1^*, B_2^*)$ in the problem of finding $\mathcal{M}(\alpha, c_1, a_1)$ where $c_1 = -4$, $a_1 = (1 - r)^2 / r$. Taking into account the change of the reduced moduli, we derive from Theorem 3.5.5 that for $\alpha \geq \frac{1-r}{2\sqrt{r}}$ the reduced moduli are given as

$$m(D_1^{z*}, 0) = \frac{1}{2\pi} \log \frac{4r}{(1+r)^2} - \frac{1}{\pi\alpha}\left(\frac{\pi}{2} - \arctan \frac{2\sqrt{r}}{1-r}\right), \qquad (3.61)$$

$$m(D_2^{z*}, 1, 1) = \frac{4}{\pi} \log \frac{4(1-r)}{1+r} + \frac{4\alpha}{\pi}\left(\frac{\pi}{2} - \arctan \frac{2\sqrt{r}}{1-r}\right). \qquad (3.62)$$

Now we consider the same problem of the extremal partition replacing $r$ by $w \in U$. Denote by $(D_1^{w*}, D_2^{w*})$ the extremal pair of domains and let $\alpha$ vary within $[0, \infty)$. By $M_w(\alpha, w)$ we denote the maximum of $\alpha^2 m(D_1^{w*}, 0) - m(D_2^{w*}, 1, 1)$. Suppose $w = x \in (0, r)$. For $\alpha \geq \frac{1-x}{2\sqrt{x}}$ we have the expressions given by (3.61), (3.62). For $0 \leq \alpha \leq \frac{1-x}{2\sqrt{x}}$ we deduce from Theorem 3.5.4

$$m(D_1^{w*}, 0) = \frac{1}{2\pi} \log \frac{4x\alpha^2}{(1-x)^2(1+\alpha^2)} - \frac{1}{\pi\alpha}\left(\frac{\pi}{2} - \arctan \frac{1}{\alpha}\right), \qquad (3.63)$$

$$m(D_2^{w*}, 1, 1) = \frac{2}{\pi} \log \frac{4(1-x)^2}{x(1+\alpha^2)} + \frac{4\alpha}{\pi}\left(\frac{\pi}{2} - \arctan \frac{1}{\alpha}\right). \qquad (3.64)$$

**Lemma 3.5.2.** 1) Let $\beta \geq 1$ and be fixed. For $\frac{1-r}{2\sqrt{r}} \leq \alpha < \infty$ and $w = x \in (0, r)$ the equation

$$m(D_2^{z*}, 1, 1) + \frac{4}{\pi} \log \beta = m(D_2^{z*}, 1, 1) \qquad (3.65)$$

defines the unique solution $x = x(\alpha)$ whose value belongs to the interval $(0, r)$ for a fixed $\alpha$. The function $x(\alpha)$ is differentiable and increases with respect to $\alpha$ from $x(\frac{1-r}{2\sqrt{r}}) =: x_1$, which is the solution of the equation

$$\frac{x}{(1-x)^2} = \frac{1}{\beta^2} \frac{r}{(1-r)^2},$$

to $\lim_{\alpha \to \infty} x(\alpha) = r$.

2) *For $\frac{1-r}{2\sqrt{r}} \leq \alpha \leq \alpha_0$ this solution $x(\alpha)$ is defined by the equation*

$$\frac{x}{(1-x)^2} =$$

$$= \frac{1}{4\beta^2}\left(\frac{1+r}{1-r}\right)^2 \frac{1}{1+\alpha^2} \exp\left[2\alpha\left(\arctan\frac{2\sqrt{r}}{1-r} - \arctan\frac{1}{\alpha}\right)\right], \qquad (3.66)$$

*as $x \in [x_1, r)$.*

*For $\alpha_0 \leq \alpha < \infty$ the solution $x(\alpha)$ is defined by the equation*

$$\log\beta\frac{1+x}{1-x}\frac{1-r}{1+r} = \alpha\left(\arctan\frac{2\sqrt{r}}{1-r} - \arctan\frac{2\sqrt{x}}{1-x}\right), \qquad (3.67)$$

*as $x \in [x_1, r)$.*

*Here $\alpha_0$ is a unique solution of the equation $P(\alpha) = 0$, where*

$$P(\alpha) := \log\frac{1-r}{1+r}\beta\frac{\sqrt{1+\alpha^2}}{\alpha} + \alpha\left(\arctan\frac{1}{\alpha} - \arctan\frac{2\sqrt{r}}{1-r}\right).$$

**Proof.** First we consider the case $\frac{1-r}{2\sqrt{r}} \leq \alpha \leq \frac{1-x}{2\sqrt{x}}$. The equation (3.65) and the formulae (3.62), (3.64) imply the equation (3.66). Since the left-hand side of (3.66) is positive, always there is a unique solution to the equation (3.66). Differentiating both sides of (3.66) with respect to $\alpha$ we obtain

$$x'(\alpha)\frac{1+x}{1-x} = \frac{1}{2\beta}\left(\frac{1+r}{1-r}\right)^2\frac{1}{1+\alpha^2}\left(\arctan\frac{2\sqrt{r}}{1-r} - \arctan\frac{1}{\alpha}\right) \times$$

$$\times \exp\left[2\alpha\left(\arctan\frac{2\sqrt{r}}{1-r} - \arctan\frac{1}{\alpha}\right)\right].$$

Therefore, $x'(\alpha) > 0$ and the function $x(\alpha)$ increases in $\frac{1-r}{2\sqrt{r}} \leq \alpha \leq \frac{1-x}{2\sqrt{x}}$.

For $\alpha = \frac{1-r}{2\sqrt{r}}$ we have $x(\frac{1-r}{2\sqrt{r}}) = x_1 < r$. For $\alpha = \frac{1-x}{2\sqrt{x}}$ the equation (3.66) is of the form

$$H(x) := \frac{1+x}{1-x} - \frac{1}{\beta}\cdot\frac{1+r}{1-r}\exp\left[\frac{1-x}{2\sqrt{x}}\left(\arctan\frac{2\sqrt{r}}{1-r} - \arctan\frac{2\sqrt{x}}{1-x}\right)\right] = 0,$$

where

$$H(r) = \frac{1+r}{1-r}\left(1 - \frac{1}{\beta}\right) \geq 0, \quad \lim_{x\to 0}H(x) = -\infty.$$

We calculate the derivative

$$H'(x) = \frac{2}{(1-x)^2} +$$

$$+ \frac{1}{2\beta}\cdot\frac{1+r}{1-r}\left(\frac{1+x}{2\sqrt{x}}\left(\arctan\frac{2\sqrt{r}}{1-r} - \arctan\frac{2\sqrt{x}}{1-x}\right) + \frac{1-x}{4x(1+x)}\right) \times$$

$$\times \exp\left[\frac{1-x}{2\sqrt{x}}\left(\arctan\frac{2\sqrt{r}}{1-r} - \arctan\frac{2\sqrt{x}}{1-x}\right)\right],$$

which is positive in so far as $x \in (0, r)$. Therefore, the equation $H(x) = 0$ has a unique solution $x = x_2 \in (0, r)$. Moreover,

$$
H(x_1) = \frac{\sqrt{(1+r)^2 - 2r(1 - \frac{1}{\beta^2})}}{1-r} -
$$

$$
- \frac{1+r}{1-r} \cdot \frac{1}{\beta} \exp\left[\frac{\beta}{2} \frac{1-r}{\sqrt{r}} \left(\arctan \frac{2\sqrt{r}}{1-r} - \arctan \frac{2\sqrt{r}}{\beta(1-r)}\right)\right] < 0,
$$

hence, $x_2 > x_1$.

Finally, the equation $\alpha = \frac{1-x(\alpha)}{2\sqrt{x(\alpha)}}$ implies the equation $P(\alpha) = 0$. In order to show that $\alpha_0$ is the unique solution of the latter equation, we point out that $P(\frac{1-r}{2\sqrt{r}}) = \log \beta > 0$ and $\lim_{\alpha \to \infty} P(\alpha) = -\infty$. Moreover, $P'(\alpha) = -\frac{1}{\alpha} - \arctan \frac{2\sqrt{r}}{1-r} + \arctan \frac{1}{\alpha} < 0$. Therefore, there is a unique solution $\alpha_0$ of the equation $P(\alpha) = 0$.

Now we consider the case $\frac{1-x}{2\sqrt{x}} \le \alpha < \infty$. The equation (3.65) and the formula (3.62) imply the equation (3.67). Denote by

$$
G(x) := \log \beta \frac{1+x}{1-x} \frac{1-r}{1+r} - \alpha \left(\arctan \frac{2\sqrt{r}}{1-r} - \arctan \frac{2\sqrt{x}}{1-x}\right).
$$

The equation (3.67) is equivalent to the equation $G(x) = 0$. Calculation gives

$$
G(x_1) = \log \frac{\sqrt{(1+r)^2 - 2r(1 - \frac{1}{\beta^2})}}{1+r} + \alpha \left(\arctan \frac{1}{\beta} \frac{2\sqrt{r}}{1-r} - \arctan \frac{2\sqrt{r}}{1-r}\right) < 0,
$$

$G(r) = \log \beta > 0$. Moreover, $G'(x) > 0$. Therefore, there is a unique solution to the equation (3.67) which we denote by $x(\alpha) \in (x_1, r)$. Calculating its derivative we obtain

$$
x'(\alpha) \left(\frac{1}{1-x^2} + \frac{\alpha}{(1+x)\sqrt{x}}\right) = \arctan \frac{2\sqrt{r}}{1-r} - \arctan \frac{2\sqrt{x}}{1-x}.
$$

Since $\alpha \ge \frac{1-x}{8\sqrt{x}}$, the function $x(\alpha)$ increases with increasing $\alpha$. The condition $\alpha = \frac{1-x(\alpha)}{2\sqrt{x(\alpha)}}$ leads to the same equation $P(\alpha) = 0$. Therefore $x(\alpha_0) = x_2$. This completes the proof. $\qquad \square$

Now we need results about some properties of the monotonicity of the modulus. Let $U' = U \setminus \{0, w\}$, $|w| < 1$ be the twice punctured unit disk. We consider on $U'$ an admissible system of curves $(\gamma_1, \gamma_2^{(n)})$, where $\gamma_1 = \{z : |z| = \varepsilon\}$ and $\gamma_2^{(n)}$ is from the countable set of arcs with certain homotopy on $U'$, with starting and ending points at $1$, enclosing $w$ such that $\gamma_1 \cap \gamma_2^{(n)} = \emptyset$. By means of $n = 1$ we assume that $\gamma_2^{(1)}$ is homotopic to the segment $[1, w]$. Here

$\varepsilon < |w|$ and sufficiently small. Let $\mathfrak{D}^{(n)}$ be the set of all pairs $(D_1^{(n)}, D_2^{(n)})$ of doubly connected parabolic domains and digons in $U'$ associated with the admissible system $(\gamma_1, \gamma_2^{(n)})$ and the vector $(\alpha, 1)$. Suppose that digons have the inner angles $\pi/2$ at the vertices and are conformal there. Then, the problem of the extremal partition of $U'$ consists of maximizing the sum $\alpha^2 m(D_1^{(n)}, 0) - m(D_2^{(n)}, 1, 1)$ over all $(D_1^{(n)}, D_2^{(n)}) \in \mathfrak{D}^{(n)}$. The maximum of this sum we denote by $M_w^{(n)}(\alpha, w)$.

**Lemma 3.5.3.** *In the family* $\mathfrak{D}^{(n)}$ *the inequality*

$$M_w^{(n)}(\alpha, w) \le M_w^{(1)}(\alpha, w)$$

*holds for all* $\alpha$ *and* $n = 2, 3, \ldots$.

*Proof.* Without loss of generality we assume $\operatorname{Im} w > 0$. Now we are going to apply the results about polarization of doubly connected domains. For this we construct the pair of doubly connected domains $(D_{1\varepsilon}^{(n)}, D_{2\varepsilon}^{(n)})$ where $D_{1\varepsilon}^{(n)}$ is the extremal circular domain $D_1^{(n)^*}$ in the above modulus problem minus the disk $|z| < \varepsilon$, and $D_{2\varepsilon}^{(n)}$ is the extremal strip domain $D_2^{(n)^*}$ minus the disk $|z - 1| < \varepsilon$ plus the symmetric image of this quadrilateral with respect to the circle $|z - 1| = \varepsilon$. Now we apply polarization to the domains $(D_{1\varepsilon}^{(n)}, D_{2\varepsilon}^{(n)})$ with respect to the real axis for $n \ge 2$. We obtain as a result the pair of non-overlapping doubly connected domains $(\tilde{D}_1^\varepsilon, \tilde{D}_2^\varepsilon)$ with the moduli $M(D_{1\varepsilon}^{(n)}) \le M(\tilde{D}_1^\varepsilon)$, $M(D_{2\varepsilon}^{(n)}) \le M(\tilde{D}_2^\varepsilon)$. Moreover, the part of $\tilde{D}_2^\varepsilon$ lying outside the disk $|z - 1| < \varepsilon$ is still symmetric to the inside one. So, the same inequality is true for the modulus of the quadrilateral $\tilde{D}_2^\varepsilon \setminus \{|z - 1| < \varepsilon\}$ inside $U$. Letting $\varepsilon \to 0$ we obtain the pair of domains $(D_1^0, D_2^0)$ and the inequality $M^{(n)}(\alpha, w) \le \alpha^2 m(D_1^0, 0) - m(D_2^0, 1, 1)$ holds for them. The pair $(D_1^0, D_2^0)$ is admissible for the family $\mathfrak{D}^{(1)}$. Therefore, $\alpha^2 m(D_1^0, 0) - m(D_2^0, 1, 1) \le \alpha^2 m(D_1^{(1)^*}, 0) - m(D_2^{(1)^*}, 1, 1)$. This completes the proof.  $\square$

Let $D = D_2^{(1)^*}$ denote the extremal strip domain from $\mathfrak{D}^{(1)}$ with two vertices $a$, $b$ (with the same support 1). Now we redefine of the quantity of the reduced modulus of digon for this particular case. Denote by $D_\varepsilon'$ the domain obtained from the digon $D$ by fixing two connected arcs $\delta_a$ and $\delta_b$ on one side starting from $a$, $b$ respectively lying on its circular boundary within the disk $\{|z - 1| < \varepsilon\}$ for a sufficiently small $\varepsilon$. Denote by $M(D_\varepsilon')$ the modulus of the family of arcs in $D_\varepsilon'$ that connect $\delta_a$ and $\delta_b$.

**Lemma 3.5.4.** *The limit*

$$\lim_{\varepsilon \to 0} \left( \frac{1}{M(D_\varepsilon')} + \left(\frac{1}{\varphi_a} + \frac{1}{\varphi_b}\right) \log \varepsilon \right) = m(D, a, b) + \frac{2}{\pi} \log 4,$$

*exists where* $\varphi_a$ *and* $\varphi_b$ *are the angles between tangent rays starting from the corresponding points* $a$ *and* $b$.

*Proof.* Since $D$ is a strip domain, it is conformal at the points $a$, $b$. Moreover, there is a conformal univalent map $f(z)$ from $D$ onto the upper half-plane $H^+$ such that

$$f(z)(z-a)^{-\frac{\pi}{\varphi_a}} = c_1 + \alpha(\varepsilon),$$

and

$$f(z)(z-b)^{\frac{\pi}{\varphi_b}} = d_1 + \beta(\varepsilon),$$

where $\alpha(\varepsilon)$, $\beta(\varepsilon)$ are infinitesimal functions. The image of the arc $\delta_a$ is the interval $(0, \Delta_1)$ such that

$$\varepsilon^{\frac{\pi}{\varphi_a}}(|c_1| - |\alpha(\varepsilon)|) \leq \Delta_1 \leq \varepsilon^{\frac{\pi}{\varphi_a}}(|c_1| + |\alpha(\varepsilon)|).$$

A similar inequality one can derive for the point $b$ and for the image $(\Delta_2, \infty)$ of the arc $\delta_b$. The modulus of the quadrilateral $D'_\varepsilon$ can be calculated as

$$M(D'_\varepsilon) = \frac{\mathbf{K}'}{\mathbf{K}}\left(\frac{\sqrt{\Delta_2 - \Delta_1}}{\sqrt{\Delta_2}}\right) = \frac{\mathbf{K}}{\mathbf{K}'}\left(\sqrt{\frac{\Delta_1}{\Delta_2}}\right),$$

where, as usual, $\mathbf{K}(k)$ and $\mathbf{K}'(k)$ stand for the complete elliptic integrals. We deduce that

$$\frac{\Delta_2}{\Delta_1} \in \left(\frac{\varepsilon^{-\frac{\pi}{\varphi_b}}(|d_1| - |\beta(\varepsilon)|)}{\varepsilon^{\frac{\pi}{\varphi_a}}(|c_1| + |\alpha(\varepsilon)|)}, \frac{\varepsilon^{-\frac{\pi}{\varphi_b}}(|d_1| + |\beta(\varepsilon)|)}{\varepsilon^{\frac{\pi}{\varphi_a}}(|c_1| - |\alpha(\varepsilon)|)}\right).$$

Moreover, we have the following asymptotic behaviour

$$\lim_{k \to 0}\left(\frac{\mathbf{K}'}{\mathbf{K}}(k) - \frac{2}{\pi}\log\frac{4}{k}\right) = 0.$$

Therefore,

$$\lim_{\varepsilon \to 0}\left(\frac{1}{M(D'_\varepsilon)} + \frac{1}{\varphi_a}\log\varepsilon + \frac{1}{\varphi_b}\log\varepsilon\right) =$$

$$= \frac{1}{\pi}\log\left|\frac{d_1}{c_1}\right| + \frac{2}{\pi}\log 4 = m(D, a, b) + \frac{2}{\pi}\log 4.$$

Taking into account that $m(H^+, 0, \infty) = 0$, we arrive at the assertion of the lemma. $\qquad\square$

**Theorem 3.5.6.** *i) The boundary curve $\Gamma^+$ of the range of the system of the functionals $(|f(r)|, |f'(0)|)$ over the class $S^1(\beta)$ that corresponds to the problem of $\max |f'(0)|$ with $|f(r)|$ fixed, consists of the points $(x, y)$. The part $\Gamma_1^+$ of $\Gamma^+$ over the segment $x \in [x_1, x_2]$ is given by the parameter $\alpha$ as $(x(\alpha), y(\alpha))$ for $\frac{1-r}{2\sqrt{r}} \leq \alpha \leq \alpha_0$, where $x(\alpha)$ is defined in Lemma 3.5.2 by the equation (3.66), $x(\frac{1-r}{2\sqrt{r}}) = x_1$, $x(\alpha_0) = x_2$, and*

$$y(\alpha) = \frac{\alpha^2(1+r)^4}{4\beta^2 r(1-r)^2(1+\alpha^2)^2} \exp\left[2(\frac{1}{\alpha} - \alpha)\left(\arctan\frac{1}{\alpha} - \arctan\frac{2\sqrt{r}}{1-r}\right)\right].$$

*The part $\Gamma_2^+$ of $\Gamma^+$ over the semi-interval $[x_2, r)$ is given explicitly by the formula*

$$y = \frac{x(1+r)^2}{r(1+x)^2} \exp\left[\frac{2\left(\arctan\frac{2\sqrt{r}}{1-r} - \arctan\frac{2\sqrt{x}}{1-x}\right)}{\log\frac{\beta(1-r)(1+x)}{(1+r)(1-x)}}\right].$$

*ii) Each point of the curve $\Gamma^+$ is given by a unique function from the class $S^1(\beta)$. The point $(x(\frac{1-r}{2\sqrt{r}}), y(\frac{1-r}{2\sqrt{r}})) = (x_1, \frac{1}{\beta^2})$ is given by the canonical map $K(z)$, satisfying the equation*

$$\frac{w}{(1-w)^2} = \frac{1}{\beta^2} \cdot \frac{z}{(1-z)^2}.$$

*iii) Each point of $\Gamma^+$ over the semi-interval $(x_1, x_2]$ is given by the function that maps the unit disk $U$ onto $U$ slit along the negative real segment $[-1, \frac{(1-c(\alpha))^2}{c(\alpha)}]$ and two analytic arcs starting from $\frac{(1-c(\alpha))^2}{c(\alpha)}$ at the angles $\frac{2\pi}{3}$. This function satisfies the differential equation*

$$\frac{(w+1)^2(w - \frac{(1-c(\alpha))^2}{c(\alpha)})(w - \frac{c(\alpha)}{(1-c(\alpha))^2})dw^2}{w^2(w-x(\alpha))(w-1/x(\alpha))(w-1)^2} = \qquad (3.68)$$

$$= \frac{(z-d(\alpha))^2(z-\overline{d(\alpha)})^2 dz^2}{z^2(z-r)(z-1/r)(z-1)^2},$$

*where $\alpha \in (\frac{1-r}{2\sqrt{r}}, \alpha_0)$. Here $d(\alpha)$ and $\overline{d(\alpha)}$ are two conjugated roots of the equation*

$$\frac{(1-d)^2}{d} = -\frac{2(1-r)}{\alpha^2\sqrt{r}},$$

*with $|d(\alpha)| = 1$, $c(\alpha) = \frac{-(1-x(\alpha))^2}{\alpha^2 x(\alpha)}$.*

*iv) Each point of $\Gamma^+$ over the interval $(x_2, r)$ is given by the function that maps the unit disk $U$ onto $U$ slit along two symmetric (with respect to the real axis) analytic arcs starting orthogonally from the points $h(\alpha)$ and $\overline{h(\alpha)}$ of $\partial U$. This function satisfies the differential equation*

$$\frac{(w-h(\alpha))^2(w-\overline{h(\alpha)})^2 dw^2}{w^2(w-x(\alpha))(w-1/x(\alpha))(w-1)^2} = \frac{(z-d(\alpha))^2(z-\overline{d(\alpha)})^2 dz^2}{z^2(z-r)(z-1/r)(z-1)^2}, \quad (3.69)$$

*where $h(\alpha)$ and $\overline{h(\alpha)}$ are two conjugated roots of the equation*

$$\frac{(1-h)^2}{h} = -\frac{2(1-x(\alpha))}{\alpha^2\sqrt{x(\alpha)}},$$

*with $|h(\alpha)| = 1$.*

*Remark 3.5.2.* 1) Since the class $S^1(\beta)$ is not compact, the point $x(\infty) = r$ is not reachable and the limiting function $f(z) \equiv z$ as $\alpha \to \infty$ is not from the class $S^1(\beta)$.

2) Theorem 3.5.6 gives the sharp lower estimate of $|f(r)|$ over the class $S^1(\beta)$ with $|f'(0)|$ fixed, $|f'(0)| \in [1/\beta^2, 1]$.

*Proof.* We start the proof constructing the extremal maps. Let $\alpha \in [\frac{1-r}{2\sqrt{r}}, \alpha_0]$. Assume $a_1 = \frac{(1-r)^2}{r}$, $c_1 = -4$, $a = \frac{(1-x(\alpha))^2}{x(\alpha)}$. Then the differential equation (3.68) is equivalent just to the equation $\psi(Z)dZ^2 = \varphi(W)dW^2$ under the transformations $Z = \frac{(1-z)^2}{z}$, $W = \frac{(1-w)^2}{w}$, where the differentials $\varphi$ and $\psi$ are defined in (3.43) and (3.55). Now we construct the functions $f_2(Z)$ and $F_2(W)$ that map the domains $B_2^*$ and $D_2^*$ respectively onto the same strip domain $\mathbb{C} \setminus [0, \infty)$. By (3.49) and (3.60), they satisfy the equations

$$\left(\frac{df_2(Z)}{f_2(Z)}\right)^2 = 4\pi^2 \varphi(Z)dZ^2, \quad \left(\frac{dF_2(W)}{F_2(W)}\right)^2 = 4\pi^2 \psi(W)dW^2.$$

Then we construct the map $w = f^*(z) \equiv W^{-1} \circ F_2^{-1} \circ f_2 \circ Z(z)$ of the domain $D_2^{z*}$ onto $D_2^{w*}$. This map can be continued analytically by the equation (3.68) into $D_1^{z*}$ through the analytic arc of the trajectory of the right-hand side differential in (3.68) connecting $d(\alpha)$ and $\overline{d(\alpha)}$. Calculating the derivatives and taking into account Lemma 3.5.2, we deduce that $|f^{*\prime}(0)| = y(\alpha)$, $|f^{*\prime}(1)| = \beta$, $|f^*(r)| = x(\alpha)$. The same we do in the case $\alpha_0 < \alpha < \infty$. In this case

$$y(\alpha) = \frac{x(\alpha)}{r} \cdot \frac{(1+r)^2}{(1+x(\alpha))^2} \cdot \exp\left[\frac{2}{\alpha}\left(\arctan \frac{2\sqrt{x(\alpha)}}{1-x(\alpha)} - \arctan \frac{2\sqrt{r}}{1-r}\right)\right],$$

and we can obtain the explicit formula substituting $\alpha$ by Lemma 3.5.2.

Let $f$ be an arbitrary map from the class $S^1(\beta)$. Then, two extremal domains $(D_1^{z*}, D_2^{z*})$ in the problem about $M_z(\alpha, r)$ are mapped onto two admissible domains $(f(D_1^{z*}), f(D_2^{z*}))$ in the problem about $M_w^{(n)}(\alpha, f(r))$ for some $(n)$, and

$$\alpha^2 m(D_1^{z*}, 0) - m(D_2^{z*}, 1, 1) + \frac{\alpha^2}{2\pi} \log |f'(0)| - \frac{4}{\pi} \log \beta = \qquad (3.70)$$
$$= \alpha^2 m(f(D_1^{z*}), 0) - m(fD_2^{z*}), 1, 1) \leq M_w^{(n)}(\alpha, f(r)).$$

Since $x(\alpha)$ increases in $\alpha \in [0, \infty)$ from $x_1$ to $r$, and $x_1 = \min\limits_{f \in S^1(\beta)} |f(r)|$, there is $\alpha^* \in [0, \infty)$ such that $x(\alpha^*) = |f(r)|$. Denote by $(\tilde{D}_1^{(n)}, \tilde{D}_2^{(n)})$ the extremal pair of domains in the problem of $M_w^{(n)}(\alpha, f(r))$ and by $(\tilde{D}_1, \tilde{D}_2)$ the extremal pair of domains in the problem of $M_w^{(1)}(\alpha, f(r))$. By Lemma 3.5.3, we have $M_w^{(n)}(\alpha, f(r)) \leq M_w^{(1)}(\alpha, f(r))$. Now we apply circular symmetrization to the domains $(\tilde{D}_1, \tilde{D}_2)$ in the following way. Denote by $\tilde{D}_1^\varepsilon$ the doubly connected domain $\tilde{D}_1 \setminus \{|w| \le \varepsilon\}$. Now we construct the quadrilateral $\tilde{D}_2^\varepsilon$ fixing two arcs of the domain $\tilde{D}_2$ that start from the point 1 lying within the disk $|w - 1| < \varepsilon$. Then, we apply circular symmetrization in the usual way to the doubly connected domain $\tilde{D}_1^\varepsilon$ with respect to the positive real axis, and to the quadrilateral $\tilde{D}_2^\varepsilon$ with respect to the negative real axis (first we have to construct the symmetric image of $\tilde{D}_2^\varepsilon$ with respect to the unit circle and then apply symmetrization to the doubly connected domain obtained). Denote by $(D_1^*, D_2^*)$ the result of this symmetrization as $\varepsilon \to 0$. Using Lemma 3.5.4, adding necessary expressions dependent on $\varepsilon$, we derive

$$\alpha^2 m(\tilde{D}_1, 0) - m(\tilde{D}_2, 1, 1) \le \alpha^2 m(\tilde{D}_1^*, 0) - m(\tilde{D}_2^*, 1, 1). \qquad (3.71)$$

In its turn, the pair $(D_1^*, D_2^*)$ is admissible in the problem about $M_w(\alpha, x(\alpha))$. Therefore,

$$\alpha^2 m(\tilde{D}_1^*, 0) - m(\tilde{D}_2^*, 1, 1) \le M_w(\alpha, x(\alpha)). \qquad (3.72)$$

Taking into account the equality

$$M_w(\alpha, x(\alpha)) = \alpha^2 m(D_1^{z*}, 0) - m(D_2^{z*}, 1, 1) + \frac{\alpha^2}{2\pi} \log y(\alpha) - \frac{4}{\pi} \log \beta,$$

the chain of inequalities (3.71–3.72) leads to the inequality $|f'(0)| \le y(\alpha)$. The uniqueness of the extremal function $f^*$ follows from the uniqueness of the extremal configuration for the maximum $M_w(\alpha, x(\alpha))$.    $\square$

The boundary curve $\Gamma^-$ of the range of the system of the functionals $(|f(r)|, |f'(0)|)$ in the class $S^1(\beta)$ that corresponds to the problem of min $|f'(0)|$ with $|f(r)|$ fixed is still unknown. The results of this sections were obtained in [168], [169]. The distortion theorems were obtained in [162]. It turns out that the minimum of $|f'(r)|$ in the class $S^1(\beta)$, $0 < r < 1$, is given by the canonical function for $r$ close to the origin whereas another function is extremal for $r \to 1$.

Recently, the hyperbolically convex univalent functions have received much attention (see [95], [99], [167]). Distortion theorems as well as two-point distortion are studied in [167] by the modulus method in this class of univalent functions.

# 4. Moduli in Extremal Problems for Quasiconformal Mapping

Classical results (P. P. Belinskiĭ [16], M. Schiffer, G. Schober [121], S. Krushkal [71]) state that in various extremal problems a special form of extremal maps (the Teichmüller maps) is used. These maps are given in terms of complex dilatation of the inverse map. For maps close to the identity this leads to complete solution of the problem. However, for the whole corresponding class of quasiconformal maps the number of unknown parameters does not allow to obtain an exact solution and gives only its qualitative characteristic, namely, this extremal mapping is a Teichmüller map. So, a major problem in the theory is this: to find a description of the extremal maps in terms of dilatation of the direct map with all parameters known. By the suggested modulus method we will obtain the value regions for systems of functionals connected with two-point distortion in basic compact classes of quasiconformal maps. Each extremal function will be naturally unique for every boundary point of the range in question.

## 4.1 General information and simple extremal problems

### 4.1.1 Quasiconformal mappings of Riemann surfaces

The starting point is quasiconformal mappings of planar domains. Let $D$ be a domain in $\overline{\mathbb{C}}$ (possibly equal to $\overline{\mathbb{C}}$) and $w = f(z)$ be a homeomorphism of $D$ onto a domain $D' \subseteq \overline{\mathbb{C}}$. We define Sobolev distributional derivatives as

$$f_{\bar{z}} := \frac{\partial f}{\partial \bar{z}} = \frac{1}{2}\left(\frac{\partial f}{\partial x} + i\frac{\partial f}{\partial y}\right); \ f_z := \frac{\partial f}{\partial z} = \frac{1}{2}\left(\frac{\partial f}{\partial x} - i\frac{\partial f}{\partial y}\right),$$

which are locally square integrable, on $D$, $z = x + iy$. A homeomorphism $f$ is said to be quasiconformal in $D$ if the complex valued function $\mu_f(z) = f_{\bar{z}}/f_z$ satisfies the inequality $|\mu_f(z)| < 1$ uniformly almost everywhere in $D$. If $\|\mu_f\|_\infty = \operatorname*{ess\,sup}_{z \in D} |\mu_f(z)| \le k < 1$, then the homeomorphism $f$ is said to be $K$-quasiconformal, $K = (1 + k)/(1 - k)$. The function $\mu_f(z)$ is called its complex characteristic or *dilatation*. A quasiconformal map is a homeomorphic generalized solution of the Beltrami equation

$$w_{\bar{z}} = \mu_f(z)w_z. \tag{4.1}$$

This solution is unique up to a conformal homeomorphism. A usual conformal normalization (for instance, three boundary fixed points for a simply connected domain) implies the uniqueness of the solution to (4.1). The detail description of properties of quasiconformal maps can be easily find in [9], [16], [43], [45], [71], [75]. Returning to the geometric definition of a quasiconformal map one can consider the notion of the modulus of a family of curves as a basis of the notion of quasiconformality. A sense preserving homeomorphism $f$ of a domain $D$ onto a domain $D'$ is said to be a $K$-*quasiconformal map* if for any doubly connected hyperbolic domain $R \subset D$ the ratio $M(f(R))/M(R)$ is bounded and the following inequality is satisfied

$$\frac{M(R)}{K} \leq M(f(R)) \leq KM(R), \tag{4.2}$$

where $M(R)$ is the modulus of the family of curves that separate the boundary components of $R$. The inequality (4.2) we call the property of *quasiinvariance* of the modulus. A quasiconformal map is conformal if and only if $K = 1$ (or $k = 0$).

We will consider basic compact classes of quasiconformal maps. They are: $Q_K-$ a class of all $K$-quasiconformal automorphisms of $\overline{\mathbb{C}}$ normalized as $f(0) = 0$, $f(1) = 1$, $f(\infty) = \infty$; $U_K-$ a class of all $K$-quasiconformal automorphisms of $U$ with the normalization $f(0) = 0$.

An important point to note here is the dependence of a quasiconformal map on its dilatation. We let the dilatation $\mu_f(z,t)$ depend on $z \in U$ and $t$ be real or complex; $\mu_f(z,\cdot)$ be a measurable function with respect to $z$, $\|\mu_f\|_\infty < 1$. If $\mu_f$ is $n$-differentiable with respect to $z$ and the $n$-th derivative is Hölder continuous of order $\alpha \in (0,1)$, then a quasiconformal solution $f^\mu \in U_K$ to the equation (4.1) is $(n+1)$-differentiable and the $(n+1)$-th derivative satisfies the same Hölder condition [16], $n \geq 1$. Thus, one could expect that $f^\mu$ possesses a continuous derivative whenever $\mu$ is continuous. However, this fails making use of the map $f(z) = z(1 - \log|z|)$, $f(0) = 0$, $z \in U$, which is even not Lipschitz continuous at $z = 0$. P. P. Belinskiĭ [16] proved that a continuous $\mu$ produces a Hölder continuous $f$ with any $0 < \alpha < 1$.

The dependence on the parameter $t$ is much easier. If $\mu(\cdot, t)$ is a differentiable or continuous function with respect to $t$ (for instance, holomorphic for complex $t$), then the function $f^\mu$ is also of the same quality.

Now we proceed with the definition of quasiconformality for Riemann surfaces. Let $S_0 = U/G_0$ and $\omega$ be a quasiconformal automorphism of $U$. Let $m_\omega$ be a dilatation of the map $\omega$ which is compatible with the Fuchsian group $G_0$, that is,

$$m_\omega(\gamma(z))\frac{\overline{\gamma'(z)}}{\gamma'(z)} = m_\omega(z), \text{ for } \gamma \in G_0.$$

We construct a Fuchsian group by the rule $G = \omega \circ G_0 \circ \omega^{-1}$. Then, the Riemann surface $S = U/G$ induces the following commutative diagram where $f = J \circ \omega \circ J_0^{-1}$ is a quasiconformal homeomorphism of $S_0$ onto $S$ and $J, J_0$

$$\begin{array}{ccc} U & \xrightarrow{\ \omega\ } & U \\ \downarrow{J_0} & & \downarrow{J} \\ S_0 & \xrightarrow{\ f\ } & S \end{array}$$

are automorphic projections. The map $f$ has the dilatation

$$\mu_f(\zeta) = m_\omega(J_0^{-1}(\zeta)) \frac{\overline{(J_0^{-1}(\zeta))'}}{(J_0^{-1}(\zeta))'},$$

in terms of a local parameter $\zeta$. The groups $G_0$ and $G$ are the groups of automorphisms for the functions $J_0$, $J$ respectively. A homeomorphism $f$ induces an isomorphism $\chi_f$ of the Fuchsian groups $G_0 \to G$. Dilatations $m_\omega$ form the unit ball $D(G_0)$ of the Beltrami differentials in the whole space of the Beltrami differentials $B(G_0)$ of finite norm $\|\cdot\|_\infty$ which are invariant with respect to the actions from $G_0$.

The *Teichmüller maps* are of particular importance in the theory of quasiconformal mapping. They are the Sobolev generalized $K$-quasiconformal homeomorphic solutions to the Beltrami equation (4.1) with the complex dilatation

$$\mu_f(\zeta) = k\frac{\overline{\varphi(\zeta)}}{|\varphi(\zeta)|}, \ \ k = \frac{K-1}{K+1}, \tag{4.3}$$

where $\varphi(\zeta)d\zeta^2$ is a holomorphic quadratic differential on $S_0$. The inverse map is also Teichmüller's, i.e., there exists a holomorphic quadratic differential $\psi(w)dw^2$ on $f(S_0)$, such that the inverse map has the dilatation $\mu_{f^{-1}}(w) = k\psi(w)/|\psi(w)|$. Teichmüller maps are locally affine and map infinitesimal circles onto infinitesimal ellipses having their big semi-axes along or orthogonal to a trajectory of the differential $\psi(w)dw^2$. The ratio of the big and small axes of ellipses is $K$ for any point of $S_0$ different from the singularities of $\varphi$.

We denote by $K(f)$ a maximal deviation of a quasiconformal map $f$ from conformality, i.e., $K(f) = (1 + \|\mu_f\|_\infty)/(1 - \|\mu_f\|_\infty)$. Let $(S, f)$ be a marked (with respect to $S_0$) Riemann surface, where $f$ is a homeomorphism $S_0 \to S$. The homeomorphism $f$ is thought of as a representative of the homotopy class of homeomorphisms of $S_0$ onto $S$. A homeomorphism $g : S \to S'$ induces a homeomorphism of the marked Riemann surface $(S, f)$ onto $(S', f')$ where $f' = g \circ f$. Let $(S, f)$ and $(S', f')$ be equally oriented marked Riemann surfaces of type $(g, n, l)$ (the orientation and type of marked Riemann surfaces are generated by those of Riemann surfaces). It is known [18] that there are quasiconformal maps among all homotopic homeomorphisms of $(S, f)$ onto $(S', f')$.

A fundamental result in the theory of quasiconformal mapping was formulated by O. Teichmüller [150] and rigorously proved by L. Ahlfors, L. Bers [11], and by another method by S. Krushkal [71]. In the homotopy class of homeomorphisms of $(S, f)$ onto $(S', f')$ there is a unique Teichmüller map which is either conformal $(k(f) = 0)$ or there are such $k$ and a quadratic differential $\varphi(\zeta)d\zeta^2$, that the complex dilatation of this map is of the form (4.3). In particular, two different Teichmüller maps are never homotopic to each other.

Another result that invokes Teichmüller maps is concerned with extremal problems for compact classes of quasiconformal maps. Let us formulate the so-called interpolation Schur-Pick-Nevanlinna problem. Let $A$ be a class of maps of a domain $D$ into $\mathbb{C}$. Let $z_1, \ldots, z_n$ be punctures of $D$, $w_1, \ldots, w_n$ be punctures of $\mathbb{C}$ and $w_1^1, \ldots, w_1^{i_1}, \ldots, w_2^1, \ldots, w_2^{i_2}, \ldots, w_n^1, \ldots, w_n^{i_n}$ be complex numbers. The problem consists of construction of a map from the class $A$, such that it admits the values $w_j$, $j = 1, \ldots, n$ at the points $z_j$ and at the points of differentiability $f^{(k_j)}(z_j) = w_j^{k_j}$ (the map is supposed to be complex differentiable at $z_j$ if $i_j \geq 1$). Let $F(f)$ be a functional (or a system of functionals) dependent on the values of a function and its derivatives at the points $z_j$. Then, finding the extremum of this functional (or the range of a system of functionals) is equivalent to the necessary condition of solvability of the above interpolation problem.

Among such problems we indicate the Bieberbach problem on estimation of the Taylor coefficients of a univalent conformal map, a mutual behaviour of the value of a univalent function and the value of its derivative at a certain point of the set of definition, Gronwall's problems in the conformal case. In the quasiconformal case, however, one finds only few completely resolved problems such as, for instance, description of the range of $f(\zeta)$ for the maps from $Q_K$, the Mori theorem, and some others. In fact, there is known only a qualitative description of the inverse maps to the extremal ones by variational methods [15], [16], [71], and optimal control methods [26].

We show such a result for one concrete problem. We consider the compact class $Q_K$ of quasiconformal automorphisms of the complex plane and denote by $F(w_1, \ldots, w_{n-3})$ a holomorphic function on $\mathbb{C}^{n-3}$. The maximum

$$\max_{f \in Q_K} \operatorname{Re} F(f(z_1), \ldots, f(z_{n-3}))$$

is to be found for $z_1, \ldots, z_{n-3}$ fixed and different from 0,1. It is known [15], [16], [71] that the extremal map is Teichmüller's and the inverse map satisfies the Beltrami equation $z_{\overline{w}} = \mu_{f^{-1}}(w)z_w$ with the dilatation

$$\mu_{f^{-1}}(w) = ke^{it}\frac{\overline{\psi(w)}}{|\psi(w)|}, \; k = \frac{K-1}{K+1},$$

$$\psi(w) = \sum_{j=1}^{n-3} \frac{\partial F}{\partial w_j} \frac{w_j(w_j - 1)}{w(w-1)(w - w_j)}, \; \text{for } w = f^*(z), w_j = f^*(z_j).$$

This Beltrami coefficient contains a lot of unknown parameters, thus, the extremal problem can be solved only qualitatively. Therefore, our main effort aims at elaboration of a method of finding representation of extremal maps by the dilatation of the direct map.

Among particular cases of this functional we consider two-point distortion under quasiconformal maps from $Q_K$. This can be reduced to estimation of functionals dependent on two fixed points from $S_0$. There are only few results devoted to such estimation. In the case of the class $Q_K$ there is a result by S. Krushkal [4]. He obtained that there is such $K_0 > 1$ that for all $f \in Q_K$, $1 \le K \le K_0$, and for fixed points $z_1, z_2 \in \mathbb{C} \setminus \{0, 1\}$, $z_1 + z_2 = 1$, the extremal map for the functional $\max_{f \in Q_K} |f(z_1) - f(z_2)|$ is the Teichmüller map with the dilatation

$$\mu_f(z) = k e^{it} \frac{\overline{\varphi(z)}}{|\varphi(z)|}, \quad k = \frac{K-1}{K+1},$$

where $\varphi(z) = z_1 z_2 [z(1-z)(z-z_1)(z-z_2)]^{-1}$.

This chapter deals with the application of the modulus method to this extremal problem in the classes $Q_K$, $U_K$. We evaluate the range of the system of functionals $(|f(r_1)|, |f(r_2)|)$ for fixed real values $r_1$ and $r_2$. From these general results we deduce some sharp estimates of functionals.

### 4.1.2 Growth and Hölder continuity

We start with two well-known results applying the modulus method. The first one is about estimation of growth of the quasiconformal automorphisms of the unit disk $U$.

**Theorem 4.1.1** (Belinskiĭ [15], [16], Hersch, Pfluger [59]). *Let $f$ be a $K$-quasiconformal homeomorphism of the unit disk with the motionless point $0$, say from the class $U_K$. Then,*

$$\Phi^{-1}(K\Phi(|z|)) \le |f(z)| \le \Phi^{-1}(\tfrac{1}{K}\Phi(|z|)),$$

*where $\Phi(t) = \mathbf{K}'(t)/2\mathbf{K}(t)$, $0 \le t < 1$.*

*The equality sign in the right-hand side inequality is attained for the function that realizes an affine map along the trajectories of the quadratic differential $\varphi(\zeta)d\zeta^2 = [\zeta(z-\zeta)(1-\zeta\bar{z})]^{-1}d\zeta^2$ (affine in terms of the local parameter $w = \int \sqrt{\varphi(\zeta)}d\zeta$). The same assertion can be made for the equality sign in the left-hand side equality for the differential $(-\varphi(\zeta)d\zeta^2)$.*

*Remark 4.1.1.* We have to clear up the notation $\Phi$ we have used. The standard notation (see [91]) for the relation

$$\mu(t) = \frac{\pi}{2} \cdot \frac{\mathbf{K}'(t)}{\mathbf{K}(t)}$$

contradicts the standard notation for the complex dilatation, so it seems to be more convenient to change it here. Denote by $p_K(t) := \Phi^{-1}(\frac{1}{K}\Phi(t))$. For this quantity the useful estimate $(p_K(t))^2 \le 16^{1-\frac{1}{k}} \cdot t^{\frac{1}{k}}$ can be found in e.g. [G].

Now we proceed with the proof of Theorem 4.1.1.

*Proof.* We suppose the contrary. Denote by $\rho_0 = p_K(|z|)$ Let $f$ be a function from the class $U_K$ with $|f(z)| \geq \rho_0$. The doubly connected domain $D_z = U \setminus [z, e^{i \arg z})$ has the modulus

$$M(D_z) = \frac{1}{2}\Phi(|z|).$$

Let $D^*$ be a result of circular symmetrization of the domain $f(D_z)$ with respect to the origin and the positive real axis. Then, $M(f(D_z)) \leq M(D_z)$. For a quasiconformal map from $U_K$ the inequality $M(f(D_z)) \geq \frac{1}{K} M(D_z)$ holds. The unique extremal map is the Teichmüller map that realizes an affine mapping along the trajectories of the quadratic differential

$$\varphi(\zeta)d\zeta^2 = \frac{d\zeta^2}{\zeta(z - \zeta)(1 - \zeta\bar{z})}.$$

Moreover, $D^* \subset \{U \setminus [0, \rho_0]\}$, and, hence, $M(D^*) \geq \frac{1}{2}\Phi(\rho_0)$. The whole chain of these inequalities is valid only in the case $\rho_0 = |f(z)|$. The uniqueness of the extremal configuration and the uniqueness of the extremal Teichmüller map imply the uniqueness of the extremal map in the theorem up to rotation.

The left-hand side inequality can be obtained starting with the quadrilateral $U'_z = U \setminus \{(-1, 0] \cup [r, 1)\}$, where $r = |z|$. Symmetrization of the domain $f(U'_z)$ yields the inequality

$$\frac{\mathbf{K}(r)}{K\mathbf{K}'(r)} \leq \frac{\mathbf{K}(|f(z)|)}{\mathbf{K}'(|f(z)|)}$$

which is equivalent to the inequality in Theorem 4.1.1.    □

The next theorem deals with the boundary tension under a quasiconformal map and can be proved similarly using the Mori domain (see Section 2.4) $\mathbb{C} \setminus \{[0, -\infty) \cup \{e^{i\theta}, -\alpha/2 \leq \theta \leq \alpha/2\}\}$.

**Theorem 4.1.2** (Belinskiĭ [16]). *Let $f$ be a $K$-quasiconformal homeomorphism of the unit disk with the motionless point 0 and let an arc of the unit circle of length $\alpha$ be mapped onto an arc of the unit circle of length $\beta$. Then,*

$$\beta \leq 4\arcsin[p_K(\sin\frac{\alpha}{4})].$$

The extremal function exists, is unique, and realizes an affine map along the trajectories of the quadratic differential with simple poles at 0 and at the limiting endpoints of the arc $\alpha$.

Now we briefly sketch some other results about the deviation of a point under a quasiconformal map to complete this subsection. Their proofs do not use the modulus method and exceed the scope of our book.

P. P. Belinskiĭ [16] has proved the sharp estimate

$$|f(z) - z| \leq \lambda^{-1}(\frac{1}{K}\lambda(|z|)) - |z|,$$

where $\lambda(t) = 2\Phi(\sqrt{t/(t+1)})$, for the functions from $U_K$.

The absolute constant for the preceding estimate was obtained by the parametric method by Shah Dao-Sing [126]. Let $f \in U_K$. Then, the sharp estimate

$$|f(z) - z| \leq \frac{\Gamma^4(1/4)}{4\pi^2} \log K,$$

holds where $\Gamma(\cdot)$ is the Euler gamma-function.

We also indicate here an asymptotic estimate by P. P. Belinskiĭ [16]. Let $f \in U_K$, $K = 1 + \varepsilon$ for a small $\varepsilon$. Then, the asymptotically sharp estimate

$$|f(z) - z| \leq \varepsilon\frac{8}{\pi}\int\limits_0^1 \mathbf{K}(t^2)dt \approx 4.5\varepsilon$$

holds.

One can find analogous results for quasiconformal homeomorphisms of the complex plane in [16], [126], [91].

We mention here some results obtained by the modulus method that are covered in [9]. We ask the reader to look through the proofs there. One of them deals with the Hölder continuity. A classical result is the Mori theorem.

**Theorem 4.1.3** (Mori [102]). *Let $f \in U_K$. Then, for any $z_1, z_2 \in U$ the following unimprovable inequality is valid*

$$|f(z_1) - f(z_2)| \leq 16|z_1 - z_2|^{1/K}.$$

One can find the *proof* by the modulus method in [9].

A topical problem is to replace the absolute constant 16 by a relative constant dependent on $K$. This means to find

$$M(K) = \sup\{\frac{|f(z_1) - f(z_2)|}{|z_1 - z_2|^{1/K}}; f \in U_K; z_1, z_2 \in U\}.$$

Obviously, $\sup\limits_{K \geq 1} M(K) = 16$. O. Lehto, K. Virtanen conjectured ([91], page 68) that $M(K) = 16^{1-1/K}$. Another problem is to establish an analog of the Mori theorem for the lower estimate of $|f(z_1) - f(z_2)|$. P. P. Belinskiĭ [16] obtained the inequality

$$|f(z_1) - f(z_2)| > \left|\frac{z_1 - z_2}{48}\right|^K.$$

Establishing refined estimates of $M(K)$ attracted attention of various specialists.

**Theorem 4.1.4** (Wang Chuang-Fang [175]). *Let $f \in U_K$. Then, the following unimprovable inequalities*

$$4^{1-K}|z|^K \leq |f(z)| \leq 4^{1-1/K}|z|^{1/K}$$

*hold.*

This estimate is also found in works by P. P. Belinskiĭ [16], O. Lehto, K. Virtanen [91] and in recent papers by V. Semenov [122], [123] as an application of the method of quasiconformal flows. He also obtained the following sharp estimate $1 - |f(z)|^2 \leq 16^{1-1/K}(1 - |z|^2)^{1/K}$.

**Theorem 4.1.5** (Semenov [122]). $16^{1-1/K} \leq M(K) \leq 124^{1-1/K}$.

### 4.1.3 Quasiconformal motion of a quadruple of points

Here we apply the modulus method to establishing geometric properties of the solution of the Teichmüller problem on the minimum of deviation of a quasiconformal image of a quadruple of punctures on the Riemann sphere.

Let $S_0 = \mathbb{C} \setminus \{0, 1, r\}$ and $S = \mathbb{C} \setminus \{0, 1, w_0\}$ where $r > 1$, $w_0 \in \mathbb{C} \setminus \{0, 1\}$. Let us map $S \to \mathbb{C}$ by the integral

$$J(w) = \int^w \frac{dw}{\sqrt{w(w-1)(w-w_0)}},$$

fixing a branch of the root. Denote by $\omega_1', \omega_2'$ the periods of the function $J^{-1}$ such that $\omega_1' > 0$, Im $\omega_2' > 0$. In the case $S = S_0$ the analogous integral $J_0(z)$ maps this surface onto the plane of the periods $\omega_1, \omega_2$ (see Section 2.3), where

$$\omega_1 = 2\mathbf{K}\left(\sqrt{\frac{r-1}{r}}\right), \quad \omega_1 = i\mathbf{K}'\left(\sqrt{\frac{r-1}{r}}\right).$$

Denote by $w = f(\zeta)$ a quasiconformal homeomorphism of $S_0$ onto $S$ that, being extended onto $\overline{S_0} = \overline{\mathbb{C}}$, keeps the points $0, 1, \infty$ motionless. Then, the function $J \circ f \circ J_0^{-1}$ maps the rectangular lattice of periods $\omega_1, \omega_2$ onto the parallelogram lattice of periods $\omega_1', \omega_2'$. Let $C_z = \mathbb{C} \setminus \{n\omega_1 + m\omega_2; n, m, \in \mathbb{Z}\}$ and $C_w = \mathbb{C} \setminus \{n\omega_1' + m\omega_2'; n, m, \in \mathbb{Z}\}$. We consider the countable set of homotopy classes $H_j$ of quasiconformal homeomorphisms $f : S_0 \to S$; $H = \cup H_j$. The index $j = 1$ refers to the class of homeomorphisms homotopic to $J^{-1} \circ g \circ J_0$, where $g$ is an affine map of the lattice of periods on $C_z$ onto the lattice of periods on $C_w$. The Teichmüller problem for a homotopy class of mappings consists of finding the infimum of the deviation $K(f)$ over all maps from this homotopy class.

We formulate a problem connected with the Teichmüller problem and the motion of a quadruple of points. Namely, for which $K$ there is a $K$-quasiconformal map of $S_0$ onto $S$? This problem was solved by L. Ahlfors by the extremal length method.

**Theorem 4.1.6** (Ahlfors [9]). *If $\frac{1}{2}\log K \geq \rho(r, w_0)$, then there exists a $K$-quasiconformal map of $S_0$ onto $S$ which is motionless in the punctures $0, 1, \infty$. Here $\rho(\cdot, \cdot)$ is the non-Euclidean metric in $\mathbb{C}\setminus\{0, 1\}$. The equality sign is attained only for the $K$-quasiconformal Teichmüller map from $H \cap Q_K$.*

For the *proof* we again refer to [9].

Generalizations of this theorem were made by S. Krushkal [75]. Now we consider a countable set of homotopy classes $\Gamma_j'$ of simple loops that separate the points $0, 1$ from $w_0, \infty$. The index $j = 1$ refers to the class of loops homotopic on $S$ to the slit along the segment $[0, 1]$. We denote by $m(\Gamma_1') \equiv m(w_0)$ the corresponding modulus. Let $\Gamma_1 := \Gamma_1'$ in the case $w_0 = r$. We define a curve on $S$ by the equation $l(R) = \{w : \frac{m(w)}{m(r)} = R\}$. The shape of this curve was shown in Section 3.4.1 (or see [78], Corollary 5.3), $l(1)$ contains the point $r$. We consider the homotopy classes $H$ of maps $S_0$ onto $S$. The following result was obtained in [151].

**Theorem 4.1.7.** *If the point $w_0$ is located to the left from the counter clockwise direction of $l(1)$, then the extremum in the Teichmüller problem is attained for the homotopy class $H_1$, i.e.,*

$$\min_{f \in H_1} K(f) = \min_{f \in H} K(f) < \min_{f \in H_j} K(f), \quad j \neq 1.$$

*Proof.* Let us consider on $S$ the family of simple loops $\gamma \in \Gamma_1'$, that separate the punctures $0, 1$ from $w_0, \infty$. This family is transformed by the mapping $J(w)$ to the family of curves connecting on $C_w$ the points $u'$ and $u' + \omega_1'$ which are homotopic on $C_w$ to the segment $[u', u' + \omega_1']$. The modulus of this family is equal to Im $\frac{\tau'}{2}$, where $\tau' = \frac{\omega_2'}{\omega_1'}$. The extremal metric is Euclidean. Consider two neighboring parallelograms of periods on $C_w$, pasted along a sloping side with identification of other sloping sides. We map the configuration obtained by the Weierstrass function $\wp(u' \mid \omega_1', \omega_2')$. As a result, we obtain a doubly connected domain $\Omega$ which is admissible for the extremal partition of the Riemann surface $S$ among the domains of the first type, associated with the homotopy class $\Gamma_1'$. Thus, $m(w_0) = M(\Omega) = m(\Gamma_1') \geq$ Im $\frac{\tau'}{2}$.

We make use of transformations $s(\tau')$ which are automorphisms of the upper half-plane Im $\tau' > 0$, and which are the covering transformations for the surface $S$. There is a one-to-one correspondence [9] between the homotopy classes of maps $H_j$ and the automorphisms $s(\tau')$. The solution of the Teichmüller problem about the minimum of deviation in a homotopy class of maps is given by the formula

$$\min_{f \in H_j} K(f) = \exp(2d(\tau, s(\tau'))),$$

where $d(\cdot, \cdot)$ is the hyperbolic Poincaré distance in the upper half-plane, and $\tau = \frac{\omega_1}{\omega_2}$. For the transformations $s(\tau')$ the inequality Im $s(\tau') \leq$ Im $\tau'$ holds and the equality sign is attained only for the identity. Obviously, for the surface $S_0$ we have $m(r) = \Phi(1/\sqrt{r})$. In this case Im $\tau/2 = m(r)$. The condition of the theorem that the point $w_0$ lies on $S$ left to the counter clock-wise direction of the curve $l(1)$ implies that the modulus $m(w_0)$ increases to $m(r)$ when we delete the part of the critical trajectory of the extremal quadratic differential for $m(w_0)$ up to the intersection with $l(1)$ leaving its connected part with the infinite endpoint. Thus, the inequality $m(r) > m(w_0)$ holds. We have the inequality Im $\tau \geq$ Im $\tau' \geq$ Im $s(\tau')$. The map $(\zeta - \tau)(\zeta - \bar{\tau})^{-1}$ geometrically shows the inequality $d(\tau, s(\tau')) \geq d(\tau, \tau')$ with the equality sign only for the identity. Thus,

$$\min_{j} \min_{f \in H_j} K(f) = \exp(2d(\tau, \tau')) = \min_{f \in H_1} K(f).$$

This completes the proof.    $\square$

The class of homeomorphisms $Q_K$ is defined and $Q_K \subset H$ if $K(f) < K$ for $f \in H$. Obviously,

$$\min_{f \in Q_K} m(f(\Gamma_1)) = \frac{1}{K}\Phi(1/\sqrt{r}), \quad \max_{f \in Q_K} m(f(\Gamma_1)) = K\Phi(1/\sqrt{r}).$$

The equality sign is attained for the function $f_1$, $f_2$ respectively satisfying the Beltrami equation with the dilatation

$$\mu_f(z) = \pm k\frac{\overline{\varphi(\zeta)}}{|\varphi(\zeta)|}, \quad \varphi(\zeta) = \frac{1}{\zeta(\zeta - 1)(\zeta - r)}.$$

These functions belong to the class $H_1$.

Denote by $T(K)$ the set of all points $w_0$ such that $\min_{f \in H} K(f) = K$, and by $T_1(K)$ the set of all points $w_0$ such that $\min_{f \in H_1} K(f) = K$.

**Proposition 4.1.1.** *The set $T(K)$ is the boundary of the range of $f(r)$ in the class $Q_K$. It is the boundary of the non-Euclidean disk $E$ in $\mathbb{C} \setminus \{0, 1\}$ centered in $r$ with the non-Euclidean radius $\frac{1}{2} \log K$.*

*Proof.* In fact, all homeomorphisms giving the points of $T(K)$ belong to $Q_K$, and $T(K) \subset E$. Let there be a point $w_0 = f_0(r)$ in $T(K)$, $w_0 \notin \partial E$. Then, there is $K_1 < K$ such that the range of $f(r)$ in the class $Q_{K_1}$ is the non-Euclidean disk $E_1 \subset E$, and $w_0 \in \partial E_1$. The boundary points of $E$ are given by the functions with $K(f) = K$. Hence, $K(f_0) < K$ and this contradicts the definition of $T(K)$. Now we prove that there is no point in $\partial E$ which is not from $T(K)$. Indeed, if there were such functions $f$ giving such points, then $K(f) > K$, and the functions would not be from the class $Q_K$. This ends the proof.    $\square$

**Theorem 4.1.8.** *The curve $T_1(K)$ has homotopy with respect to $K$ to the point $T_1(1) = r$, it is located between the curves $l(K)$ and $l(1/K)$. $T_1(K)$ has a unique common point with $l(K)$ in the real axis to the right from $r$. $T_1(K)$ and $T(K)$ have a unique common point with $l(1/K)$ in the real axis between 1 and $r$.*

*Proof.* a) $T_1(K)$ and $l(K)$ are intersected at a unique point. Indeed, let there be at least two points $w_1$ and $w_2$ of the intersection. But $\max\limits_{f \in Q_K} m(f(\Gamma_1)) = Km(r)$ and there exist two functions such that $f_1(r) = w_1 \in T_1(K)$, $f_2(r) = w_2 \in T_1(K)$, $f_1, f_2 \in H_1$, and, therefore, $f_1, f_2 \in Q_K$. At the same time $w_1, w_2 \in l(K)$ and $m(w_1) = m(w_2) = Km(r)$ that contradicts the uniqueness of the extremal map with respect to the modulus.

b) Now we prove that $\text{Im}\,(T_1(K) \cap l(K)) = 0$. The extremal map (for the modulus $m(w_0)$) belongs to the class $H_1$, and transforms the point $r$ into the point $w_0$ that lies on the real axis and $l(K)$. At the same time, if it were not a point from $T_1(K)$, then there would be such $K_1 < K$ that $w_0 \in T_1(K_1)$, and $T_1(K_1) \subset \text{int}\,T_1(K)$. Then, $T_1(K)$ would have more than one point of intersection with $l(K)$. This contradicts (a).

Analogously one can prove the assertion of the theorem for the point of intersection of $T_1(K)$ with $l(1/K)$. Theorem 4.1.7 yields the conclusion about $T_1(K)$.                                                                        $\square$

**Corollary 4.1.1.** *Solve the Teichmüller problem for the class $H_j$ and let $K^j_{w_0} = \min\limits_{f \in H_j} K(f)$, $f(r) = w_0$. Then, $\min\limits_{w_0 \in l(1/K)} K^1_{w_0} = \min\limits_{w_0 \in l(K)} K^1_{w_0} = K$ with the equality sign only for the intersections of $T_1(K)$ with $l(K)$ and $l(1/K)$.*

**Corollary 4.1.2.** $\min\limits_{j} K^j_{w_0} = K^1_{w_0}$ *for* $w_0 \in l(1/K)$. $\min\limits_{f \in H, w_0 \in l(1/K)} K(f) = K$.

**Corollary 4.1.3.** *If $|w_0| \geq r > 1$, then $K^1_{w_0} \geq K^1_{|w_0|}$ with the equality sign only for $w_0 = |w_0|$.*

## 4.2 Two-point distortion for quasiconformal maps of the plane

The main results of this section we compile in two theorems.

**Theorem 4.2.1.** *Let $r_1$ and $r_2$ be fixed real points, $f \in Q_K$. Then,*
*(i) The unique extremal map $f^*$ gives the maximum of $|f(r_2)| - |f(-r_1)|$, $r_1 > 0$, $r_2 > 1$ in the class $Q_K$. This map satisfies the Beltrami equation (4.1) with the Beltrami coefficient (4.3) where*

$$\varphi(z) = \frac{c - z}{z(z-1)(z+r_1)(z-r_2)},$$

*and*

$$c = \frac{r_1 r_2 (r_1 - r_2 + 2)}{r_2 (r_2 - 1) + r_1 (1 + r_1)}.$$

*(ii) The unique extremal map $f^{**}$ gives the maximum of $|f(r_2) - f(-r_1)|$ and $|f(r_2)| + |f(-r_1)|$, $r_1 > 0$, $r_2 > 1$ in the class $Q_K$. This map satisfies the Beltrami equation (4.1) with the Beltrami coefficient (4.3) where*

$$\varphi(z) = \frac{c - z}{z(z - 1)(z + r_1)(z - r_2)}, \quad \text{for } r_2 - r_1 > 1,$$

$$\varphi(z) = \frac{z - c}{z(z - 1)(z + r_1)(z - r_2)}, \quad \text{for } r_2 - r_1 < 1,$$

$$\varphi(z) = \frac{-1}{z(z - 1)(z + r_1)(z - r_2)}, \quad \text{for } r_2 - r_1 = 1,$$

*and*

$$c = \frac{r_1 r_2}{1 + r_1 - r_2}.$$

*(iii) The unique extremal map $f^*$ gives the maximum of $|f(r_2) + f(r_1)|$ and $|f(r_2)| + |f(r_1)|$, $1 < r_1 < r_2$ in the class $Q_K$. This map satisfies the Beltrami equation (4.1) with the Beltrami coefficient (4.3) with*

$$\varphi(z) = \frac{c - z}{z(z - 1)(z - r_1)(z - r_2)}$$

*and*

$$c = \frac{r_1 r_2 (r_1 + r_2 - 2)}{r_2 (r_2 - 1) + r_1 (r_1 - 1)}.$$

*(iv) The unique extremal map $f^{**}$ gives the maximum of $|f(r_2)| - |f(r_1)|$, $1 < r_1 < r_2$ in the class $Q_K$. This map satisfies the Beltrami equation (4.1) with the Beltrami coefficient (4.3) with $\varphi(z)$ as in (iii) and*

$$c = \frac{r_1 r_2}{r_1 + r_2 - 1}.$$

**Theorem 4.2.2.** *(i) The unique map $f^*$ gives the absolute maximum of $|f(r_2)/f(-r_1)|$, $r_1 > 0$, $r_2 > 1$ in the class $Q_K$. This map satisfies the Beltrami equation (4.1) with the Beltrami coefficient (4.3) where*

$$\varphi(z) = \frac{-1}{z(z + r_1)(z - r_2)}.$$

*Moreover, the following sharp estimate*

$$\frac{|f(r_2)|}{|f(-r_1)|} \leq \frac{u^2}{1 - u^2} \leq \frac{16^{1 - \frac{1}{k}} \cdot r_2^{\frac{1}{2k}}}{(r_1 + r_2)^{\frac{1}{2k}} - 16^{1 - \frac{1}{k}} \cdot r_2^{\frac{1}{2k}}},$$

$$u = p_K \left( \sqrt{\frac{r_2}{r_2 + r_1}} \right),$$

holds[1] where $p_K(t)$ is defined in the Remark after Theorem 4.1.1.

(ii) The unique map $f^{**}$ gives the absolute maximum of $|f(r_2)/f(r_1)|$, $1 < r_1 < r_2$ in the class $Q_K$. This map satisfies the Beltrami equation (4.1) with the Beltrami coefficient (4.3) where

$$\varphi(z) = \frac{-1}{z(z - r_1)(z - r_2)}.$$

Moreover, the following sharp estimate is valid

$$\frac{|f(r_2)|}{|f(r_1)|} \leq \frac{1}{1 - \left( p_K \left( \sqrt{\frac{r_2 - r_1}{r_2}} \right) \right)^2}.$$

Other locations of real $r_1, r_2$ can be also considered similarly.

### 4.2.1 Special differentials and extremal partitions

Suppose $1 < r_1 < r_2$. We consider some special modulus problems and extremal partitions generated by certain quadratic differentials. Set $P(z) = z(z-1)(z-r_1)(z-r_2)$. Let us consider the following one-parametric families of holomorphic quadratic differentials on $S_0 = \mathbb{C} \setminus \{0, 1, r_1, r_2\}$.

$$\varphi_1(z)dz^2 = A_1(\alpha)(z - c_1(\alpha))P^{-1}(z)dz^2, \tag{4.4}$$

where $c_1(\alpha) \in [-\infty, 0] \cup [r_2, \infty]$, $A_1(\alpha) > 0$. These values are calculated by the equations

$$\int_0^1 \sqrt{\varphi_1(x)}dx = \frac{1}{2}, \quad \int_{r_1}^{r_2} \sqrt{\varphi_1(x)}dx = \frac{\alpha}{2};$$

$\alpha$ is a fixed number from the segment $[\alpha_0, \alpha_1]$, where

$$\alpha_0 = \frac{F\left( \arcsin \sqrt{\frac{r_2 - r_1}{r_2 - 1}}, \frac{1}{\sqrt{r_1}} \right)}{K\left( 1/\sqrt{r_1} \right)},$$

$F(\cdot, \cdot)$ is the elliptic integral of the first kind,

$$\alpha_1 = \frac{K\left( \sqrt{\frac{r_2 - r_1}{r_2 - 1}} \right)}{F\left( \arcsin \frac{1}{\sqrt{r_1}}, \sqrt{\frac{r_2 - r_1}{r_2 - 1}} \right)}.$$

---

[1] In Theorem 4.2.2 the inequality in (i) follows also from [2]

$$\varphi_2(z)dz^2 = A_2(\alpha)(z - c_2(\alpha))P^{-1}(z)dz^2, \qquad (4.5)$$

where $c_2(\alpha) \in [0, r_1]$, $A_2(\alpha) < 0$. These values are calculated by the equations

$$\int\limits_{-\infty}^{0} \sqrt{\varphi_2(x)}dx = \frac{1}{2}, \quad \int\limits_{r_1}^{r_2} \sqrt{\varphi_2(x)}dx = \frac{\alpha}{2};$$

$\alpha$ is a fixed number from the segment $[\alpha_2, \alpha_3]$, where

$$\alpha_2 = \frac{\mathbf{F}\left(\arcsin\sqrt{\frac{r_2 - r_1}{r_2 - 1}}, \sqrt{1 - /r_2}\right)}{\mathbf{K}\left(\sqrt{1 - 1/r_2}\right)}, \quad \alpha_3 = \frac{\mathbf{K}\left(\sqrt{\frac{r_2 - r_1}{r_2 - 1}}\right)}{\mathbf{F}\left(\arcsin\sqrt{1 - 1/r_2}, \sqrt{\frac{r_2 - r_1}{r_2 - 1}}\right)}.$$

$$\varphi_3(z)dz^2 = A_3(\alpha)(z - c_3(\alpha))P^{-1}(z)dz^2, \qquad (4.6)$$

where $c_3(\alpha) > r_1$, $A_3(\alpha) < 0$. These values are calculated by the equations

$$\int\limits_{-\infty}^{0} \sqrt{\varphi_3(x)}dx = \frac{1}{2}, \quad \int\limits_{1}^{r_1} \sqrt{\varphi_3(x)}dx = \frac{\alpha}{2};$$

$\alpha$ is a fixed number from the segment $[\alpha_4, 1]$, where

$$\alpha_4 = \frac{\mathbf{F}\left(\arcsin\sqrt{\frac{r_2(r_1 - 1)}{r_1(r_2 - 1)}}, \sqrt{1 - 1/r_2}\right)}{\mathbf{K}\left(\sqrt{1 - 1/r_2}\right)}.$$

From (4.4–4.6) one can learn the dynamics of zeros and trajectories with respect to the parameter $\alpha$. The values $\alpha_0, \ldots, \alpha_4$ of the parameter $\alpha$ correspond to degeneracy of $D_1^*$ or $D_2^*$ in the trajectory structure of the differentials (4.4–4.6). The differentials are of finite $L_1$-norm and have finite trajectories. For each of them one can define the modulus where the corresponding differential is extremal. This means that the admissible system of curves consists at most of two non-homotopic non-critical trajectories of the differential with the weight vector $(1, \alpha)$.

### 4.2.2 Quasisymmetric functions and the extremal maps

Let $f(x)$ be a continuous strictly increasing function with respect to a real $x$ defined in the axis $\mathbb{R}$. This function is said to be *quasisymmetric* on $\mathbb{R}$ if there exists a positive constant $M$ such that

$$\frac{1}{M} \le \frac{f(x + t) - f(x)}{f(x) - f(x - t)} \le M$$

for any real $t > 0$. A finite real function $f$ is quasisymmetric on the real axis if and only if there exists a quasiconformal continuation of this function onto the upper (or, which is the same, lower) half-plane (see e.g. [91] Section 7). We consider the compact class of all quasisymmetric functions that admit $K$-quasiconformal continuation normalized by $f(0) = 0$, $f(1) = 1$. For us it will be convenient to consider functions extended into the upper half-plane $\mathbb{H}^+$. We denote such class of extended normalized maps by $H_K$. Then, we extend them onto the whole complex plane by the rule $f(\bar{z}) = \overline{f(z)}$. So, the class $H_K$ is supposed to be a subclass of $Q_K$. First, we will solve the extremal problem on the range of the system of functionals $I(f) = (f(r_1), f(r_2))$, $1 < r_1 < r_2$, in the class $H_K$.

We construct the Teichmüller maps $f_j$ defined in the upper half-plane $\mathbb{H}^+$ satisfying the Beltrami equation (4.1) with the complex dilatation

$$\mu_{f_j}(\zeta) = k\frac{\overline{\varphi_j(\zeta)}}{|\varphi_j(\zeta)|}, \quad k = \frac{K-1}{K+1}, \quad j = 1, 2, 3,$$

keeping the points $z = 0, 1, \infty$ motionless. These maps can be extended into the whole plane $\mathbb{C}$ by the rule $f(\bar{z}) = \overline{f(z)}$. We can do this because the Beltrami coefficients are symmetric. Then, the maps become extremal with respect to the following modulus problems. Let us consider the differential $\varphi_1(z)dz^2$, $\alpha \in (\alpha_0, \alpha_1)$ and its two non-homotopic in $S_0$ trajectories as an admissible system of curves $\gamma = \{\gamma_1, \gamma_2\}$ on $S_0$. One may define the problem of the extremal partition for this system and the vector $(1, \alpha)$, $\alpha \in [\alpha_0, \alpha_1]$. Let $m = m(S, \Gamma, \alpha)$ be the modulus in this problem. For $f$ which is an arbitrary quasiconformal homeomorphism from $H_K$ we define the modulus $m_f = m(f(S), f(\Gamma), \alpha)$. The Teichmüller map $f_1$ is the unique extremal homeomorphism in the problem of $\min_{f \in H_K} m_f$. More in details, $m_f \geq m/K \equiv m(S, \Gamma, \alpha)/K$ and the extremal mapping exists. Consider the map $f_1$. In the trajectory structure of $\varphi_1(z)dz^2$ there are two ring domains $D_1$ and $D_2$. The domain $D_1$ is bounded by the segment $[0, 1]$, the ray $[c_1, \infty)$, and the connected arc $\delta$ of the critical trajectory starting and ending at $c_1$, $D_2$ is bounded by the segment $[r_1, r_2]$ and the arc $\delta$. The Teichmüller homeomorphism $f_1$ maps $D_1$ and $D_2$ onto a couple of ring domains in the trajectory structure of a quadratic differential $\psi(w)dw^2$ on $\mathbb{C}$ with singularities relevant to those of $\varphi_1(z)dz^2$. Let us introduce the local parameters $\zeta = \exp(\int^z \sqrt{\varphi_1(z)}dz)$ and $\zeta' = \exp(\int^w \sqrt{\psi(w)}dw)$. In each domain $D_j$ the map $f_1$, being considered in $\mathbb{C}$, acts affinely in these coordinates $\zeta' = \zeta + k\bar{\zeta}$. The ratio $\alpha \in [\alpha_0, \alpha_1]$ of the length of trajectories remains the same for $\varphi_1$ and for $\psi$, therefore, the normalization of $\psi$ implies $m_{f_1} = M(f_1(R_1)) + \alpha^2 M(f_1(R_2)) = m/K$. The uniqueness follows from [63]. The same statements one can deduce for the differentials $\varphi_j$ and the maps $f_j$ for $j = 2, 3$.

**Proposition 4.2.1.** *The unique map* $f_1\big|_{\alpha=\alpha_0}$ *gives the absolute minimum and* $f_3\big|_{\alpha=1}$ *gives the absolute maximum to* $f(r_1)$ *in the class* $H_K$.

*Proof.* The result is not difficult but we give the proof for the completeness and to show the general approach to the problem in question. We prove Proposition 4.2.1 for $f_3$. If $\alpha = 1$, then $c_3(1) = r_1$ and the admissible system of curves consists of a curve $\gamma_1$ that separates the points $1, r_1$ from $0$ and $\infty$ which is homotopic on $S_0$ to the slit along the segment $[1, r_1]$. We suppose the contrary. Let there be a map $f \in H_K$ such that $f \neq f_3$ for $\alpha = 1$ and $f(r_1) \geq f_3(r_1)\big|_{\alpha=1}$. Then, $m_f > m/K$. The maps from $H_K$ are homeomorphic on $\mathbb{R}$, therefore, $f(\gamma_1)$ is homotopic on $\mathbb{C} \setminus \{0, 1, f(r_1)\}$ to the slit along the segment $[1, f(r_1)]$. We have

$$m/K \leq M(\mathbb{C} \setminus \{(-\infty, 0] \cup [1, f(r_1)]\}) = m_f$$
$$\leq M(\mathbb{C} \setminus \{(-\infty, 0] \cup [1, f_3(r_1)\big|_{\alpha=1}]\}) = m/K.$$

This chain of inequalities is valid only for $f \equiv f_3\big|_{\alpha=1}$. This contradiction proves Proposition 4.2.1. For the map $f_1\big|_{\alpha=\alpha_0}$ the proof is similar.    □

**Proposition 4.2.2.** *For any real $c$ with*

$$\min_{f \in H_K} f(r_1) = f_1(r_1)\big|_{\alpha=\alpha_0} \leq c \leq f_3(r_1)\big|_{\alpha=1} = \max_{f \in H_K} f(r_1)$$

*there are such $j$ and $\tilde{\alpha}$ that $f_j(r_1)\big|_{\alpha=\tilde{\alpha}} = c$.*

*Proof.* Theorems 2.7.1, 2.7.2 imply that the differentials $\varphi_1, \ldots, \varphi_3$ continuously depend on $\alpha$ and, hence, the same for $f_1, \ldots, f_3$. Moreover, $f_1\big|_{\alpha=\alpha_1} = f_2\big|_{\alpha=\alpha_3}$, $f_2\big|_{\alpha=\alpha_2} = f_3\big|_{\alpha=\alpha_4}$ that implies the assertion of Proposition 4.2.2.
□

**Theorem 4.2.3.** *For any $f \in H_K$ there are a unique $\tilde{\alpha}$ and $j = 1, \ldots, 3$ such that with $f(r_1) = f_j(r_1)\big|_{\alpha=\tilde{\alpha}} = c$ the inequality $f(r_2) \leq f_j(r_2)\big|_{\alpha=\tilde{\alpha}}$ holds. The equality is attained only for $f = f_j\big|_{\alpha=\tilde{\alpha}}$.*

*Proof.* By Proposition 4.2.2 we choose $j$ and $\tilde{\alpha}$, such that $f(r_1) = f_j(r_1)\big|_{\alpha=\tilde{\alpha}} = c$ for a function $f \in H_K$. Assume, for instance, $j = 3$, $\tilde{\alpha} \in [\alpha_4, 1]$. Then, we can generate the problem of the extremal partition on $S_0$ by the differential $\varphi_3(z)dz^2$ with the modulus $m = m(S, \Gamma, \tilde{\alpha})$ for the admissible system of curves $\gamma = (\gamma_1, \gamma_2)$ so that $\gamma_2$ is freely homotopic on $S_0$ to the slit along $[1, r_1]$. Now we assume the contrary. Let $f \in H_K$ and suppose that with $f_3$ as above $f(r_2) \geq f_3(r_2)\big|_{\alpha=\tilde{\alpha}}$ but that $f(z)$ is not identical with $f_3$. We define the problem of the extremal partition in $S' = \mathbb{C} \setminus \{0, 1, f(r_1), f(r_2)\}$ for the admissible system $\gamma'$ of two curves $(\gamma_1', \gamma_2')$ and the vector $(1, \tilde{\alpha})$. A simple loop $\gamma_1'$ separates the points $\infty, 0$ from $1, f(r_1), f(r_2)$, and is homotopic on $S'$ to the slit along $[-\infty, 0]$. The curve $\gamma_2'$ separates the points $\infty, 0, f(r_2)$ from $1, f(r_1)$ and is homotopic on $S'$ to the slit along $[1, f(r_1)]$. Since the domains $f(D_1)$, $f(D_2)$ are admissible, associated with the system $(\gamma_1', \gamma_2')$ the following chain of inequalities is valid

$$\frac{1}{K}m < M(f(D_1)) + \tilde{\alpha}^2 M(f(D_2)) \leq m(S', \Gamma', \tilde{\alpha}). \qquad (4.7)$$

The strict inequality sign results from the uniqueness of $f_3$ in the extremal problem of $\min m_f$. The extremal quadratic differential in the latter problem of the extremal partition is

$$\psi(w)dw^2 = A' \frac{w - c'}{w(w-1)(w - f(r_1))(w - f(r_2))} dw^2,$$

where the constants $A' < 0$, $c' \in [f(r_1), f(r_2)]$ are defined by the equations

$$\int\limits_{-\infty}^{0} \sqrt{\psi(x)}dx = \frac{1}{2}, \quad \int\limits_{1}^{f(r_1)} \sqrt{\psi(x)}dx = \frac{\alpha}{2}.$$

Since the critical trajectory of the differential $\psi$ starting from $f(r_2)$ has the negative direction, the corresponding modulus increases in so far as we move the point $f(r_2)$ in this direction. Taking into account the extremal properties of $f_3(z)$ with respect to the problem of the extremal partition, we have the inequality $m(S', \gamma', \tilde{\alpha}) \leq m_{f_3} = m/K$. This contradicts to (4.7) and ends the proof of the inequality in Theorem 4.2.3. The cases $j = 1, 2$ can be considered analogously.

Now we can prove the uniqueness of the choice of $\tilde{\alpha}$ and $j = 1, \ldots, 3$. For this we assume that there are two different pairs of these parameters that lead us to the same point of the boundary of the range of $I(f)$. One of these pairs we choose as a basic pair for the Teichmüller map (analog of $f_3$) and the Teichmüller map defined by the other one we denote by $f$. Then, we can repeat the preceding proof and arrive at the same contradiction. This ends the whole proof.     □

### 4.2.3 Boundary parameterization

We set

$$w_k(t) = f_1(r_k)\Big|_{\alpha = 3(\alpha_1 - \alpha_0)t + \alpha_0} \qquad \text{for} \quad t \in [0, 1/3],$$

$$w_k(t) = f_2(r_k)\Big|_{\alpha = 3(\alpha_2 - \alpha_3)t + 2\alpha_3 - \alpha_2} \qquad \text{for} \quad t \in [1/3, 2/3],$$

$$w_k(t) = f_3(r_k)\Big|_{\alpha = 3(1 - \alpha_4)t + 3\alpha_4 - 2} \qquad \text{for} \quad t \in [2/3, 1]$$

$k = 1, 2.$

**Theorem 4.2.4.** *Let $f \in H_K$. Then, the upper boundary curve for the range of the functional $I(f) = (f(r_1), f(r_2))$, $1 < r_1 < r_2$, i.e., the curve of $\max\limits_{f \in H_K} f(r_2)$ for $f(r_1)$ fixed, is assigned parameterically by $(w_1(t), w_2(t))$, $t \in [0,1]$. This curve is smooth and, being considered in the plane $(w_1, w_2) \in \mathbb{R}^2$, increases in $t \in [0, 2/3]$ and decreases in $t \in [2/3, 1]$. The normal vector to this curve at the point $t = 2/3$ is vertical and at the points $t = 0, 1$ is horizontal.*

*Proof.* By (4.4-4.6), the functions $A_j(\alpha)$ and $c_j(\alpha)$ for $j = 1, 2, 3$ are differentiable with respect to $\alpha$ in the corresponding intervals, so the same for $\varphi_j$, $f_j$. Therefore, the functions $w_1(t)$, $w_2(t)$ are piecewise differentiable and they have left-side and right-side derivatives at the points $1/3$, $2/3$. We consider the interval $(2/3, 1)$. We have in this interval the equality $m_{f_3} \equiv m/K$ where the modulus $m$ is defined by the extremal differential $\varphi_3$. Theorem 2.7.2 states that

$$\frac{dm}{d\alpha} = 2\alpha M(D_2); \quad \frac{\partial m_{f_3}}{\partial \alpha} = 2\alpha M(f_3(D_2)); \quad \frac{\partial m_{f_3}}{\partial w_k} = \pi \mathrm{Res}\,_{w=w_k} \psi(w).$$

Each of $w_k$ is a simple pole of the extremal differential $\psi$. By this we come to the equality

$$\frac{\partial m_{f_3}}{\partial \alpha} + \frac{\partial m_{f_3}}{\partial w_1}\frac{dw_1}{d\alpha} + \frac{\partial m_{f_3}}{\partial w_2}\frac{dw_2}{d\alpha} = \frac{1}{K}\frac{dm}{d\alpha}.$$

The map $f_3$ is extremal for the problem of the extremal partition with the modulus $m_f$ so it maps the extremal configuration in the trajectory structure of the differential $\varphi_3$ onto the extremal configuration in the trajectory structure of the differential $\psi$. Therefore,

$$\frac{\partial m_{f_3}}{\partial \alpha} = \frac{1}{K}\frac{dm}{d\alpha}.$$

The direct differentiation leads us to the derivative $dw_2/dw_1$ which exists and is negative in all points of $(2/3,1)$. Moreover, $dw_2/dw_1 \to 0$ as $t \to 2/3 + 0$ and $dw_2/dw_1 \to -\infty$ as $t \to 1 - 0$. By the analogy with the above, we prove the cases $j = 1, 2$. This ends the proof.  $\square$

The lower boundary is obtained in the same way. We omit the awkward statements of these results because one can learn them by Theorems 4.2.3, 4.2.4. Here we present the quadratic differentials that lead us to the extremal functions for the lower boundary.

$$\varphi_4(z)dz^2 = A_4(\alpha)(z - c_4(\alpha))P^{-1}(z)dz^2,$$

where $c_4(\alpha) \in [r_2, \infty]$, $A_4(\alpha) < 0$. These values are calculated by the equations

$$\int\limits_{-\infty}^{0} \sqrt{\varphi_4(x)}\,dx = \frac{1}{2}, \quad \int\limits_{1}^{r_1} \sqrt{\varphi_4(x)}\,dx = \frac{\alpha}{2},$$

$\alpha$ is a fixed number from the segment $[1, \alpha_5]$.

$$\varphi_5(z)dz^2 = A_5(\alpha)(z - c_5(\alpha))P^{-1}(z)dz^2,$$

where $c_5(\alpha) \in (-\infty, 1]$, $A_5(\alpha) > 0$. These values are calculated by the equations

$$\int\limits_{r_2}^{\infty} \sqrt{\varphi_5(x)}\,dx = \frac{1}{2}, \quad \int\limits_{1}^{r_1} \sqrt{\varphi_5(x)}\,dx = \frac{\alpha}{2},$$

$\alpha$ is a fixed number from the segment $[\alpha_6, \alpha_7]$.

$$\varphi_6(z)dz^2 = A_6(\alpha)(z - c_6(\alpha))P^{-1}(z)dz^2,$$

where $c_6(\alpha) \in [1, r_2]$, $A_6(\alpha) > 0$. These values are calculated by the equations

$$\int\limits_{r_2}^{\infty} \sqrt{\varphi_6(x)}\,dx = \frac{1}{2}, \quad \int\limits_{0}^{1} \sqrt{\varphi_6(x)}\,dx = \frac{\alpha}{2},$$

$\alpha$ is a fixed number from the segment $[\alpha_8, \alpha_9]$.

Here all boundary values for the parameter $\alpha$ can be obtained by integration along trajectories of the differentials in case of degeneracy of one of the ring domains in their trajectory structure.

Theorems 4.2.3, 4.2.4 define the range of $I(f)$ in the class of quasisymmetric functions.

### 4.2.4 The class $Q_K$. Estimations of functionals

Here we apply symmetrization to describe the part of the boundary of the range of $I(f)$ for the class $Q_K$ and obtain some non-trivial estimates of functionals dependent on $|f(r_1)|$ and $|f(r_2)|$.

**Proposition 4.2.3.** *The unique map* $f_3\big|_{\alpha=1}$ *gives the absolute maximum to* $|f(r_1)|$ *in the class* $Q_K$.

*Proof.* If $\alpha = 1$, then $c_3(1) = r_1$ and the admissible system of curves consists of a curve $\gamma$ that separates the points $1, r_1$ from $0, \infty$, and is homotopic on $S_0$ to the slit along the segment $[1, r_1]$ as for Proposition 4.2.1. We suppose the contrary to the assertion of Proposition 4.2.3. Let there be a map $f \in Q_K$ such that for $f \neq f_3$, $\alpha = 1$ the reverse inequality $|f(r_1)| \geq f_3(r_1)\big|_{\alpha=1}$ holds. We define the problem of the extremal partition of $S' = \mathbb{C} \setminus \{0, 1, |f(r_1)|\}$ for an admissible curve $\gamma'$ which is a simple loop separating the points $\infty, 0$

from $1, |f(r_1)|$, homotopic on $S'$ to the slit along $[-\infty, 0]$. Denote by $D = \mathbb{C} \setminus \{(-\infty, 0] \cup [1, r_1]\}$. For the domain $f(D)$ we have the inequality

$$\frac{1}{K}m = m_{f_3} \leq M(f(D)). \tag{4.8}$$

Let $D(x) = \mathbb{C} \setminus \{(-\infty, 0] \cup [1, x]\}$ and the domain $D' = D(|f(r_1)|)$ be extremal for $m(S', \Gamma')$, say $m(S', \Gamma') = M(D')$. We apply circular symmetrization to the domain $f(D)$ with the center at the origin and with the direction along the positive real axis. Denote by $D^*$ the resulting domain of this symmetrization. This domain is admissible in the problem of the extremal partition of $S'$ for the admissible curve $\gamma'$. Since $D' \subset D(|f_3(r_1)|)\Big|_{\alpha=1}$, the inequality

$$M(f(D)) \leq M(D^*) \leq M(D') \leq M(D(|f_3(r_1)|)\Big|_{\alpha=1}) = m/K \tag{4.9}$$

holds. The first inequality admits the equality only in the case $D^* = f(D)$, and the chain of inequalities (4.8–4.9) is valid only if $f \equiv f_3\Big|_{\alpha=1}$. This contradiction ends the proof. $\qquad \square$

**Theorem 4.2.5.** *Let $f \in Q_K$. Then, the upper boundary curve of the range of the system of functionals $I(f) = (|f(r_1)|, |f(r_2)|)$, $1 < r_1 < r_2$ in the class $Q_K$ coincides with that for the class $H_K$ for $|f(r_1)| \geq f_2(r_1)\Big|_{\alpha=1}$.*

*Proof.* Let $f^* \in Q_K$ be a function that maximizes $\max\limits_{f \in Q_K} |f(r_2)|$ for a fixed value of $|f(r_1)| \geq f_2(r_1)\Big|_{\alpha=1}$. Then, by Propositions 4.2.1, 4.2.3, 4.2.3, there is a function $f_j$ for some $j = 2, 3$ and $\tilde{\alpha}$ such that $|f^*(r_1)| = f_j(r_1)$. For this function the inequality $f_j(r_1) < f_j(r_2) \leq |f^*(r_2)|$ holds. Therefore, we can choose extremal functions among those satisfying the inequality $|f(r_1)| < |f(r_2)|$.

We prove Theorem 4.2.5 for $j = 3$. For $j = 2$ the proof is similar. We start with the problem of the extremal partition defined by the differential $\varphi_3$. By Proposition 4.2.2, we choose $\tilde{\alpha}$ such that $|f(r_1)| = f_3(r_1)\Big|_{\alpha=\tilde{\alpha}} = c$ for a function $f \in Q_K$ and $|f(r_1)| \geq f_3(r_1)\Big|_{\alpha=\alpha_4}$, $\tilde{\alpha} \in [\alpha_4, 1]$. Then, we introduce the problem on the extremal partition of $S_0$ by the differential $\varphi_3(z)dz^2$ with the modulus $m = m(S, \Gamma, \tilde{\alpha})$ for the admissible system of curves $\gamma = (\gamma_1, \gamma_2)$ as in Theorem 4.2.3. We assume the contrary. Let $f \in Q_K$ and suppose that with $f_3$ as above $|f(r_2)| \geq f_3(r_2)\Big|_{\alpha=\tilde{\alpha}}$, but that $f(z)$ is not identical with $f_3$. Now we define the problem of the extremal partition of $S' = \mathbb{C} \setminus \{0, 1, |f(r_1)|, |f(r_2)|\}$ by the admissible system $\gamma'$ of two curves $(\gamma_1', \gamma_2')$ and the vector $(1, \tilde{\alpha})$ as above. The curve $\gamma_1'$ is a simple loop that separate the points $\infty, 0$ from $1, |f(r_1)|, |f(r_2)|$, homotopic on $S'$ to the slit along $[-\infty, 0]$. The curve $\gamma_2'$ separates the punctures $\infty, 0, |f(r_2)|$ from $1, |f(r_1)|$ with the homotopy defined by $\gamma_1'$. The domains $f(D_1)$, $f(D_2)$ satisfy the following inequality

$$\frac{1}{K}m \leq M(f(D_1)) + \tilde{\alpha}^2 M(f(D_2)). \tag{4.10}$$

Let $D_1'$ and $D_2'$ be the extremal pair of ring domains for $m(S', \Gamma', \tilde{\alpha})$. Then, $m(S', \Gamma', \tilde{\alpha}) = M(D_1') + \tilde{\alpha}^2 M(D_2')$. Now we apply circular symmetrization to the domains $f(D_1)$, $f(D_2)$ with the center at the origin and with the direction of the positive and negative real axes respectively. Denote by $D_1^*$ and $D_2^*$ the resulting domains of this symmetrization. This pair of domains is admissible in the problem of the extremal partition of $S'$ for the admissible system $\gamma'$ and the vector $(1, \tilde{\alpha})$ as above. Then,

$$m(S', \Gamma', \tilde{\alpha}) \geq M(D_1^*) + \tilde{\alpha}^2 M(D_2^*) \geq M(f(D_1)) + \tilde{\alpha}^2 M(f(D_2)). \tag{4.11}$$

The inequality can be replaced by equality only in the case $D_1^* = f(D_1)$ and $D_2^* = f(D_2)$. Since the trajectory of the extremal differential in the latter problem of the extremal partition starting from $|f(r_2)|$ has negative direction to $|f(r_1)|$, the gradient of $m(S', \Gamma', \tilde{\alpha})$ at the point $|f(r_2)|$ has the same direction and the modulus increases in so far as we move the point $|f(r_2)|$ to $|f_3(r_2)|_{\alpha=\tilde{\alpha}}$. Taking into account the extremal properties of $f_3(z)$ with respect to the problem of the extremal partition we arrive at the inequality $m(S', \Gamma', \tilde{\alpha}) \leq m_{f_3} = m/K$. This contradicts the chain of inequalities (4.10–4.11) and ends the proof.                                                            □

For the rest of the boundary of the range of the system of functionals $I(f) = (|f(r_1)|, |f(r_2)|)$ symmetrization can not be applied because symmetrized domains are not admissible in the corresponding problems of the extremal partition.

**Theorem 4.2.6.** *The unique map $f_2\Big|_{\alpha=1}$ gives the absolute maximum to the functional $|f(r_2)/f(r_1)|$, $1 < r_1 < r_2$ in the classes $Q_K$ and $H_K$. Moreover, the following sharp estimate is valid*

$$\frac{|f(r_2)|}{|f(r_1)|} \leq \frac{1}{1 - \left( p_K \left( \sqrt{1 - \frac{r_1}{r_2}} \right) \right)^2}.$$

*Proof.* If $\alpha = 1$, then $c_2(1) = 1$ and the admissible system of curves consists of a curve $\gamma$ that separates the points $r_1, r_2$ from $0$ and $\infty$ and homotopic on $S$ to the slit along the segment $[r_1, r_2]$. We assume the contrary. Let $f \in Q_K$ and suppose that with $f_2$ as above $|f(r_2)/f(r_1)| \geq f_2(r_2)/f_2(r_1)\Big|_{\alpha=1}$, but that $f(z)$ is not identical with $f_2$. We define the problem of the extremal partition of $S' = \mathbb{C} \setminus \{0, 1, |f(r_2)/f(r_1)|\}$ for the admissible curve $\gamma'$ which is a simple loop separating the punctures $\infty, 0$ from $1, f(r_2)/f(r_1)$, homotopic on $S'$ to the slit along $[-\infty, 0]$. Let $D = \mathbb{C} \setminus \{(-\infty, 0] \cup [r_1, r_2]\}$. We consider the map $w = f(z)/f(r_1)$. The domain $f(D)/f(r_1)$ satisfies the following inequality

$$\frac{1}{K}M(D) = \frac{1}{K}m \le M(f(D)/f(r_1)). \tag{4.12}$$

Let $D(x) = \mathbb{C} \setminus \{(-\infty, 0] \cup [1, x]\}$ and $D' = D(|f(r_2)/f(r_1)|)$ be the extremal ring domain for $m(S', \Gamma')$, say $m(S', \Gamma') = M(D')$. Now we apply circular symmetrization to the domain $f(D)/f(r_1)$ with the center at origin and with the direction of the positive real axis. Denote by $D^*$ the resulting domain of this symmetrization. This domain is admissible in the problem of the extremal partition of $S'$ for the admissible curve $\gamma'$. Since $D' \subset D(f_2(r_2)/f_2(r_1))\big|_{\alpha=1}$, the inequality

$$M(D^*) \le M(D') \le M(D(f_2(r_2)/f_2(r_1))\big|_{\alpha=1}) = m/K \tag{4.13}$$

holds. The chain of inequalities (4.12–4.13) is valid only for $f \equiv f_2\big|_{\alpha=1}$. The explicit calculation of the moduli and the relation $m/K \le M(D(|f(r_2)/f(r_1)|))$ gives us the inequality of Theorem 4.2.6 for the class $Q_K$. Since the extremal function is from $H_K$ the same result is true for the class $H_K$. This ends the proof. □

The next corollary immediately follows from the inequality $\frac{K'(k)}{K(k)} \le \frac{1}{\pi}\log 4/k$.

**Corollary 4.2.1.** *Let $f(z)$ be a function from $Q_K$ or $H_K$, $1 < r_1 < r_2$. Then,*

$$\frac{|f(r_2)|}{|f(r_1)|} \le \frac{16}{16 - \exp\frac{2\pi}{K}\frac{K'(\sqrt{1-r_1/r_2})}{K(\sqrt{1-r_1/r_2})}}.$$

By the analogy with Theorem 4.2.6, we obtain the next theorem and corollary. We omit their proofs but show the initial problems of the extremal partition in the $(z)$-plane. For Theorem 4.2.7 the admissible curve is a simple loop that separates the punctures $0, r_1$ from $r_2, \infty$. It is homotopic on $S_0 = \mathbb{C} \setminus \{0, r_1, r_2\}$ to the slit along the segment $[0, r_1]$.

**Theorem 4.2.7.** *Let $f \in H_K$, $1 < r_1 < r_2$. Then, the following sharp estimate*

$$\frac{|f(r_2)|}{|f(r_1)|} \ge \left(p_K\left(\sqrt{\frac{r_1}{r_2}}\right)\right)^{-2}$$

*holds. The extremal function is the Teichmüller map is induced by the differential $\varphi_5\big|_{\alpha=\alpha_7}$.*

**Corollary 4.2.2.** *Let $f(z)$ be a function from $H_K$, $1 < r_1 < r_2$. Then,*

$$\frac{|f(r_2)|}{|f(r_1)|} \ge \frac{1}{16}\exp\frac{2\pi}{K}\frac{K'(\sqrt{r_1/r_2})}{K(\sqrt{r_1/r_2})}.$$

For the forthcoming Theorem 4.2.8 (right-hand side inequality) the problem of the extremal partition is defined by the admissible curve which is a simple loop that separates the punctures $0, 1$ from $r_1, r_2$. It is homotopic on $S_0 = \overline{\mathbb{C}} \setminus \{0, 1, r_1, r_2\}$ to the slit along the segment $[0, 1]$. For the left-hand side inequality we consider the problem of the extremal partition for the admissible curve which is a simple loop that separates the punctures $1, r_1$ from $0, r_2$. It is homotopic on the same $S_0$ to the slit along the segment $[1, r_1]$.

**Theorem 4.2.8.** *Let $f \in H_K$, $1 < r_1 < r_2$. Then, the following sharp estimates*

$$Q^2(r_1, r_2) \le \frac{1 - 1/|f(r_1)|}{1 - 1/|f(r_2)|} \le \frac{1}{Q^2(r_1, r_2)}$$

*hold where*

$$Q(r_1, r_2) = p_K\left(\sqrt{\frac{r_2(r_1 - 1)}{r_1(r_2 - 1)}}\right).$$

*The extremal functions are the Teichmüller maps generated by the extremal differentials in the above problems of the extremal partition.*

More difficult estimates are valid for the functionals $|f(r_1) \pm f(r_2)|$ and $|f(r_1)| \pm |f(r_2)|$.

**Theorem 4.2.9.** *The unique extremal map $f^*$ gives the maximum to $f(r_2) - f(r_1)$ in the class $H_K$ and the maximum to $|f(r_2)| - |f(r_1)|$, $1 < r_1 < r_2$ in the class $Q_K$. This map satisfies the Beltrami equation (4.1) with the Beltrami coefficient (4.2) with*

$$\varphi(z) = \frac{c - z}{z(z - 1)(z - r_1)(z - r_2)},$$

*where*

$$c = \frac{r_1 r_2}{r_1 + r_2 - 1}.$$

*Proof.* The proof falls into three steps.

**1.** We are looking for the extremal functions among those which give the points $(w_1(t), w_2(t))$ of the upper boundary curve $\Gamma^+$ of the range of the system of functionals $I(f) = (|f(r_1)|, |f(r_2)|)$.

Consider the point $(w_1(1/3), w_2(1/3))$ of $\Gamma^+$. Theorems 4.2.4, 4.2.6 imply that

$$\beta = \arctan \frac{w_2(1/3)}{w_1(1/3)} = \arctan \frac{w_2'(1/3)}{w_1'(1/3)} > \frac{\pi}{4}.$$

**2.** Then, the extremal function is $f_2$ for some $\alpha^* \in (1, \alpha_2)$ that satisfies the necessary condition of extremality for the functional given

$$\frac{w_2'(t^*)}{w_1'(t^*)} = 1, \quad \alpha^* - 3t^*(\alpha_2 \quad \alpha_3) + 2\alpha_3 - \alpha_2, \quad t^* \in (1/3, 2/3).$$

For the function $f_2$ we have

$$\nabla_{w_1} m_{f_2} = \pi \overline{\text{Res}_{w=w_1} \psi_2(w)}; \quad \nabla_{w_2} m_{f_2} = \pi \overline{\text{Res}_{w=w_2} \psi_2(w)}. \tag{4.14}$$

The notation $\nabla_{w_j}$ stands for the gradient which is taken at the point $w_j$. Each of $w_1, w_2$ is a simple pole of the extremal differential $\psi_2$,

$$\psi_2(w)dw^2 = B \frac{w-C}{w(w-1)(w-w_1)(w-w_2)} dw^2,$$

$C \in [1, w_1]$ and $B < 0$. Hence, we obtain as in Theorem 4.2.4, the equality

$$\frac{\partial m_{f_2}}{\partial w_1} \frac{dw_1}{dt} + \frac{\partial m_{f_2}}{\partial w_2} \frac{dw_2}{dt} = 0.$$

From (4.14) we deduce that for all points $\alpha^*$ and, consequently $t^*$, satisfying the necessary condition we have

$$\frac{w_2'(t^*)}{w_1'(t^*)} = \frac{w_2(w_2-1)(w_1-C)}{w_1(w_1-1)(w_2-C)} = 1.$$

or equivalently,

$$C = \frac{w_1 w_2}{w_1 + w_2 - 1}, \tag{4.15}$$

where $C = C(\alpha^*, w_1, w_2)$ is the function defined by the conditions for the differential $\psi_2$,

$$\int_{-\infty}^{0} \sqrt{\psi_2(s)} ds = 1/2, \quad \int_{w_1}^{w_2} \sqrt{\psi_2(s)} ds = \alpha^*/2.$$

**3.** Now we claim that

$$c(\alpha^*) = \frac{r_1 r_2}{r_1 + r_2 - 1}. \tag{4.16}$$

This implies that there exists a unique $\alpha^*$ and $t^*$ satisfying the necessary condition that, therefore, becomes sufficient, because the extremal function giving a point of $\Gamma^+$ is unique for each point of $\Gamma^+$. To prove this we need more refined observations.

For a fixed $\alpha^*$ we consider the quadratic differential

$$\widetilde{\varphi}_2(z)dz^2 = \tilde{A}_2 \frac{z - \tilde{c}_2(u,v)}{z(z-1)(z-u)(z-v)} dz^2, \tag{4.17}$$

where $\tilde{c}_2 \in [1, u]$, $\tilde{A}_2 < 0$. These values are calculated by the equations

$$\int_{-\infty}^{0} \sqrt{\widetilde{\varphi}_2(s)} ds = 1/2, \quad \int_{u}^{v} \sqrt{\widetilde{\varphi}_2(s)} ds = \alpha^*/2,$$

where the real-valued differentiable functions $u = u(\mu)$, $v = v(\mu)$ accept their values from some neighbourhood of $(r_1, r_2)$, $\mu \in (-\varepsilon, \varepsilon)$, $u(0) = r_1$, $v(0) = r_2$. We construct the Teichmüller map $\tilde{f}$ as the solution of the Beltrami equation (4.1) with the Beltrami coefficient (4.2) for the quadratic differential (4.17) with the normalization of the class $H_K$. Denote by $w_1(\alpha^*, u, v) = \tilde{f}(r_1)$, $w_2(\alpha^*, u, v) = \tilde{f}(r_2)$. These functions are differentiable with respect to $u$ and $v$ because $\tilde{f}$ is the solution to the Beltrami equation with the Beltrami coefficient which is differentiable with respect to $u$ and $v$. Moreover, we have that $\tilde{c}_1(r_1, r_2) = c_1(\alpha^*)$, $w_1(\alpha^*) = w_1(\alpha^*, r_1, r_2)$, $w_2(\alpha^*) = y(\alpha^*, r_1, r_2)$. Since $\tilde{f} \in H_K$, the points $(w_1(\alpha^*, u(\mu), v(\mu)), w_2(\alpha^*, u(\mu), v(\mu)))$ form a curve, parameterized by $\mu$, that touches the boundary curve $\Gamma^+$ at the point $(w_1(\alpha^*), w_2(\alpha^*))$.

Denote by $m(u, v)$ the modulus defined by the differential (4.17), and by $m_{\tilde{f}}$ the modulus defined by the Teichmüller map $\tilde{f}$. Of course, $m_{\tilde{f}} = \frac{1}{K} m(u, v)$. Now we differentiate this equality with respect to $u$ and $v$ and obtain

$$\frac{1}{K} \frac{\partial m(u,v)}{\partial u} = \frac{\partial m_{\tilde{f}}}{\partial w_1} \cdot \frac{\partial w_1(\alpha^*,u,v)}{\partial u} + \frac{\partial m_{\tilde{f}}}{\partial w_2} \cdot \frac{\partial w_2(\alpha^*,u,v)}{\partial u},$$
$$\frac{1}{K} \frac{\partial m(u,v)}{\partial v} = \frac{\partial m_{\tilde{f}}}{\partial w_1} \cdot \frac{\partial w_1(\alpha^*,u,v)}{\partial v} + \frac{\partial m_{\tilde{f}}}{\partial w_2} \cdot \frac{\partial w_2(\alpha^*,u,v)}{\partial v}. \qquad (4.18)$$

Here in the right-hand sides of these equalities the partial derivatives are taken over the simple poles $w_1$, $w_2$ of the extremal differential for $m_{\tilde{f}}$. Taking into account the rule of differentiation for the modulus we obtain that

$$\frac{u(u-1)(v - \tilde{c}_2(u,v))}{v(v-1)(u - \tilde{c}_2(u,v))} = \frac{\frac{\partial m_{\tilde{f}}}{\partial w_1} \frac{\partial w_1(\alpha^*,u,v)}{\partial v} + \frac{\partial m_{\tilde{f}}}{\partial w_2} \frac{\partial w_2(\alpha^*,u,v)}{\partial v}}{\frac{\partial m_{\tilde{f}}}{\partial w_1} \frac{\partial w_1(\alpha^*,u,v)}{\partial u} + \frac{\partial m_{\tilde{f}}}{\partial w_2} \frac{\partial w_2(\alpha^*,u,v)}{\partial u}}. \qquad (4.19)$$

Observe, that

$$\left. \frac{\partial m_{\tilde{f}}}{\partial w_1} \right|_{u=r_1, v=r_2} = \left. \frac{\partial m_{f_2}}{\partial w_1} \right|_{\alpha=\alpha^*}, \qquad \left. \frac{\partial m_{\tilde{f}}}{\partial w_2} \right|_{u=r_1, v=r_2} = \left. \frac{\partial m_{f_2}}{\partial w_2} \right|_{\alpha=\alpha^*}.$$

Therefore, the equalities (4.14), (4.15) imply

$$\left. \frac{\partial m_{\tilde{f}}}{\partial w_1} \right|_{u=r_1, v=r_2} = - \left. \frac{\partial m_{\tilde{f}}}{\partial w_2} \right|_{u=r_1, v=r_2}.$$

Now we choose the parameterization $u = r_1 + \mu/2$, $v = r_2 - \mu/2$. Since the curve $(w_1(\alpha^*, u(\mu), v(\mu)), w_2(\alpha^*, u(\mu), v(\mu)))$, given with the parameter $\mu \in (-\varepsilon, \varepsilon)$, touches the boundary curve of Theorem 4.2.4 at the point $\mu = 0$,

$$\frac{d(w_2(\alpha^*, u, v) - w_1(\alpha^*, u, v))}{d\mu}\bigg|_{\mu=0} =$$

$$= \frac{1}{2} \left( \frac{\partial(w_2(\alpha^*, u, v) - w_1(\alpha^*, u, v))}{\partial u} - \right.$$

$$\left. - \frac{\partial(w_2(\alpha^*, u, v) - w_1(\alpha^*, u, v))}{\partial v} \right)_{\mu=0} = 0,$$

and the right-hand side of the equality (4.19) reduces to 1 at $\mu = 0$, $(u, v) = (r_1, r_2)$. This leads to the value $\tilde{c}_2(r_1, r_2) = c(\alpha^*)$ given by (4.16).

Thus, the extremal function $f^*$ is $f_2$ defined by the Beltrami coefficient given in the formulation of the theorem. This completes the proof.    □

By the analogy with the preceding theorem we arrive at the following statement.

**Theorem 4.2.10.** *The unique extremal map $f^*$ gives the maximum to $|f(r_2) + f(r_1)|$, $1 < r_1 < r_2$ in the classes $H_K$, $Q_K$. This map satisfies the Beltrami equation (4.1) with the Beltrami coefficient (4.2) with*

$$\varphi(z) = \frac{c - z}{z(z - 1)(z - r_1)(z - r_2)},$$

*where*

$$c = \frac{r_1 r_2 (r_1 + r_2 - 2)}{r_2(r_2 - 1) + r_1(r_1 - 1)}.$$

The extremal function in this theorem is $f_3$ for some $\alpha = \alpha^*$.

The case of a negative $r_1$ is obtained by the analogy with the above. Moreover, for $r_1 > 0$, $r_2 > 1$ an analog of Theorem 4.2.5 is stronger. It turns out that the upper boundary curve of the range of the system of functionals $I(f) = (|f(-r_1)|, |f(r_2)|)$ in the class $Q_K$ coincides with that for the class $H_K$ completely.

## 4.2.5 Conclusions and unsolved problems

We start with the conformal case. An earlier result by J. Jenkins [64] asserts that the upper boundary curve of the range of the system of functionals $(|f(-r_1)|, |f(r_2)|)$, $0 < r_1, r_2 < 1$ in the class $S$ coincides with that in the class $S_R$. By the words "upper" and "lower" we mean, as before, that the points $(|f(-r_1)|, |f(r_2)|)$ are considered as points of the real plane $\mathbb{R}^2$. In Jenkins' proof it was important that the fixed points $(-r_1)$ and $r_2$ were situated in different legs of the real diameter. Later on, the author and S. Fedorov [163] have reached to a similar result for the system of functionals $(|f(r_1)|, |f(r_2)|)$, $0 < r_1 < r_2 < 1$. As for the lower boundary curves for these ranges, they are different in the classes $S$ and $S_R$. In $S_R$ the points of the lower boundary curves are given by the simple function $z(1 - uz + z^2)^{-1}$, $-2 \le u \le 2$. The points of the upper boundary curves are given by functions with more complicated structure. They have been described in [64], [163].

In the quasiconformal case we introduce a subclass $H_K$ of the class $Q_K$ of functions satisfying the same symmetry condition $\overline{f(z)} = f(\bar{z})$. This class plays the same role for the class $Q_K$ as the subclass $S_R$ for the class $S$. One can easily see that the upper boundary curve of the range of the system of functionals $(|f(-r_1)|, |f(r_2)|)$, $r_1 > 0, r_2 > 1$ over the class $Q_K$ coincides with that over the class $Q_K^R$. Hence, this result is close to [64] and [163]. Of course, the method of proof is completely different. The extra normalization $f(1) = 1$ causes another difficulty. So, we are not able right now to prove an analogous result for the system of functionals $(|f(-r_1)|, |f(r_2)|)$, $r_1 > 0, 0 < r_2 < 1$ or for the system of functionals $(|f(r_1)|, |f(r_2)|)$, $0 < r_1 < 1, r_2 > 1$. But for the system of functionals $(|f(r_1)|, |f(r_2)|)$, $1 < r_1 < r_2$ this is the case as the author announced at the International Congress of Mathematicians, Berlin, 1998 [159]. In [156] the author has shown this for $K$-quasiconformal homeomorphisms of $U$. Corresponding theorem we will present in the next section.

From these results we deduce the sharp estimates for the functionals $|f(r_2)| \pm |f(\pm r_1)|$ and $|f(r_2) \pm f(\pm r_1)|$. The proofs open a way to present extremal functions in terms of Beltrami coefficients of direct mapping. But the solution of the problem for $|f(r_2) \pm f(\mp r_1)|$ is still unknown. The same holds for the lower estimates of the functional $|f(r_2)|/|f(r_1)|$.

Hereby we consider the following problems as interesting and difficult to solve:

**1)** Find the lower boundary curve of the range of the system of functionals $(|f(r_1)|, |f(r_2)|)$ in the class $Q_K$ for different real $r_1, r_2$.

**2)** Find the upper boundary curve of the range of the system of functionals $(|f(r_1)|, |f(r_2)|)$ in the class $Q_K$ for different $r_1, r_2$, so that one of $r_1, r_2$ lies in the segment $(0, 1)$ (for $r_1, r_2 \in (0, 1)$ this can be obtained in the same way as presented in this work and, therefore, it is not so interesting).

**3)** Obtain the sharp estimates of $|f(r_2) - f(r_1)|$ and $|f(r_2) + f(-r_1)|$ for $r_2, r_1 > 0$ in terms of the Beltrami coefficient of the direct mapping.

**4)** Obtain the lower sharp estimates of $|f(r_2)|/|f(\pm r_1)|$ for different real values of $r_1$ and $r_2$.

Here we do not speak about non-real fixed points because our method is based on symmetric structures.

## 4.3 Two-point distortion for quasiconformal maps of the unit disk

We obtain here analogous results for the range of the system of functionals $(|f(r_1)|, |f(r_2)|)$ for quasiconformal homeomorphisms of the unit disk $U$. Denote by $U_K$ the class of all $K$-quasiconformal automorphisms $w = f(z)$ of the disk $U$, $K \geq 1$, with the normalization $f(0) = 0$, $f(1) = 1$. A subclass $U_K^R \subset U_K$ consists of the functions satisfying the condition $f(\bar{z}) = \overline{f(z)}$. Here we will need some special tools to symmetrize domains.

### 4.3.1 Special differentials and extremal partitions

We consider the following one-parametric families of holomorphic differentials in $U' = U \setminus \{0, r_1, r_2\}$. Firstly,

$$\varphi_1(z)dz^2 = A_1(\alpha) \frac{(z - c_1(\alpha))(1 - c_1(\alpha)z)}{z(z - r_1)(z - r_2)(1 - r_1 z)(1 - r_2 z)} dz^2, \qquad (4.20)$$

where $c_1(\alpha) \in [r_2, 1]$, $A_1(\alpha) > 0$. These values are calculated by the equations

$$\int_{r_1}^{r_2} \sqrt{\varphi_1(x)}dx = \frac{\alpha}{2}, \quad \int_{-1}^{0} \sqrt{\varphi_1(x)}dx = \frac{1}{2};$$

$\alpha$ is a fixed number from the segment $[\alpha_0, \alpha_1]$, where

$$\alpha_0 = \int_{r_1}^{r_2} p_1(x)dx \left( \int_{-1}^{0} p_1(x)dx \right)^{-1} = \frac{\mathbf{F}(\varkappa, p)}{\mathbf{F}(\beta, p)}, \quad \text{where } p_1(x) := \sqrt{\varphi_1(x)}\Big|_{c_1(\alpha)=r_2},$$

$\varkappa = \arcsin \sqrt{\frac{r_2 - r_1}{r_2(1 - r_1^2)}}$, $\beta = \arcsin \frac{1}{\sqrt{1 + r_1}}$, $p = \sqrt{1 - r_1^2}$, $\mathbf{F}(\cdot, \cdot)$ is the elliptic integral of the first kind, and

$$\alpha_1 = \int_{r_1}^{r_2} p_2(x)dx \left( \int_{-1}^{0} p_2(x)dx \right)^{-1}, \quad \text{where } p_2(x) := \sqrt{\varphi_1(x)}\Big|_{c_1(\alpha)=1}.$$

Secondly,

$$\varphi_2(z)dz^2 = A_2(\alpha) \frac{(z - e^{i\beta(\alpha)})(z - e^{-i\beta(\alpha)})}{z(z - r_1)(z - r_2)(1 - r_1 z)(1 - r_2 z)} dz^2, \qquad (4.21)$$

where $\beta(\alpha) \in [0, \pi]$, $A_2(\alpha) < 0$. These values are calculated by the equations

$$\int_{r_1}^{r_2} \sqrt{\varphi_2(x)}dx = \frac{\alpha}{2}, \quad \int_{-1}^{0} \sqrt{\varphi_2(x)}dx = \frac{1}{2};$$

$\alpha$ is a fixed number from the segment $[\alpha_1, \alpha_2]$, where

$$\alpha_2 = \int_{r_1}^{r_2} p_3(x)dx \left(\int_{-1}^{0} p_3(x)dx\right)^{-1}, \text{ where } p_3(x) := \sqrt{\varphi_2(x)}\Big|_{\beta(\alpha)=\pi}.$$

Thirdly,

$$\varphi_3(z)dz^2 = A_3(\alpha)\frac{(z - c_3(\alpha))(1 - c_3(\alpha)z)}{z(z - r_1)(z - r_2)(1 - r_1 z)(1 - r_2 z)}dz^2, \qquad (4.22)$$

where $c_3(\alpha) \in [-1, r_1]$, $A_3(\alpha) < 0$. These values are calculated by the equations

$$\int_{r_1}^{r_2} \sqrt{\varphi_3(x)}dx = \frac{\alpha}{2}, \quad \int_{0}^{2\pi} \sqrt{-\varphi_3(e^{i\eta})}e^{i\eta}d\eta = 1;$$

$\alpha$ is a fixed number from the segment $[\alpha_2, \alpha_3]$, where

$$\alpha_3 = \frac{\int_{r_1}^{r_2} (x(x - r_1)(1 - r_1 x))^{-1/2}dx}{\int_0^{2\pi} (1 + r_1^2 - 2r_1 \cos\eta)^{-1/2}d\eta} = \frac{F(\varkappa, p)}{2K(r_1)},$$

where $K(\cdot)$ is the complete elliptic integral.

Finally,

$$\varphi_4(z)dz^2 = A_4(\alpha)\frac{(z - c_4(\alpha))(1 - c_4(\alpha)z)}{z(z - r_1)(z - r_2)(1 - r_1 z)(1 - r_2 z)}dz^2, \qquad (4.23)$$

where $c_4(\alpha) \in [r_1, r_2]$, $A_4(\alpha) < 0$. These values are calculated by the equations

$$\int_{0}^{r_1} \sqrt{\varphi_4(x)}dx = \frac{\alpha}{2}, \quad \int_{0}^{2\pi} \sqrt{-\varphi_4(e^{i\eta})}e^{i\eta}d\eta = 1;$$

$\alpha$ is a fixed number from the segment $[\alpha_4, 1]$, where

$$\alpha_4 = \frac{\int_0^{r_1} (x(x - r_2)(r_2 x - 1))^{-1/2}dx}{\int_0^{2\pi} (1 + r_1^2 - 2r_1 \cos\eta)^{-1/2}d\eta} = \frac{F(\gamma, r_2)}{2K(r_1)},$$

where $\gamma = \arcsin \sqrt{r_1/r_2}$.

From the construction of the differentials one can learn the dynamics singularities and the trajectory changes depending on the parameter $\alpha$. The differentials $\varphi_1$, $\varphi_2$, $\varphi_3$, and $\varphi_4$ have finite norms and finite trajectories. For each of them one can define the problem of the extremal partition the admissible system of curves that consists of at most two elements which are non-homotopic non-critical trajectories of the corresponding differential. We pose now more general problems.

I) Let $U' = U \setminus \{0, A, B\}$, $|A| < |B| < 1$. We consider the admissible system of two curves $\gamma^{(j)} = (\gamma_1^{(j)}, \gamma_2^{(j)})$ where $\gamma_1^{(j)}$ is a countable set of curves of the 2-nd type containing arcs with endpoints on $\partial U$ that separate 0 from $A$, $B$; $\gamma_2^{(j)}$ is a countable set of simple loops of the 1-st type with the homotopy defined by $\gamma_1^{(j)}$, and separating $A$, $B$ from 0, $\partial U$. Assume $j = 1$ if $\gamma_1^{(j)}$ is homotopic on $U'$ to the slit along $[A, B]$, $0 \notin [A, B]$, and to the slit along the broken line $[A, \varepsilon] \bigcup [\varepsilon, B]$ for some complex $\varepsilon$ with a small $|\dot\varepsilon|$ if $0 \in [A, B]$.

We pose the problem of the extremal partition of $U'$ for the admissible system $\gamma^{(j)}$ and a vector $\alpha \equiv (1, \alpha)$. The extremal metric is defined by the quadratic differential $\psi_1(w) dw^2$,

$$\psi_1(w) = A_1(\alpha) \frac{(w - d_1(\alpha))(w - 1/\bar{d}_1(\alpha))}{w(w - A)(w - B)(w - 1/\bar{A})(w - 1/\bar{B})},$$

where $A_1(\alpha)$ and $d_1(\alpha)$ are defined by the $\psi_1$-lengths of non-critical trajectories under the assumption that two characteristic domains do not degenerate. If $j = 1$, then the equality $\arg A_1(\alpha) = \pi - \arg(AB/d_1(\alpha))$ is satisfied.

The extremal differential can also have the form $\psi_2(w) dw^2$,

$$\psi_2(w) = A_2(\alpha) \frac{(w - e^{i\beta(\alpha)})(w - e^{i\varkappa(\alpha)})}{w(w - A)(w - B)(w - 1/\bar{A})(w - 1/\bar{B})},$$

where $A_2(\alpha)$, $\beta(\alpha)$, $\varkappa(\alpha)$ are defined by $\psi_2$-lengths of non-critical trajectories and $\beta$, $\varkappa$ are connected by the following equalities

$$\int_\beta^\varkappa \sqrt{\psi_2(e^{i\eta})} e^{i\eta} d\eta = 0, \quad \arg A_2(\alpha) = -\frac{\varkappa + \beta}{2}.$$

II) On $U'$ we consider the admissible system of curves $\gamma^{(j)} = (\gamma_1, \gamma_2^{(j)})$ where $\gamma_1$ is the simple loop of the 1-st type separating $\partial U$ from 0, $A$, $B$; $\Gamma_2^{(j)}$ is defined in the same way as the above in I). The equality $j = 1$ is thought of as before.

The extremal quadratic differential $\psi_3(w) dw^2$ in the corresponding problem of the extremal partition has the form

$$\psi_3(w) = A_1(\alpha) \frac{(w - d_3(\alpha))(w - 1/\bar{d}_3(\alpha))}{w(w - A)(w - B)(w - 1/\bar{A})(w - 1/\bar{B})},$$

where $A_3(\alpha)$ and $d_3(\alpha)$ are defined in the same way as in I). If $j = 1$, then the equality $\arg A_3(\alpha) = \arg(AB/d_3(\alpha))$ is satisfied.

III) Consider on $U'$ the admissible system of curves the first type $\gamma^{(j)} = \{\gamma_1, \gamma_2^{(j)}\}$ where $\gamma_1$ is the same as in II) and $\gamma_2^{(j)}$ is a simple loop that separates 0, $A$ from $B$, $\partial U$. The homotopy $j = 1$ is defined by the slit along $[0, A]$, if $B \notin [0, A]$, and by the slit along the broken line $[0, B + \varepsilon] \bigcup [B + \varepsilon, A]$ if $B \in [0, A]$.

The extremal quadratic differential $\psi_4(w)dw^2$ is given by

$$\psi_4(w) = A_4(\alpha)\frac{(w - d_4(\alpha))(w - 1/\bar{d}_4(\alpha))}{w(w - A)(w - B)(w - 1/\bar{A})(w - 1/\bar{B})},$$

where $A_4(\alpha)$ and $d_4(\alpha)$ are defined as in I), II). If $j = 1$, then the equality $\arg A_4(\alpha) = \arg(AB/d_4(\alpha))$ is satisfied. If $A = r_1$, $B = r_2$, then $\psi_k \equiv \varphi_k$ with an appropriate $\alpha$. The limiting values of $\alpha$ correspond to degeneracy of a characteristic domain and one can calculate them integrating along critical trajectories.

Further we need some statements which we formulate as propositions. Without loss of generality we assume $A > 0$.

**Proposition 4.3.1.** *In each of the problems of the extremal partition I), II), III), the modulus satisfies the inequality*

$$m(U', \Gamma^{(j)}, \alpha) \leq m(U', \Gamma^{(1)}, \alpha).$$

**Proposition 4.3.2.** *Let $j = 1$. In the problem III) the simple zero $d_4(\alpha)$ and the arc of the critical trajectory of the differential $\psi_4(w)dw^2$ connecting $d_4(\alpha)$ and $B$ are located within the triangle with the vertices $0$, $A$, $B$, when $\alpha < 1$, and in the sector $\{w : |B| \leq |w| < 1, \arg B \leq \arg(w - B) \leq \arg(B - A)\}$, when $\alpha > 1$.*

*Proof of Propositions 4.3.1, 4.3.2* are based on the use of polarization. Let $0 \leq \alpha < 1$, $\text{Im } B > 0$. We consider first the problem III). We apply the polarization transform with respect to the real axis to the extremal doubly connected domains $D_1$ and $D_2$ associated with $\gamma_1$, $\gamma_2^{(j)}$. As a result, we obtain the pair of non-overlapping doubly connected domains $D_1^0$, $D_2^0$ admissible with respect to the problem associated with $\gamma_1$, $\gamma_2^{(1)}$. Moreover $M(D_k) < M(D_k^0)$ if $D_k \neq D_k^0$. Thus, the statement of Proposition 4.3.1 follows from the extremality of the pair of the domains. The problems I), II) are considered analogously.

Besides, the trajectories of $\psi_4$ are analytic. By the definition of polarization the arc of the trajectory connecting $B$ and $d_4$ is located in the upper half-plane. The analogous application of polarization with respect to the lines $(0, B)$, $(A, B)$ leads to the statement of Proposition 4.3.2. $\qquad\square$

Denote by $m_I(B) = m(U', \Gamma^{(1)}, \alpha) = \|\psi_1\|$ (or $\|\psi_2\|$), $m_{II}(B) = \|\psi_3\|$, $m_{III}(B) = \|\psi_4\|$ emphasizing the dependence of the modulus on the simple pole of the corresponding differential in $U$.

**Proposition 4.3.3.** *If $0 < \alpha < 1$, then $m_{III}(|B|e^{it})$ is a decreasing function in $t \in [0, \pi]$ and an increasing function in $t \in [-\pi, 0]$.*

*Proof.* We prove the proposition for non-negative values of $t$. It follows from Theorem 2.7.2 that the modulus is a differentiable function with respect to

$B$ and that the vector $\nabla m(B)$ is directed tangentially to the trajectory of the differential $\psi_4$ at the point $B$. By Proposition 4.3.2 this vector is directed into the half-plane $\operatorname{Im} w\bar{B} < 0$. The tangent vector to the circle $|B|e^{it}$ at the point $B$ is directed into the half-plane $\operatorname{Im} w\bar{B} > 0$. Therefore, their scalar product is negative. This leads to the statement of Proposition 4.3.3.    □

The following Propositions can be proved analogously.

**Proposition 4.3.4.** *In the problem I), if $j = 1$, then the arc of the critical trajectories of the differentials $\psi_1(w)dw^2$, $\psi_2(w)dw^2$ connecting $A$ and $B$ are situated in the half-plane $\operatorname{Im} w\bar{B} < 0$.*

**Proposition 4.3.5.** *In the problem I), if $\alpha > 0$, then the modulus $m_k(|B|e^{it})$ is a decreasing function in $t \in [0, \pi]$ and an increasing function in $t \in [-\pi, 0]$, $k = I, II$.*

### 4.3.2 Extremal problems

We construct the Teichmüller maps $f_j$ by the differentials $\psi_1, \ldots, \psi_4$ as in Section 4.2.2, i.e., $K$-quasiconformal automorphisms defined on $U$ satisfying the Beltrami equation 5.1 with the complex dilatation

$$\mu_{f_j}(\zeta) = k\frac{\overline{\varphi_j(\zeta)}}{|\varphi_j(\zeta)|}, \; k = \frac{K-1}{K+1}, \; j = 1, 2, 3, 4,$$

keeping the points $z = 0, 1$ motionless. Obviously, $f_j \in U_K$. These maps are extremal with respect to the problems of the extremal partition established by the differentials (4.20–4.23).

The following remark concerns with the continuous dependence of $f_1, \ldots, f_4$ on the parameter $\alpha$. We have $f_1|_{\alpha=\alpha_1} = f_2|_{\alpha=\alpha_1}$, $f_2|_{\alpha=\alpha_2} = f_3|_{\alpha=\alpha_2}$, $f_3|_{\alpha=\alpha_3} = f_4|_{\alpha=\alpha_3}$.

Theorem 4.1.1 yields that the unique map $f_1\Big|_{\alpha=\alpha_0}$ gives the absolute minimum and $f_4\Big|_{\alpha=1}$ gives the absolute maximum $|f(r_1)|$ in the class $U_K$.

The continuous properties of $f_1, \ldots, f_4$ on $\alpha$ imply that for any real $c$ with

$$\min_{f \in Q_K} |f(r_1)| = |f_1(r_1)|_{\alpha=\alpha_0} \leq c \leq |f_4(r_1)|_{\alpha=1} = \max_{f \in Q_K} |f(r_1)|$$

there are such $j$ and $\tilde{\alpha}$ that $|f_j(r_1)|_{\alpha=\tilde{\alpha}} = c$. Moreover, the Teichmüller maps $f_j$ act homeomorphically in the segment $[-1, 1]$. Hence, $|f_j(r_1)| = f_j(r_1) < |f_j(r_2)| = f_j(r_2)$, $j = 1, \ldots, 4$.

**Theorem 4.3.1.** *For any $f \in U_K$ there are unique pair $\tilde{\alpha}$ and $j = 1, \ldots, 4$ such that*

$$|f(r_1)| = f_j(r_1)\Big|_{\alpha=\tilde{\alpha}}, \text{ and } |f(r_2)| \leq f_j(r_2)\Big|_{\alpha=\tilde{\alpha}}.$$

*The equality is attained only for $f = f_j\Big|_{\alpha=\tilde{\alpha}}$.*

*Proof.* We choose $j$ and $\tilde{\alpha}$ such that $|f(r_1)| = f_j(r_1)\Big|_{\alpha=\tilde{\alpha}} = c$ for a function $f \in U_K$. Let, for instance, $j = 4$, $\tilde{\alpha} \in [\alpha_4, 1]$. Then, we generate the problem of the extremal partition induced by the differential $\varphi_4(z)dz^2$ with the modulus $m$ for the admissible system of curves $\gamma = (\gamma_1, \dot{\gamma_2})$ with the homotopy defined by the slit along $[0, r_1]$. Now we suppose the contrary. Let $f \in U_K$ and suppose that with $f_4$ as above $|f(r_2)| \geq f_4(r_2)\Big|_{\alpha=\tilde{\alpha}}$, but that $f(z)$ is not identical with $f_4$. The map $e^{-i\arg f(r_1)}f$ defines a system of curves $e^{-i\arg f(r_1)}f(\gamma) = (e^{-i\arg f(r_1)}f(\gamma_1), e^{-i\arg f(r_1)}f(\gamma_2))$ which is an admissible system $\gamma^{(k)} = f(\gamma)$ on $U' = U \setminus \{c, f(r_2)e^{-i\arg f(r_1)}, 0\}$ for some $k$. Then, $m/K < m_f < m(U', \Gamma^{(k)}, \tilde{\alpha})$, and by Proposition 4.3.1, $m/K < m_f < m(U', \Gamma^{(1)}, \tilde{\alpha})$. By Proposition 4.3.3 we have the inequality

$$m(U', \Gamma^{(1)}, \tilde{\alpha}) \leq m(U \setminus \{c, |f(r_2)|, 0\}, \Gamma^{(1)}, \tilde{\alpha}).$$

In the latter modulus its gradient at the point $|f(r_2)|$ is directed to $0$ and

$$m(U \setminus \{c, |f(r_2)|, 0\}, \Gamma^{(1)}, \tilde{\alpha}) \leq m(U \setminus \{c, |f_4(r_2)|_{\alpha=\tilde{\alpha}}, 0\}, \Gamma^{(1)}, \tilde{\alpha}).$$

For $j = 1, 2$ we use Propositions 4.3.4 and for j=3 Proposition 4.3.5. The uniqueness of the pair $(j, \tilde{\alpha})$ follows from the uniqueness of the extremal map for quasiconformal distortion of the modulus as in the proof of Theorem 4.2.3. Thus, the theorem is proved completely.     □

Assume $w_k(t) = f_j(r_k)\Big|_{\alpha=t}$ where $j = 1$ for $t \in [\alpha_0, \alpha_1]$, $j = 2$ for $t \in [\alpha_1, \alpha_2]$, $j = 3$ for $t \in [\alpha_2, \alpha_3]$, $w_k(t) = f_4(r_k)\Big|_{\alpha=t-\alpha_3+\alpha_4}$ for $t \in [\alpha_3 + 1 - \alpha_4]$; $k = 1, 2$.

**Theorem 4.3.2.** *Let $f \in U_K$. The upper boundary curve for the range of the functional $I(f) = (|f(r_1)|, |f(r_2)|)$, $0 < r_1 < r_2 < 1$, i.e., the curve of $\max_{f \in U_K} |f(r_2)|$ for $|f(r_1)|$ fixed, is parameterized by $(w_1(t), w_2(t))$ in $t \in [\alpha_0, \alpha_3 + 1 - \alpha_4]$. This curve is smooth and, being considered in $\mathbb{R}^2$, increases in $t \in [\alpha_0, \alpha_3]$ and decreases in $t \in [\alpha_3, \alpha_3 + 1 - \alpha_4]$. The normal vector to this curve is vertical at the point $t = \alpha_3$ and is horizontal at the points $t = \alpha_0, \alpha_3 + 1 - \alpha_4$.*

The proof of this theorem is similar to the proof of Theorem 4.2.4. Also, the following estimates are valid.

**Theorem 4.3.3.** *Let $f \in U_K$, $0 < r_1 < r_2 < 1$. If $|f(r_1)| < |f(r_2)|$, then*

$$K\Phi\left(\frac{1 - |f(r_2)|}{1 - |f(r_1)|}\sqrt{\frac{|f(r_1)|}{|f(r_2)|}}\right) \geq \Phi\left(\frac{1 - r_2}{1 - r_1}\sqrt{\frac{r_1}{r_2}}\right).$$

*If $|f(r_1)| > |f(r_2)|$, then*

$$K\Phi\left(\frac{1-|f(r_1)|}{1-|f(r_2)|}\sqrt{\frac{|f(r_2)|}{|f(r_1)|}}\right) \geq \Phi\left(\frac{1-r_1}{1-r_2}\sqrt{\frac{r_2}{r_1}}\right).$$

**Theorem 4.3.4.** *Let* $f \in U_K$, $0 < r_1 < r_2 < 1$. *If* $|f(r_1)| < |f(r_2)|$, *then*

$$K\Phi\left(\frac{1+|f(r_2)|}{1+|f(r_1)|}\sqrt{\frac{|f(r_1)|}{|f(r_2)|}}\right) \geq \Phi\left(\frac{1+r_2}{1+r_1}\sqrt{\frac{r_1}{r_2}}\right).$$

Results of this chapter were obtained in [156], [157], [159], [161].

# 5. Moduli on Teichmüller Spaces

The main purpose of this chapter is to study the moduli on a Riemann surface varying the conformal structure of this surface. Thus, we will consider the modulus as a functional on a Teichmüller space. The variational formulae will be derived. By this variations we will deduce harmonic properties of the modulus with respect to the complex parameters of the Teichmüller space.

## 5.1 Some information on Teichmüller spaces

Let $g, n$ be non-negative integers and $S_0$ be a Riemann surface of finite type $(g, n)$, i.e., it is of genus $g$ with $n$ possible punctures. If $3g - 3 + n > 0$, then $S_0$ is of hyperbolic conformal type and its universal cover is conformally equivalent to the unit disk $U$. The covering maps replacing the sheets of the universal covering induce a corresponding Fuchsian group $G_0$ of Möbius automorphisms of $U$. One says that the Fuchsian group $G_0$ uniformizes the Riemann surface $S_0$ and $S_0 = U/G_0$.

The Teichmüller space $T(S_0)$ is the space of analytically finite conformal structures on a topological surface $S_0$ where two are equivalent if there is a conformal map between them which is homotopic to the identity. In other words, it is a quotient space of marked (with respect to the initial surface $S_0$) Riemann surfaces of the same orientation and analytically finite conformal type. Let $S$ be a Riemann surface and $f$ be a homeomorphism of the initial Riemann surface $S_0$ onto $S$. We consider the pair $(S, [f])$ as a *marked Riemann surface* where $[f]$ is the class of all homeomorphisms homotopic to $f$. For the convenience we will denote the marked Riemann surface by $(S, f)$ where $f$ represents $[f]$. The marked Riemann surfaces $(S_1, f_1)$ and $(S_2, f_2)$ are said to be *equivalent* if there is a conformal homeomorphism $h : S_1 \to S_2$ such that the map $f_2^{-1} \circ h \circ f_1$ is homotopic to the identity map on $S_0$. We call an equivalence class of the marked Riemann surfaces the *point of the Teichmüller space* and the collection of all equivalence classes the *Teichmüller space* $T \equiv T(S_0)$ (with respect to the initial surface $S_0$). The equivalence class represented by the marked Riemann surface $(S_0, id)$ is said to be the initial point $0$ of $T(S_0)$. The Teichmüller space $T(S_0)$ is a complex analytic manifold of complex dimension $3g - 3 + n$. For those results we refer the reader to L. V. Ahlfors [8], [9], L. Bers [17] who proved the existence of global co-ordinates (moduli) in the Teichmüller space (see also [71]). For the detail information we also refer to [1], [90], [104].

The Teichmüller space $T$ is a metric, connected, simply connected space with the natural Teichmüller metric $\tau_T(x, y)$. Assume $(S_1, f_1)$ and $(S_2, f_2)$ to be the marked Riemann surfaces. The homeomorphism $f_2 \circ f_1^{-1} : S_1 \to S_2$ generates a homeomorphism of marked Riemann surfaces. There are quasiconformal maps (see e.g. [18]) among all homeomorphisms homotopic to $f_2 \circ f_1^{-1}$. Let $(S_1, f_1)$ represent a point $x \in T(S_0)$ and $(S_2, f_2)$ represent a point $y \in T(S_0)$. Then,

$$\tau_T(x, y) := \frac{1}{2} \inf_f \log \frac{1 + \|\mu_f\|_\infty}{1 - \|\mu_f\|_\infty},$$

where we take $f$ over all quasiconformal homeomorphisms which are homotopic to $f_2 \circ f_1^{-1}$ and $\|\mu_f\|_\infty = \text{ess} \sup_{z \in S_1} |\mu_f(z)| < 1$.

Let $S_0 = U/G_0$ and let $B(G_0)$ be the Banach space of Beltrami differentials $\mu(z)d\bar{z}/dz$ which are invariant with respect to actions from $G_0$, i.e., $\mu(z)$ is a measurable function in $U$ with $\|\mu_f\|_\infty < \infty$ and $\mu(\gamma(z))\overline{\gamma'(z)}/\gamma'(z) = \mu(z)$, $\gamma \in G_0$. Let $D(G_0)$ be the unit ball in $B(G_0)$: $\|\mu_f\|_\infty < 1$. Each $\mu \in D(G_0)$ defines by means of the Beltrami equation a unique quasiconformal automorphism $\omega^\mu : U \to U$ of the unit disk with the normalization $\omega^\mu(\pm 1) = \pm 1$, $\omega^\mu(i) = i$. We denote by $I(G_0)$ the space consisting of the differentials $\mu \in D(G_0)$ so that the map $\omega^\mu$ acts identically on $\partial U$ and call it the *space of trivial Beltrami differentials*. The Teichmüller space $T(S_0)$ can be realized as the Teichmüller space of Fuchsian groups $T(G_0) = D(G_0)/I(G_0)$. Riemann surfaces of the same conformal type $(g, n)$ generate isomorphic Teichmüller spaces, so they can be thought of as a single Teichmüller space $T(g, n)$. However, for convenience we will link the Teichmüller space with some initial Riemann surface. There is a natural holomorphic projection $\Phi : D(G_0) \to T(G_0)$.

Denote by $H^{2,0}(G_0)$ the space of all quadratic holomorphic differentials $q(z)dz^2$ in $U$, invariant with respect to $G_0$: $q(\gamma(z))(\gamma'(z))^2 = q(z)$, $\gamma \in G_0$, say the space of all parabolic 2-forms of weight (-4) for the group $G_0$ with the finite norm $\|q\| := \sup_U (1 - |z|^2)^{-2}|q(z)|$. The projection of $H^{2,0}(G_0)$ onto $S_0$ is the space of quadratic holomorphic differentials $\varphi$ on $S_0$ and in the case of finite Riemann surfaces they have finite integral norm

$$\iint_{S_0} |\varphi(\zeta)| d\sigma_\zeta < \infty.$$

On $B(G_0) \times H^{2,0}(G_0)$ one can define a coupling

$$A(\mu) = \iint_{U/G_0} \mu q \, d\sigma_z.$$

Denote by $N(G_0)$ the space of *locally trivial Beltrami differentials*, i.e., the subspace of $B(G_0)$ that forms the kernel of the operator $A(\mu)$. Then, one can identify the tangent space to $T(G_0)$ at the initial point with the space $H(G_0) := B(G_0)/N(G_0)$. It is natural to relate it to a subspace of $B(G_0)$. Define the Bergmann integral $\Lambda_\mu : \mu \mapsto \Lambda_\mu(z) \in H^{2,0}(G_0)$,

$$\Lambda_\mu = \frac{12}{\pi} \iint\limits_{U/G_0} \frac{\overline{\mu(\zeta)}}{(\zeta - z)^4} d\sigma_\zeta, \quad \mu \in B(G_0).$$

The space $N(G_0)$ is also the kernel of the operator $\Lambda_\mu$. Define the operator $\Lambda_q^* : q \mapsto \Lambda_q^*(z) \in B(G_0)$ as

$$\Lambda_q^*(z) = \frac{\overline{q(z)}}{(1 - |z|^2)^2}, \quad q \in H^{2,0}(G_0).$$

The operator $\Lambda_\mu \circ \Lambda_q^*$ acts identically on $H^{2,0}(G_0)$. Thus, the operator $\Lambda_q^*$ splits the following exact sequence

$$0 \longrightarrow N(G_0) \hookrightarrow B(G_0) \xrightarrow{\Lambda_\mu} H^{2,0}(G_0) \longrightarrow 0.$$

Then, $H(G_0) = \Lambda_q^*(H^{2,0}(G_0)) \cong B(G_0)/N(G_0)$. At any other point of $T(G_0)$ one can obtain the tangent space $H(G)$ by the same observations with respect to the group $G = \chi_\mu(G_0)$ where $\chi_\mu$ is the group isomorphism defined by $\chi_\mu : G_0 \to \omega^\mu \circ G_0 \circ (\omega^\mu)^{-1}$. The coupling $\langle \mu, q \rangle := A(\mu)$ defines the cotangent space $H^{2,0}(G_0)$ where the Hermitian product of Petersson is defined as

$$(q_1, q_2) = \iint\limits_{U/G_0} q_1(z) \frac{\overline{q_2(z)}}{(1 - |z|^2)^2} d\sigma_z.$$

The Kählerian Weil-Petersson metric (see the paper by S. Wolpert [177]) $\{\mu_1, \mu_2\} = \langle \mu_1, \Lambda_{\mu_2} \rangle$ can be defined on the tangent space to $T(G_0)$ that gives a Kählerian manifold structure to $T(G_0)$.

## 5.2 Moduli on Teichmüller spaces

On a Riemann surface $S_0$ of type $(g, n, l)$ (in this section we let $l \neq 0$) we define an admissible system $\gamma = (\gamma_1, \ldots, \gamma_m)$ of curves of type I, II. This system induces a free family $\Gamma$ of the homotopy classes of curves generated by the curves of the admissible system. A major problem considered in this section is this: *given a fixed vector $\alpha$, how the modulus $m(S_0, \Gamma, \alpha)$ changes under the variation of the conformal structure of $S_0$?*

### 5.2.1 Variational formulae

Now our attempt is to vary the complex structure of $S_0$ to observe what happens with the modulus $m(S_0, \Gamma, \alpha)$, i.e., we are going to extend the modulus onto the Teichmüller space $T(S_0)$. Let us define the modulus problem (2.12–2.13) for the free family $\Gamma$ on $S_0$ and a weight vector $\alpha$ fixed. Denote the modulus in this problem by $m(S_0, \Gamma, \alpha)$. Consider a quasiconformal map $w = f(\zeta) : S_0 \to S$ with a complex Beltrami dilatation $\mu_f = f_{\bar\zeta}/f_\zeta$ such that the marked Riemann surface $(S, f)$ represents a point $x$ of the Teichmüller space $T(S_0)$. Now let us define the modulus problem (2.12–2.13) for the admissible system of curves $\Gamma_f = (f(\gamma_1), \ldots, f(\gamma_m))$ with the same weight vector $\alpha$ on the Riemann surface $S$. Define $m_0 = m(S_0, \Gamma, \alpha)$, $m_f = m(S, \Gamma_f, \alpha)$. Let $\varphi_\mu(w)dw^2$ ($w$ is a local parameter on $S$) be the extremal quadratic differential in the modulus problem on $S$, $m_f = \iint_S |\varphi_\mu| d\sigma_w$. Since the homotopy type of the maps from the same equivalence class $x \in T(S_0)$ induces the homotopy type of curves on $S$, the conformal invariance of the modulus implies that the value of $m_f$ does not depend on the choice of $f$. It depends only on the point $x \in T(S_0)$ and we denote it by $m(x)$ where $x$ varies over the whole space $T(S_0)$ for the admissible system $\Gamma$ and the vector $\alpha$ fixed, $m(0) = m_0$. If $\Phi(\mu_1) = \Phi(\mu_2) = x$, then the differential $\varphi_{\mu_1}$ can be obtained from the differential $\varphi_{\mu_2}$ by a conformal transform. The quadratic differential is invariant under a conformal change of parameters. Therefore, the differentials $\varphi_{\mu_1}$, $\varphi_{\mu_2}$ represent one and the same differential $\varphi_x$, $\varphi_0 \equiv \varphi$.

Set

$$\psi(\zeta) = \varphi_\mu(f(\zeta)) f_\zeta^2 \left(1 + \mu(\zeta) \frac{\varphi(\zeta)}{|\varphi(\zeta)|}\right)^2.$$

Then, the differential $\psi(\zeta)d\zeta^2$ is locally integrable on $S_0$ and $\|\psi\|_{L_1} < \infty$.

**Lemma 5.2.1.** *For the parameters introduced above, we have*

$$m_0 \leq \iint_{S_0} |\psi(\zeta)|^{1/2} |\varphi(\zeta)|^{1/2} d\sigma_\zeta.$$

*Proof.* The critical trajectories of the extremal quadratic differential $\varphi(\zeta)d\zeta^2$ split $S_0$ into a family of non-overlapping ring domains $\{D_j\}_{j=1}^s$ and quadrangles $\{D_j\}_{j=s+1}^m$ on $S_0$ ($\cup_{j=1}^m D_j = \overline{S_0}$). These domains are associated with the homotopy classes $\{\Gamma_j\}_{j=1}^s$ of the first type and with the homotopy classes $\{\Gamma_j\}_{j=s+1}^m$ of the second type, respectively. We prove that

$$\iint_{D_j} |\varphi(\zeta)| d\sigma_\zeta \leq \iint_{D_j} |\psi(\zeta)|^{1/2} |\varphi(\zeta)|^{1/2} d\sigma_\zeta.$$

Let $\gamma \in \Gamma_j$ be a trajectory of the differential $\varphi(\zeta)d\zeta^2$; we assume that $\Gamma_j$ is of the first type, i.e., $1 \leq j \leq s$. If $\zeta \in \gamma$, then $\varphi(\zeta)d\zeta^2 \geq 0$ and $\varphi(\zeta)/|\varphi(\zeta)| =$

$d\bar{\zeta}/d\zeta$ for non-singular $\zeta$. The closed curve $f(\gamma)$ is admissible in the modulus problem for $m_f$, and we have the following chain of relations:

$$\alpha_j = \int_\gamma \sqrt{|\varphi(\zeta)|}|d\zeta| \le \int_{f(\gamma)} \sqrt{|\varphi_\mu(w)|}|dw| =$$

$$= \int_\gamma \sqrt{|\varphi_\mu(f(\zeta))|}|f_\zeta d\zeta + f_{\bar{\zeta}}d\bar{\zeta}| =$$

$$= \int_\gamma \sqrt{|\varphi_\mu(f(\zeta))|}|f_\zeta|\left|1 + \mu(\zeta)\frac{d\bar{\zeta}}{d\zeta}\right| = \int_\gamma \sqrt{|\psi(\zeta)|}|d\zeta|.$$

We slit the doubly connected domain $D_j$ along a critical orthogonal trajectory going from a simple pole of the differential to the exterior boundary component of $D_j$ and denote by $\tilde{D}_j$ the simply connected domain obtained. The parameter

$$z = \xi + i\eta = \frac{1}{\alpha_j}\int^\zeta \sqrt{\varphi(\zeta)}d\zeta$$

is natural (with respect to the quadratic differential $\varphi$). After an appropriate normalization, $z = z(\zeta)$ conformally maps the domain $\tilde{D}_j$ onto a rectangle $R$ in the $z$-plane; this rectangle is of length 1 and of height $M(D_j)$, where $M(D_j)$ is the modulus of the ring domain $D_j$ with respect to the family of curves that separate the boundary components of $D_j$. The differentials $\varphi(\zeta)d\zeta^2$ and $\psi(\zeta)d\zeta^2$ are invariant under the change of the local parameter on the surface $S_0$. Consequently, $\varphi(\zeta)d\zeta^2 = \varphi(z)dz^2$, $\psi(\zeta)d\zeta^2 = \psi(z)dz^2$, and in terms of the local parameter $z$ we have $\varphi(z) \equiv \alpha_j^2$. Thus,

$$\iint_{D_j} |\varphi(\zeta)|d\sigma_\zeta = \alpha_j^2 \int_0^{M(D_j)} d\eta \int_0^1 d\xi \le \alpha_j^2 \int_0^{M(D_j)} \left(\int_0^1 \frac{\sqrt{|\psi(z)|}}{\alpha_j}d\xi\right) d\eta.$$

Passing to the double integral, we obtain

$$\iint_{D_j} |\varphi(\zeta)|d\sigma_\zeta \le \alpha_j \iint_R \sqrt{|\psi(z)|}d\xi d\eta.$$

Changing the local parameter $z$ and using the identity

$$\sqrt{|\psi(z(\zeta))|}\,|z_\zeta'| = \sqrt{|\psi(\zeta)|},$$

we conclude that

$$\iint\limits_{D_j} |\varphi(\zeta)| d\sigma_\zeta \le \alpha_j \iint\limits_{D_j} \sqrt{|\psi(z(\zeta))|}\, |z'_\zeta|^2 d\sigma_\zeta =$$

$$= \alpha_j \iint\limits_{D_j} \sqrt{|\psi(\zeta)|}\, |z'_\zeta| d\sigma_\zeta =$$

$$= \iint\limits_{D_j} |\psi(\zeta)|^{1/2} |\varphi(\zeta)|^{1/2} d\sigma_\zeta.$$

If $j = (s+1), \dots, m$, the above arguments remain valid, and the domain $D_j$ need not be slit. Summing the resulting inequalities from 1 to $m$, we prove the lemma. $\qquad\square$

We remind that $\Phi$ is a projection $D(G_0) \to T(G_0) = T(S_0)$ and $H(G_0)$ is a tangent space to $T(S_0)$ at the initial point. The following theorem gives us a variational formula for the functional $m(x)$– the modulus extended to $T(S_0)$.

**Theorem 5.2.1.** *Let $x$ be a point of $T(S_0)$, $t = \inf \|\mu\|_\infty$ for all $\mu \in D(G_0)$ such that $\Phi(\mu) = x$. Then*

$$m(x) = m(0) - 2\mathrm{Re} \iint\limits_{U/G_0} \mu^*(z) q(z) d\sigma_z + o(t),$$

*where $\mu^* \in H(G_0)$, $\|\mu^*\|_\infty = t$; $q(z)dz^2$ is a pull-back of the extremal quadratic differential $\varphi(\zeta)d\zeta^2$ for the modulus $m(0)$ onto the universal covering $U$.*

*Proof.* By the well-known Teichmüller theorem on the minimum of the maximal deviation from a conformal map (see Section 5.1), there exists a quasiconformal map $f$ such that $\|f_{\bar\zeta}/f_\zeta\|_\infty = t$. The preceding lemma shows that

$$m(0) = \iint\limits_{S_0} |\varphi(\zeta)| d\sigma_\zeta \le \iint\limits_{S_0} \sqrt{|\varphi(\zeta)\psi(\zeta)|}\, d\sigma_\zeta =$$

$$= \iint\limits_{S_0} \sqrt{|\varphi_\mu(f(\zeta))|}|f_\zeta|\left|1 + \mu\frac{\varphi}{|\varphi|}\right|\sqrt{|\varphi|}\, d\sigma_\zeta,$$

and we use the Schwarz inequality to obtain

$$m(0) \le \left( \iint\limits_{S_0} |\varphi_\mu(\omega)| d\sigma_\omega \right)^{1/2} \left( \iint\limits_{S_0} |\varphi| \frac{\left|1 + \mu\frac{\varphi}{|\varphi|}\right|^2}{1 - |\mu|^2} d\sigma_\zeta \right)^{1/2}.$$

Squaring the two sides yields the relation

$$m(0) \leq m(x) \iint_{U/G_0} |q(z)| \frac{\left|1 + \mu(z)\frac{q(z)}{|q(z)|}\right|^2}{1 - |\mu(z)|^2} d\sigma_z,$$

where $z \in U$ (for convenience, we use the same notation for the complex dilatation on the surface and on its universal covering). Here we have made the change of variables $\mu(z) = \mu(J(z))\overline{J'(z)}/J'(z)$, $q(z) = \varphi(J(z))J'^2(z)$, where $\zeta = J(z)$ is a global uniformizing function, i.e., an automorphic function with the automorphism group $G_0$ (see Section 4.1.1). Dividing two sides by the double integral and expanding in powers of $\mu$, we see that

$$m(x) \geq m(0) - 2\mathrm{Re} \iint_{U/G_0} \mu^*(z)q(z)\, d\sigma_z + O(t^2). \tag{5.1}$$

To obtain the reverse inequality, we apply the same arguments to the inverse function $\zeta = f^{-1}(w)$. Calculation of the dilatation $\mu_1 = -\mu \cdot f_\zeta/\overline{f_\zeta}$ of $f^{-1}$ gives the inequality

$$m(0) \geq m(x) - 2\mathrm{Re} \iint_{S} \varphi_\mu\, \mu_1^*\, d\sigma_w + O(t^2),$$

or

$$m(x) \leq m(0) - 2\mathrm{Re} \iint_{S_0} \varphi_\mu(f(\zeta))f_\zeta^2\, \mu^* d\sigma_\zeta + O(t^2). \tag{5.2}$$

In order to show that

$$\mathrm{Re} \iint_{S_0} (\varphi_\mu(f(\zeta))f_\zeta^2 - \varphi(\zeta))\mu^*\, d\sigma_\zeta = o(t), \tag{5.3}$$

we construct a linear functional on $B(G_0)$ by using the locally integrable quadratic differential

$$q_\mu(\omega(z))\omega_z^2 - q(z) = \hat{q}, \quad z \in U,$$

where $\omega(z)$ is a quasiconformal automorphism of $U$ with the dilatation

$$\mu(z) = \frac{f_\zeta}{f_\zeta} \circ (J(z))\overline{J'(z)}/J'(z)$$

and the normalization $\omega(\pm 1) = \pm 1$, $\omega(i) = i$. For every $\nu \in B(G_0)$ and its differential element $\nu^* \in H(G_0)$ we have

$$\mathrm{Re} \iint_{U/G_0} \hat{q}\, \nu\, d\sigma_z = \mathrm{Re} \iint_{U/G_0} \hat{q}\, \nu^*\, d\sigma_z + o(\|\nu\|_\infty).$$

Setting $\nu = \mu$, we see that (5.3) will be proved if we show that

$$\lim_{\|\mu\| \to 0} \left| \iint\limits_{U/G_0} (q_\mu(\omega(z))\omega_z^2 - q(z))\mu^* d\sigma_z \right| = 0.$$

This relation is a consequence of the absolute continuity of the Lebesgue integral and the following facts: $\omega(z)$ tends to $z$ locally uniformly in $U$; $\omega_z$ tends to 1 locally in $L^2$; $q_\mu$ tends to $q$ locally uniformly in $U$ (by virtue of the convergence $\|\varphi_\mu\| \to \|\varphi\|$ and the uniqueness of the limit differential).

Obviously, the assertion of Theorem 5.2.1 follows from (5.1–5.3), that completes the proof. □

The following corollary is a differential formula for the modulus of a family of curves with respect to its parameters; this formula was proved earlier and by different methods by A. Solynin and E. Emel'yanov (see Theorem 2.7.2).

**Corollary 5.2.1.** *Let $g = 0$, $l = 0$, $w_j = f^\mu(\zeta_j)$, $j = 1, \ldots, n$; further, let $\zeta_{n-2} = w_{n-2} = 0$, $\zeta_{n-1} = w_{n-1} = 1$, $\zeta_n = w_n = \infty$. Then*

$$\left. \frac{\partial m_\mu}{\partial w_j} \right|_{\mu=0} = \pi \operatorname{Res}_{\zeta = \zeta_j} \varphi(\zeta), \quad j = 1, \ldots, (n-3).$$

*Proof.* If $\mu^* \in H(G_0)$, then there exists a unique cusp 2-form $q^*(z) \in H^{2,0}(G_0)$ of weight (-4) such that $\mu^* = (1 - |z|^2)^{-2}\overline{q^*(z)}$. Hence,

$$\iint\limits_{U/G_0} \mu^* q \, d\sigma_z = \iint\limits_{U/G_0} q \, (1 - |z|^2)^{-2}\overline{q^*(z)} \, d\sigma_z = (q, q^*),$$

where $(\cdot, \cdot)$ is the Petersson scalar product on $H^{2,0}(G_0)$. Let $R_j$, $j = 1, \ldots, (n-3)$, be the pull-backs to $U$ of the linearly independent vectors

$$\frac{1}{\pi} \frac{\zeta_j(\zeta_j - 1)}{\zeta(\zeta - 1)(\zeta - \zeta_j)}.$$

In $H^{2,0}(G_0)$, we consider the basis $\{q_j\}_{j=1}^{n-3}$ biorthogonal with $\{R_j\}$ relative to the Petersson scalar product, i.e.,

$$(q_j, R_k) = \begin{cases} 0, & \text{for } j \neq k, \\ 1, & \text{for } j = k. \end{cases}$$

Then $\{\nu_j : \nu_j = (1 - |z|^2)^2 \overline{q_j}\}$ is the Bers basis in $H(G_0)$. The punctures $w_j$ can be regarded as local parameters on $T(0, n)$. Now the assertion follows from the Petersson differentiation formulae $\partial/\partial w_j = (\cdot, q_j)$ and $(q, q_j) = -\pi \operatorname{Res}_{w=w_j} \varphi(w)$ (see, e.g., [89]). □

The problem of the extremal partition of $S_0$ according to an admissible system of curves $\gamma = (\gamma_1, \ldots, \gamma_m)$ leads to the functional $m(S_0, \Gamma, \alpha)$ for each vector $\alpha$ given. Given a vector $b = (b_1, \ldots, b_m)$ we consider another problem of the extremal partition following K. Strebel (Propositions 2.7.2, 2.7.3) where we fix the widths of the domains swept by the curves of the corresponding homotopy classes. This implies that an extremal domain never degenerates for a positive width.

Now we construct the holomorphic quadratic differential $\varphi(\zeta)d\zeta^2$ of finite $L^1$ norm with finite trajectories on $S_0$. Let $(D_1, \ldots, D_m)$ be its characteristic system of domains. Let us consider on $S$ the family of non-overlapping doubly connected domains and quadrilaterals $(f(D_1), \ldots, f(D_m))$. The moduli of all components are bounded due to the obvious inequality $M(f(D_j)) \leq KM(D_j)$ where $K = \exp(2\tau_T(0, x))$. In each doubly connected domain $f(D_j)$ we select a simple loop $\gamma_j^x$, separating the boundary components of $f(D_j)$ and in each quadrilateral $f(D_j)$ we select a simple arc connecting its opposite sides on $\partial S$. Then $(\gamma_1^x, \ldots, \gamma_m^x)$ is an admissible system of curves on $S$. By Proposition 2.7.3 for the same vector $b = (b_1, \ldots, b_m)$ we construct a quadratic differential $\varphi(\zeta, x)d\zeta^2$, $\varphi(\zeta, 0) = \varphi(\zeta)$ with finite trajectories, of homotopy type $(\gamma_1^x, \ldots, \gamma_m^x)$. Analogously to $m(x)$ the differential $\varphi(\zeta, x)d\zeta^2$ does not depend on concrete $\mu$ or $f$, but on the point $x \in T(S_0)$. Denote by $s(x) = s_\varphi(x) = \|\varphi(\zeta, x)\|_{L_1}$. We deduce the properties of this functional.

**Proposition 5.2.1.** *Let* $x \in T(S_0)$, $K = \exp(2\tau_T(0, x))$. *Then* $s(0)/K \leq s(x) \leq Ks(0)$.

*Proof.* Let $(D_1^x, \ldots, D_m^x)$ be the characteristic family of domains for the differential $\varphi(\zeta, x)d\zeta^2$, and let $(D_1, \ldots, D_m) := (D_1^0, \ldots, D_m^0)$. Since $D_j^x$ and $f^x(D_j)$ have the same homotopic type, Proposition 2.7.2 implies the inequality

$$s(x) = \|\varphi(\cdot, x)\| = \sum_{j=1}^m \frac{b_j^2}{M(D_j^x)} \leq \sum_{j=1}^m \frac{b_j^2}{M(f^x(D_j))}.$$

On the other hand, $M(f^x(D_j)) \geq K^{-1}M(D_j)$, so

$$\sum_{j=1}^m \frac{b_j^2}{M(f^x(D_j))} \leq K \sum_{j=1}^m \frac{b_j^2}{M(D_j)} = Ks(0).$$

Using the inverse map, we obtain the left-hand side inequality of the proposition. This completes the proof. □

The following theorem is proved much as Lemma 4.1 and Theorem 4.1, by using the corresponding integration along the orthogonal trajectories. For this reason, here we only give the statement.

**Theorem 5.2.2.** *Let* $x$ *be a point of* $T(S_0)$, $t = \inf \|\mu\|_\infty$ *for all* $\mu \in D(G_0)$ *such that* $\Phi(\mu) = x$. *Then,*

$$s(x) = s(0) + 2\mathrm{Re} \iint_{U/G_0} \mu^*(z)q(z)d\sigma_z + o(t),$$

where $\mu^* \in H(G_0)$, $\|\mu^*\|_\infty = t$; $q(z)dz^2$ is a pull-back of the quadratic differential $\varphi(\zeta)d\zeta^2$ for $s(0)$ onto the universal covering $U$.

Theorems 5.2.1, 5.2.2, in particular, yield that $m(x)$ and $s(x)$ are locally harmonic in a neighbourhood of $x$ if and only if the extremal quadratic differential $\varphi_x$ is locally holomorphic in this neighbourhood.

## 5.2.2 Three lemmas

Here and further on we assume the initial Riemann surface to be of type $(g,n)$. First of all, we summarize the relevant information about the spaces of differentials on Riemann surfaces. We refer the reader to [8], [19], [71], [104] for the case of compact Riemann surfaces. Denote by $J : U \to S_0$ the $G_0$-automorphic function, say the global uniformization of $S_0 = U/G_0$, and let $p$ be an integer number. Then, every holomorphic $p$-differential $\varphi(\zeta)d\zeta^p$ on $S_0$ can be lifted to the universal cover $U$ by the rule

$$\varphi(\zeta)d\zeta^p \to q(z)dz^p,$$

where $q(z) = (\varphi \circ J(z))\,(J'(z))^p$.

For simplicity, we denote the linear vector spaces of the differentials indicated on $S_0$ and $U$ by the same symbol $H^{p,0}(S_0)$.

Classical results ([38], [140]) state for $n = 0$ that

$$\dim{}_{\mathbb{C}} H^{p,0}(S_0) = \begin{cases} 0, & \text{for } p < 0, \\ 1, & \text{for } p = 0, \\ g, & \text{for } p = 1, \\ (2p-1)(g-1), & \text{for } p > 1. \end{cases}$$

**Lemma 5.2.2.** For $p \geq 1$ and $n > \max[3 - 3g, 2(p-1)(g-1)]$, the complex dimension of the space $H^{p,0}(S_0)$ is given by the formula

$$\dim{}_{\mathbb{C}} H^{p,0}(S_0) = (2p-1)(g-1) + n.$$

*Proof.* The lemma is a consequence of the Riemann-Roch theorem. Suppose $\mathfrak{a}$ is a divisor on a compact Riemann surface $S$ of genus $g > 1$, i.e., $\mathfrak{a}$ is a finite 0-chain on $S$: $\mathfrak{a} = \sum_{j=1}^k \alpha_j p_j$, where $p_j$ are points of $S$ and $\alpha_j$ are integers. The order of the divisor at $p_j$ is $\mathrm{ord}_{p_j}\mathfrak{a} = \alpha_j$ (note that $\mathrm{ord}_p\mathfrak{a} = 0$ if $p \neq p_j$), and the degree is $\deg \mathfrak{a} = \sum_{j=1}^k \alpha_j$. The sum of two divisors $\mathfrak{a} = \sum_{j=1}^k \alpha_j p_j$ and $\mathfrak{b} = \sum_{j=1}^l \beta_j q_j$ is the 0-chain defined by $\mathfrak{a} + \mathfrak{b} = \sum_{j=1}^k \alpha_j p_j + \sum_{j=1}^l \beta_j q_j$. If $p_i = q_j$, then the coefficients are added. The set of all divisors on $S$ is an Abelian group with respect to the above operation of addition; the unit element is the divisor with zero coefficients, and the inverse element is $(-\mathfrak{a}) = \sum_{j=1}^{k}(-\alpha_j)p_j$. A divisor $\mathfrak{a}$ is called non-negative if $\alpha_j \geq 0$ for all $j = 1, \ldots, k$.

If $\varphi(\zeta)d\zeta^p$ is a holomorphic or meromorphic differential on $S$, then $(\varphi)$ denotes its divisor,

$$(\varphi) = \sum_{r \in S} r \cdot \operatorname{ord}_r\varphi, \quad \deg(\varphi) = \sum_{r \in S} \operatorname{ord}_r\varphi = N - P,$$

where $N$ is the number of zeros and $P$ the number of poles of the differential in question. For $p = 0$ we have $\deg(\varphi) = 0$. Let $[\mathfrak{a}]$ denote the equivalence class of a divisor $\mathfrak{a}$; by definition, $\mathfrak{a} \sim \mathfrak{b}$ if $\deg(\mathfrak{a} - \mathfrak{b}) = 0$, i.e., if there exists a meromorphic function $f$ such that $\mathfrak{a} - \mathfrak{b} = (f)$. Since, obviously, all divisors in $[\mathfrak{a}]$ have equal degrees, we can set $\deg[\mathfrak{a}] := \deg\mathfrak{a}$. Furthermore, let $D^p(\mathfrak{a})$ denote the complex linear space of meromorphic differentials $\varphi(\zeta)d\zeta^p$ on $S$ such that $(\varphi) - \mathfrak{a} \geq 0$. The fundamental Riemann-Roch theorem says that for $p \geq 1$ we have

$$\dim_{\mathbb{C}} D^p(\mathfrak{a}) = \dim_{\mathbb{C}} D^{1-p}(-\mathfrak{a}) + (2p - 1)(g - 1) - \deg[\mathfrak{a}], \quad g \geq 0.$$

Let $S$ be the closure $\overline{S_0}$ of the surface $S_0$, and let $\mathfrak{a} = \sum_{j=1}^{n}(-\zeta_j)$, where $\zeta_j$ are the punctures on $S_0$ and $\sum$ is understood as a formal sum. In this case, $D^p(\mathfrak{a})$ coincides with the space $H^{p,0}(S_0)$. We have $\dim_{\mathbb{C}} D^{1-p}(-\mathfrak{a}) > 0$ if and only if $\deg\mathfrak{a} > 2(p-1)(g-1)$ (see, e.g., [131]). It follows that under the condition imposed on $n$ we have $\dim_{\mathbb{C}} D^p(\mathfrak{a}) = \dim_{\mathbb{C}} H^{p,0}(S_0) = (2p - 1)(g - 1) + n$, that proves the lemma. $\qquad\square$

**Lemma 5.2.3.** *Let* $\varphi(\zeta)d\zeta^p \in H^{p,0}(S_0)$, $\zeta \in S_0$. *Let* $\nu_j$ *denote the order of the zero of* $\varphi$ *or the order of the pole of* $\varphi$ *at* $\zeta_j$ *taken with the sign (-) if* $\zeta_j$ *is a pole of* $\varphi$. *Then the number* $N$ *of zeros of* $\varphi$ *on* $S_0$ *counted with multiplicities is given by*

$$N = -\sum_{j=1}^{n} \nu_j + 2p(g - 1).$$

*Proof.* This readily follows from the fact that the differential $\varphi$ has degree $\deg\varphi = 2p(g - 1)$ on $\overline{S_0}$. Carrying the orders of the differential at the punctures of $S_0$ to the right-hand side, we obtain the assertion of the lemma. $\square$

Again, we consider $S_0 = U/G_0$, where $G_0$ is the uniformizing Fuchsian group with the unit disk $U$ as a connected component of the set of discontinuity. Any fundamental polygon $P$ in $U$ is bounded by segments of non-Euclidean lines and is convex in the non-Euclidean plane, its sides in $U$ are pairwise equivalent. Let $p_1, \ldots, p_r$ be the $J$-projection of the elliptic and parabolic fixed points of $G_0$. The universal covering is branched over $S_0$ at the elliptic points. With each point $p_k$ we associate the number $m_k$ here $(m_k - 1)$ is the order of branching at $p_k$ in the elliptic case or is $\infty$ in the parabolic case. The vector $(g, r; m_1, \ldots, m_r)$ is called the *signature* of the group $G_0$. There are $2g + r$ generators of $G_0$

$$A_1, B_1, \ldots, A_g, B_g, C_1, \ldots, C_r,$$

where $A_j, B_j$ are the hyperbolic generators and $C_k$ the elliptic and parabolic ones of the group $G_0$. They satisfy the relation

$$\begin{cases} C_r \circ \cdots \circ C_1 \circ B_g^{-1} \circ A_g^{-1} \circ B_g \circ A_g \circ \cdots \\ \cdots \circ B_1^{-1} \circ A_1^{-1} \circ B_1 \circ A_1 = \mathrm{id}, \\ C_j^{m_j} = \mathrm{id}, \quad j = 1, \ldots, r \quad \text{for all } m_j \neq \infty. \end{cases} \qquad (5.4)$$

**Lemma 5.2.4.** *For every point $x \in T(S_0)$, there exist a Jordan domain $U(x)$ and $2g + r$ transformations $z \mapsto A_j(z, x)$, $z \mapsto B_j(z, x)$, $z \mapsto C_k(z, x)$, $j = 1, \ldots, g$, $k = 1, \ldots, r$ of $U$ such that*

*1) $\partial U(x)$ is parameterized via a mapping $z = \sigma(e^{i\theta}, x)$, where $0 \leq \theta \leq 2\pi$; $\sigma(e^{i\theta}, 0) = e^{i\theta}$; $\sigma(\cdot, x)$ depends holomorphically on $x$;*

*2) $A_j(\cdot, x)$, $B_j(\cdot, x)$, and $C_k(\cdot, x)$ satisfy the relation (5.4) and depend holomorphically on $z$ and $x$; they generate a quasi-Fuchsian group $G(x)$, so that $G(0) = G_0$ and $S(x) = U(x)/G(x)$ is a Riemann surface;*

*2) every Beltrami differential $\mu$ such that $\Phi(\mu) = x$, induces a quasiconformal map $f^x : S_0 \to S(x)$ such that the equivalence class represented by the marked Riemann surface $(S(x), f^x)$ is the element $x$ of the Teichmüller space $T(G_0)$.*

*Proof.* Let $\mu_x(z) = \Phi^{-1}(x)$ be a Beltrami differential which is compatible with $G_0$. There is a quasiconformal automorphism $w = \Omega_m(z)$ on $\overline{\mathbb{C}}$ with the fixed points $-1, 1$, and $i$, satisfying the Beltrami equation $w_{\bar z} = m_x w_z$, where

$$m_x(z) := \begin{cases} \mu_x(z), \text{ for } z \in U, \\ 0, \quad \text{for } z \in \overline{\mathbb{C}} \setminus U. \end{cases}$$

Set $U(x) := \Omega_m(U)$. If we parameterize $\partial U$ by $(e^{i\theta}, 0 \leq \theta < 2\pi)$, then

$$\partial U(x) = \{w : w = \Omega_m(e^{i\theta}) =: \sigma(e^{i\theta}, x), 0 \leq \theta < 2\pi\}.$$

Since $m_x$ depends holomorphically on $x$, it follows that $\omega_m$ and, hence, $\sigma(e^{i\theta}, x)$ also depens holomorphically on $x$. The mapping $\Omega_m \circ A_j$ satisfies the Beltrami equation

$$(\Omega_m \circ A_j)_{\bar z} = m_x(z)\,(\Omega_m \circ A_j)_z.$$

Therefore, there is a mapping $A_j(z, x)$ of the domain $U(x)$ onto itself that depends holomorphically on $z$ and is such that $A_j(\cdot, x) \circ \Omega_m \equiv \Omega_m \circ A_j(\cdot)$. Similarly, starting with $B_j$ and $C_j$, we obtain transformations $B_j(z, x)$, $C_k(z, x)$. These transformations form a quasi-Fuchsian group $G(x) = \Omega_m \circ G_0 \circ \Omega_m^{-1}$ acting on $U(x)$. The hyperbolic, elliptic, and parabolic elements of $G_0$ are transformed into hyperbolic, elliptic, and parabolic elements of $G(x)$ (see [96]). Obviously, $A_j(z,x)$, $B_j(z,x)$, and $C_k(z,x)$ satisfy the relation (5.4) and depend holomorphically on $x$. The quotient $U(x)/G(x)$ is a Riemann surface $S(x)$.

Let $J^x$ stand for the projection $U(x) \to S(x)$. We define a quasiconformal map $f^x : S_0 \to S(x)$ by the formula $f^x(\zeta) := J \circ \Omega_m \circ (J^x)^{-1}(\zeta)$, $\zeta \in S_0$. Two different Beltrami differentials $\mu_1$ and $\mu_2$, for which $\Phi(\mu_1) = \Phi(\mu_2) = x$, define the maps $\Omega_1 = \Omega_{m_1}$ and $\Omega_2 = \Omega_{m_2}$, the corresponding domains $U_1(x)$ and $U_2(x)$, the quasi-Fuchsian groups $G_1(x)$ and $G_2(x)$ acting on these domains, and the marked Riemann surfaces $(S_1(x), [f_1^x])$ and $(S_2(x), [f_2^x])$. We consider the conformal map given by $H : U_1(x) \to U_2(x)$ with the fixed boundary points $-1$, $i$, and $1$. The definition of $T(S_0)$ implies that the continuous extension of the map $\Omega_2^{-1} \circ H \circ \Omega_1$ onto $\partial U$ is identical on $\partial U$. This means that, for the conformal map $h := J_2^x \circ H \circ (J_1^x)^{-1}: S_1(x) \to S_2(x)$, the map $(f_2^x)^{-1} \circ h \circ f_1^x$ is homotopic to the identity on of $S_0$, and the marked Riemann surfaces $(S_1(x), [f_1^x])$ and $(S_2(x), [f_2^x])$ are equivalent and represent the same $x$. This completes the proof of Lemma 5.2.4.    □

We will need the following statement.

**Proposition 5.2.2** (Gardiner [43], Lemma 1, p.192; Strebel [141], Theorem 2.6, page 8). *The maximal number of elements in an admissible system of curves of type I on a Riemann surface of finite conformal type $(g, n)$ is $3g - 3 + n$.*

## 5.3 Harmonic properties of the moduli

Suppose $g = 0$, i.e., we consider $S_0$ to be the $n$-punctured Riemann sphere, $n \geq 4$ (of course, one can achieve the normalization $S_0 = \mathbb{C} \setminus \{0, 1, \zeta_1, \ldots, \zeta_{n-3}\}$ by a Möbius transform). An interesting question arises when we move locally the punctures of $S_0$. Namely, *how the value of the modulus depends on the values of these punctures.* Generally $(g \geq 0)$ *how the modulus $m(S_0, \Gamma, \alpha)$ changes under the variation of the conformal structure of $S_0$.* It is a simple consequence from ([78], Theorem 5.2) that if $g = 0$, $n = 4$, then the modulus is a continuous and differentiable function of $\zeta_1$. Then, A. Yu. Solynin and E. G. Emel'yanov (see Theorems 2.7.1, 2.7.2 and references thereafter) proved that for any $n$ $(g = 0)$ the modulus is a differentiable function of the parameters $(\zeta_1, \ldots, \zeta_{n-3})$.

The next step has been made by the author [155] who proved that for $n = 5$ the modulus is a locally pluriharmonic function whenever the extremal differential $\varphi$ has exactly two non-degenerate ring domains in its trajectory structure. In the case of one degenerate ring domain the counterexample has been constructed by A. Yu. Solynin (see his construction in [151], and Theorem 5.3.1, the example thereafter). In [158] we conjectured that for any finite $g, n$, $m = 3g - 3 + n > 0$ the modulus is a locally harmonic function with respect to the parameters of conformal structures of $S_0$ when the extremal differential $\varphi$ has the maximal $(3g - 3 + n)$ number of non-degenerate ring domains in its trajectory structure. Later, this has been proved by E. G. Emel'yanov [37] and by another method in [170].

We are aimed at local harmonic properties of $m(x)$ on the Teichmüller space $T(S_0)$ with the initial Riemann surface $S_0$ of genus $g$ with $n$ possible punctures such that $3g + n - 3 > 0$.

**Theorem 5.3.1.** *Let* $m = 3g-3+n$. *Fix an admissible system of* $m$ *curves* $\Gamma$ *and a weight vector* $\alpha = (\alpha_1, \ldots, \alpha_m)$ *with positive coordinates* $\alpha_j$. *Construct the modulus* $m(S_0, \Gamma, \alpha)$. *Suppose that the extremal holomorphic quadratic differential* $\varphi_0$ *has exactly* $m$ *non-degenerate ring domains in its trajectory structure. Then, there exists a neighbourhood* $\Omega$ *of the initial point of the Teichmüller space* $T(S_0)$ *such that the extension of the modulus* $m(x)$ *is harmonic in* $\Omega$.

*Proof.* The differential $\varphi_0$ has exactly $m$ non-degenerate ring domains in its trajectory structure that is the maximal number due to Proposition 5.3.1. There is a neighbourhood $\Omega$ of the origin of $T(S_0)$ such that for all $x \in \Omega$ the differential $\varphi_x$ also has $m$ non-degenerate ring domains in its trajectory structure. As it has been remarked at the end of Section 5.2.1, it suffices to prove that the extremal differential $\varphi_x$ depends holomorphically on $x$ in $\Omega$. Let us denote by $q(x, z)dz^2$ the pull-back of the differential $\varphi_x(\zeta)d\zeta^2$ onto the universal cover $U(x)$ of the surface $S(x)$ (here we use notations introduced in Lemma 5.2.4). Lemma 5.2.2 implies that this differential can be represented as

$$q(x, z)dz^2 = \sum_{j=1}^{m} \beta_j(x)p_j(x, z)dz^2, \qquad (5.5)$$

where the differentials $p_j(x, z)dz^2$, $j = 1, \ldots, m$ form the basis of $H^{2,0}(S(x))$, and they are holomorphic with respect to $x$. Thus, it suffices to prove that the coefficients $\beta_j(x)$ are also holomorphic with respect to $x$. We denote by $q(z) \equiv q(z, 0)$, $p_j(z) \equiv p_j(z, 0)$, $D_k^* \equiv D_k^*(0)$. We know (see Section 2.7) that in each non-degenerate ring domain $D_k^*(x)$, $k = 1, \ldots, m$, in the trajectory structure of the differential $\varphi_x(\zeta')(d\zeta')^2$, this differential is represented as

$$\varphi_x(\zeta')(d\zeta')^2 = -\frac{\alpha_k^2 dw^2}{4\pi^2 w^2}, \qquad (5.6)$$

where the parameter $w$ is defined in the annulus $K_k(x) = \{1 < |w| < R_k(x)\}$, $M(D_k^*(x)) = \frac{1}{2\pi} \log R_k(x)$, and the parameter $\zeta'$ is defined in $D_k^*(x)$. Now we consider the equality (5.6) at the initial point $x = 0$ and rewrite it as

$$\varphi(\zeta)d\zeta^2 = -\frac{\alpha_k^2 dv^2}{4\pi^2 v^2}, \qquad (5.7)$$

where the parameter $v$ is defined in the annulus $K_k = \{1 < |v| < R_k\}$, $M(D_k^*) = \frac{1}{2\pi} \log R_k$, and the parameter $\zeta$ is defined in $D_k^*$. Among all homotopic homeomorphisms $f : S_0 \to S(x)$, that represent the point $x \in T(S_0)$, there is one that maps each $D_k^*$ onto $D_k^*(x)$. We denote it by $f^*$. Its dilatation $\mu_x^*$ holomorphically depends on $x$. Then the superposition $w \circ f^* \circ \zeta(v)$ maps the ring $K_k$ onto the ring $K_k(x)$ as $K_k \to D_k^* \to D_k^*(x) \to K_k(x)$. It satisfies in $K_k$ the differential equation $dw^2/w^2 = dv^2/v^2$. Changing variables by (5.6, 5.7) and applying the representation (5.5) we have

$$\sum_{j=1}^m \beta_j(x) p_j(x,z) dz^2 = -\frac{\alpha_k^2 w_\zeta^2}{4\pi^2 w^2} \left(1 + \mu_x^*(\zeta) \frac{d\bar\zeta}{d\zeta}\right)^2 d\zeta^2$$

in each domain $D_k^*$.

Now we fix $3g - 3 + n$ points $v_k$ in annuli $K_k$ and obtain $3g - 3 + n$ images $z_k(x)$ in $U(x)$ which are holomorphic with respect to $x$. We choose the points $v_k$ so that their images $z_k(x)$ are different from the Weierstrass points of order 2 (in Petersson's sense, see [19]) of the surface $S(x)$. All functions $d\zeta/dz$, $w \circ v(\zeta)$, $w_\zeta$ are holomorphic with respect to $x$. Therefore, we have obtained a system of $m$ equations with respect to $m$ unknowns $\beta_j(x)$ with coefficients which are holomorphic with respect to $x$.

In order to prove that the obtained system has a unique solution, we note that since the points $z_k(x)$ are different from the Weierstrass points, the Wronskian for the basis of $H^{2,0}(S(x))$ with respect to these points does not vanish and the columns

$$[p_j(z_1(x), x), \ldots, p_j(z_m(x), x)]', \quad j = 1, \ldots, m$$

are linearly independent. Therefore, the system has a unique solution for each $x$ which is holomorphic in $x \in \Omega$. This ends the proof.     $\square$

Remark 5.3.1. Under the conditions of Theorem 5.3.1 the modulus $m(x)$ is locally harmonic in all points of $T(S_0)$ whenever the extremal differential $\varphi_x$ has $3g - 3$ (the maximal number by Proposition 5.2.2) non-degenerate ring domains in its trajectory structure.

However, we can not speak about global pluriharmonicity of $m(x)$. A counterexample by A. Solynin shows this.

Counterexample. Let $\Gamma$ be a homotopy class of simple loops that separate the points $0, \infty$ and $a, \exp(i\beta), \exp(-i\beta)$ where $|a - 1| < \varepsilon$, $\arccos(1 - \varepsilon^2/2) < \beta < \pi/2$, that are homotopic on $\mathbb{C} \backslash \{a, \exp(i\beta), \exp(-i\beta)\}$ to the slit along the ray $[0, \infty]$. Varying $a$ in the neighbourhood of 1 we see that the modulus of $\Gamma$ attains its maximum at the whole arc of the unit circle $\exp(it)$, $|t| < \arccos(1 - \varepsilon^2/2)$ (the Mori domain). This contradicts the maximum modulus property for harmonic functions.

Now let $\Gamma = (\gamma_1, \ldots, \gamma_m)$ be an admissible system of $m = 3g - 3 + n$ curves on a Riemann surface $S_0$ and we choose a vector $\alpha = (\alpha_1, \ldots, \alpha_m)$ so that the extremal quadratic differential $\varphi(\zeta)d\zeta^2$ contains $m$ non-degenerate ring domains in its trajectory structure. We denote by $m_0 = m(S_0, \Gamma, \alpha)$ the modulus in the corresponding modulus problem and let $m(x)$ be its extension onto the Teichmüller space $T_0$ which is harmonic in a neighbourhood $\Omega$ of the origin of $T(S_0)$. Now we define another admissible system of curves $\widetilde{\Gamma} = (\gamma_1, \ldots, \gamma_k) \subset \Gamma$, $k < m$, and with the vector $\widetilde{\alpha} = (\alpha_1, \ldots, \alpha_k)$ let us define the modulus problem and denote by $\widetilde{m}(x)$ the extension of the modulus $\widetilde{m}_0 = m(S_0, \widetilde{\Gamma}, \widetilde{\alpha})$. Then the metric $\sqrt{|\varphi(\zeta)|}|d\zeta|$ is admissible in the latter modulus problem with the modulus $\widetilde{m}_0$ as well as the metric $\sqrt{|\varphi_x(\zeta')|}|d\zeta'|$ is admissible in the modulus problem with the modulus $\widetilde{m}(x)$. Therefore, the inequality $\widetilde{m}(x) \leq m(x)$ holds in $\Omega$. Reciprocally, if an admissible system is incomplete, then we always can construct a complete one so that the corresponding modulus for the complete system is harmonic. Thus we come to the following result.

**Corollary 5.3.1.** *The extension $m(x)$ of the modulus $m_0 = m(S_0, \Gamma, \alpha)$ is a subharmonic function in the whole Teichmüller space $T(S_0)$ for any admissible system $\Gamma$ and a weight vector $\alpha$.*

*Remark 5.3.2.* Let us fix an admissible system $\Gamma$ and a weight vector $\alpha$. The maximum principle for subharmonic functions asserts that $m(x)$ never attains its maximum in $T(S_0)$. We observe now that $m(x)$ has no local extrema in $T(S_0)$. Indeed, let us consider the coupling $\langle \mu^*, q \rangle$ in the variational formula in Theorem 2.2. If we fix $\mu^* \in L^\infty(U, G_0)$ and $\langle \mu^*, q \rangle = 0$ for all $q \in Q(G_0)$, then $\mu^* \in N(G_0)$ and represents the origin of $H(G_0) = L^\infty(U, G_0)/N(G_0)$, $\dim_{\mathbb{C}} H(G_0) = 3g - 3 + n$. The complex dimension of the cotangent space $Q(G_0)$ is also $3g - 3 + n$. This means, that if we fix $q \in Q(G_0)$ and $\langle \mu^*, q \rangle = 0$ for all $\mu^* \in H(G_0)$, then $q \equiv 0$. But $q(z)dz^2$ is the pull-back of the extremal quadratic differential $\varphi(\zeta)d\zeta^2$, in other words, the linear term in the variational formula is never 0.

## 5.4 Descriptions of the Teichmüller metric

Now let $S_0$ be again a hyperbolic Riemann surface of finite type $(g, n, l)$ with boundary when $l \neq 0$. If $f_1$ and $f_2$ are homeomorphisms of $S_0$ onto itself and $f_1$ is homotopic to $f_2$, then they map any simple loop $\gamma$ on $S_0$ onto $\gamma_1 = f_1(\gamma)$ and $\gamma_2 = f_2(\gamma)$ and $\gamma_1$ is homotopic to $\gamma_2$ on $S_0$. The reverse statement is false. A simple example shows this: $S_0 = \{z : r \leq |z| \leq 1/r,\ 0 < r < 1\}$, $f_1 = z$, $f_2 = 1/z$, $z \in S_0$. In this section we show that we can use only homotopy of curves instead of homotopy of homeomorphisms for the definition of the Teichmüller metric. Given an admissible system $\gamma$ of curves of type I,II on $S_0$

and the weight vector $\alpha$ we define the modulus $m(S_0, \Gamma, \alpha)$. Let $m(x)$ stand for this modulus extended onto the Teichmüller space $T(S_0)$.

**Theorem 5.4.1.** *The Teichmüller metric $\tau_T(x, y)$ on the Teichmüller space $T(S_0)$ is described as*

$$\tau_T(x, y) = \sup_{\Gamma, \alpha} \frac{1}{2} \left| \log \frac{m(x)}{m(y)} \right| = \sup_{\Gamma, \alpha} \frac{1}{2} \left| \log \frac{m(f_1(S_0), f_1(\Gamma), \alpha)}{m(f_2(S_0), f_2(\Gamma), \alpha)} \right|,$$

*where $(f_1(S_0), f_1)$ and $(f_2(S_0), f_2)$ are the marked Riemann surfaces representing $x$ and $y$ respectively.*

We remark here that a similar description was obtained by S. Kerckhoff [69] for compact and punctured Riemann surfaces in terms of the Ahlfors-Beurling extremal length, so we will better prove the following theorem giving the description in terms of the functional $s(x)$.

Let $\varphi \in Q(S_0)$ be the space of all holomorphic quadratic differentials with finite trajectories on $S_0$. Then, each $\varphi$ defines the vector $b = (b_1, \ldots, b_m)$ with positive coordinates, and the system $\gamma = (\gamma_1, \ldots, \gamma_m)$ of its non-critical trajectories is an admissible system of curves on $S_0$. We define the functional $s_\varphi(x)$, $x \in T(S_0)$, as in Theorem 5.2.2.

**Theorem 5.4.2.** *Let $\varphi \in Q(S_0)$. Then the Teichmüller metric $\tau_T(x, y)$ on the Teichmüller space $T(S_0)$ is described as*

$$\tau_T(x, y) = \sup_{\varphi \in Q(S_0)} \frac{1}{2} \left| \log \frac{s_\varphi(x)}{s_\varphi(y)} \right|.$$

*Proof.* Without loss of generality one can assume $x = 0$. We denote by

$$s(0, x) := \sup_{\varphi \in Q(S_0)} \frac{1}{2} \left| \log \frac{s_\varphi(0)}{s_\varphi(x)} \right|. \tag{5.8}$$

First we will show that $s(0, x)$ is a metric and then, that $\tau_T(0, x) = s(0, x)$.

1) *There exists a finite supremum in (5.8)*. This follows from Proposition 5.2.1, because the inequality there does not depend on $\varphi$. So for any fixed $x$ the value $s(0, x)$ is uniquely defined and finite.

2) *The obvious triangle inequality $s(0, x) + s(x, y) \geq s(0, y)$ is satisfied*. Moreover, if $x = 0$, then $f^\mu$, where $\Phi(\mu) = 0$, is homotopic to the identity, $\mu \in I(G_0)$ and $s(0, 0) = 0$.

3) *If $s(0, 0) = 0$, then $x = 0$* (so we prove that $s(x, y)$ is a metric). We suppose the contrary, i.e., $x \neq 0$. Then there exists a Teichmüller differential $\mu = t|\psi|/\psi$, such that the equivalence class with the representative $\mu$ is the point $x \in T(S_0)$, and where $0 < t < 1$, $\psi(\zeta) d\zeta^2$ is a holomorphic quadratic differential on $S_0$ with $\|\psi\| < \infty$.

Proposition 2.6.3 implies that for any $\varepsilon > 0$ there exists a holomorphic quadratic differential $\varphi$ with finite trajectories and of finite norm such that

$\|\mu - \nu\|_\infty < \varepsilon$ where $\nu = t|\varphi|/\varphi$. The differential $\varphi \in Q(S_0)$ defines a vector $b = (b_1, \ldots, b_m)$ with positive coordinates and the admissible system of curves $\gamma = (\gamma_1, \ldots, \gamma_m)$. Let $S = f^\nu(S_0)$, $y = \Phi(\nu)$. Then, the map $f^\nu$ is affine in the local coordinates $z = \int \sqrt{\varphi}$ in each ring domain $D_j$ in the trajectory structure of $\varphi$ on $S_0$. Hence, this map realizes the equality signs in Proposition 5.2.1 and $s_\varphi(y) = K s_\varphi(0)$, where $K = (1 + t)/(1 - t)$. Since $s(0, x) = 0$, for any quadratic differential $\varphi \in Q(S_0)$

$$s_\varphi(x) = s_\varphi(0). \tag{5.9}$$

Now we construct the mapping $F : S \to S_0 \to f^\mu(S_0)$. Its complex dilatation is

$$\mu_F = \frac{\mu - \nu}{1 - \bar{\mu}\nu} \frac{f_\zeta^\nu}{\overline{f_\zeta^\nu}}$$

and $\|\mu_F\|_\infty < \varepsilon/(1 - t^2)$. Now we find in the equivalence class $[\mu_F]$ the Teichmüller differential $\mu_0$ which has the norm $\|\mu_0\| = \varepsilon_1 \leq \|\mu_F\|$ due to its minimal property.

The mappings $F \circ f^\nu$, $f^\mu$ form the same quadratic differential $\varphi(\zeta, x)d\zeta^2$ from the initial differential $\varphi(\zeta)d\zeta^2$. The mapping $F$ generates this differential as an image of the differential $\varphi(\zeta, y)d\zeta^2$ on the surface $S$.

The variational formula of Theorem 5.2.2 yields

$$s_\varphi(x) = s_\varphi(y) + 2\varepsilon_1 \mathrm{Re} \iint_S \mu^* q^\nu d\sigma_\zeta + o(\varepsilon_1),$$

where $\mu^*$ is a unit vector of the tangent space to $T(S_0)$ at the point $y$ and $q^\nu$ is a pull-back of the differential $\varphi(\zeta, y)d\zeta^2$. This implies

$$s_\varphi(x) - s_\varphi(0) = \frac{2t}{1 - t} s_\varphi(0) + 2\varepsilon_1 \mathrm{Re} \iint_S \mu^* q^\nu d\sigma_\zeta + o(\varepsilon_1). \tag{5.10}$$

The right-hand side is strictly positive for a sufficiently small $\varepsilon$ and, hence, $\varepsilon_1$. This contradicts (5.9) and completes the proof of 3).

4) *Prove that the obtained metric $s(0, x)$ is equal to $\tau_T(0, x)$.*

Let $x \in T(S_0)$ be different from the origin, $\mu = t|\psi|/\psi$ be a Teichmüller differential such that $\Phi(\mu) = x$. If $\psi$ is a differential with finite trajectories, then we consider it's non-freely homotopic non- critical trajectories $(\gamma_1, \ldots, \gamma_m)$ as an admissible system of curves. There are exactly $m$ characteristic ring domains $(D_1, \ldots, D_m)$ in its trajectory structure. Assume $(b_1, \ldots, b_m)$ the vector of the length of connected arcs of the orthogonal trajectories of $\psi$ situated in each $D_j$. By these data we construct the functional $s_\psi(x)$. The Teichmüller mapping $f^\mu$ acts affinely in each characteristic domain $D_j$, and therefore, either $s_\psi(x) = K s_\psi(0)$ or $s_\psi(x) = s_\psi(x)/K$. This means that $s(0, x) = \frac{1}{2} \log \frac{1+t}{1-t} = \tau_T(0, x)$.

If $\psi$ is an arbitrary holomorphic quadratic differential on $S_0$, then for any $\varepsilon$ we can approximate it by differentials $\varphi_\varepsilon$ with finite trajectories as in 3). The variation (5.10) implies

$$\tau_T(0,x) - \varepsilon \le \frac{1}{2}\left|\log\frac{s_{\varphi_\varepsilon}(x)}{s_{\varphi_\varepsilon}(0)}\right| < s(0,x) \le \tau_T(0,x).$$

Thus, $s(0,x) = \tau_T(0,x)$, that finishes the proof. □

## 5.5 Invariant metrics

The Teichmüller space $T(g,n)$ is a $(3g-3+n)$-dimensional hyperbolic complex analytic manifold and one can define on it biholomorphically invariant metrics that generalize the Poincaré hyperbolic metric. Namely, we will consider the Kobayashi and Carathéodory metrics $k_T(x,y)$ and $c_T(x,y)$, $x,y \in T(g,n)$ metrics. For their definitions and properties we refer to [43], [75], [113].

Let $S_0 = U/G_0$ and $q(z)dz^2$ be a holomorphic quadratic differential in $U$, $q(\gamma(z))(\gamma'(z))^2 = q(z)$ for any $\gamma \in G_0$. A geodesic in the Teichmüller metric disk which is the embedding of the unit disk in the Teichmüller space $T(S_0)$ by $\Delta_q := \{\Phi(t\bar{q}/|q|), t \in U\}$ is called the *Teichmüller disk*.

Denote by $d(\cdot,\cdot)$ the usual hyperbolic Poincaré metric in the unit disk $U$:

$$d(z_1,z_2) = \frac{1}{2}\log\frac{1 + \left|\frac{z_1-z_2}{1-z_1\bar{z}_2}\right|}{1 - \left|\frac{z_1-z_2}{1-z_1\bar{z}_2}\right|}.$$

Let $X,A,B$ be complex analytic manifolds and $\mathrm{Hol}(A,B)$ be the class of all holomorphic mappings from $A$ to $B$. The Kobayashi semimetric $k_X(x,y)$ on $X$ is the biggest among the semimetrics $\rho$ in $X$ satisfying the inequality $\rho(f(z_1),f(z_2)) \le d(z_1,z_2)$ for any $f \in \mathrm{Hol}(U,X)$; $z_1,z_2 \in U$. The Carathéodory semimetric $c_X(x,y)$ is the smallest among the semimetrics $\rho$ in $X$ satisfying the inequality $\rho(x,y) \ge d(f(x),f(y))$ for any $f \in \mathrm{Hol}(X,U)$; $x,y \in T(g,n)$. Both semimetrics on $X$ satisfy the main properties of the hyperbolic Poincaré metric: holomorphic contractibility (analogously to the Schwarz lemma) and biholomorphic invariance. In particular, if $X$ is a submanifold of $T(g,n)$, then $c_X(x,y) \ge c_T(x,y)$. On the Teichmüller space and on its submanifolds both semimetrics are metrics. The Kobayashi metric is inner, i.e., it can be restored by its infinitesimal generator, but the Carathéodory metric is not. This seems to be the main difference between the metrics. Besides, $k_X(x,y) \ge c_X(x,y)$.

The Kobayashi metric was completely described by H. Royden [117] in 1971 for $T(g,0)$ and by C. Earle, I. Kra [32], [20] in 1974 for $T(g,n)$. They showed that $k_T(x,y) = \tau_T(x,y)$. For the Teichmüller space $T(0,4)$ the metrics $k_T(x,y)$, $c_T(x,y)$ are the same because this space is holomorphically equivalent to a hyperbolic domain in $\mathbb{C}$. In [20] W. Abikoff, C. Earle, I. Kra posed a problem whether these metrics are the same on any Teichmüller space. In 1981 S. Krushkal [72] (see also [75]) constructed an example showing that for $3g - 3 + n > 2$, $n \geq 1$, $g > 2$ and for every $x \in T(g, n)$ there is a point $y \in T(g, n)$ such that $k_X(x, y) > c_X(x, y)$. In the same year I. Kra [70] obtained an affirmative result in this direction. Namely, he proved that in the Teichmüller disks $\Delta_q$, such that $q$ is a square of an Abelian holomorphic differential on $S_0$, the metrics coincide. Then there appeared some works of negative sense. In particular, it was shown that the Bers embedding into $T(g, n)$ (see [11]) is not geodesic in the Carathéodory metric (S. Nag [103]). Then the analogous result was obtained by S. Krushkal [73] for the universal Teichmüller space and for the general case of infinite dimensional Teichmüller space it was proved in the Ph.D. thesis of L. Liu [93]. For the universal Teichmüller space H. Shiga and H. Tonigawa [127] have presented an example of a non-Abelian Teichmüller disk where the metrics are the same. This result was obtained by a quasiconformal variation of Grunski coefficients.

The following results apply harmonic properties of the functionals $m(x)$, $s(x)$ and variational formulae from Theorems 5.2.1, 5.2.2 for sufficient conditions for Teichmüller disks or their subdisks to be geodesic in the Carathéodory metric.

**Theorem 5.5.1.** *Let $X_r = \{x : x \in T(g, n), \tau_T(0, x) < r\}$. If $\varphi$ is an extremal quadratic differential for the functional $m(x)$ at the point $x = 0$, $q$ its pull-back onto the universal covering $U$, and $m(x)$ is pluriharmonic on $X_r$, then on $\Delta_q \cap X_r$ the Teichmüller-Kobayashi metric $\tau_T$ and the Carathéodory metric $c_{X_r}$ are the same. In particular, this statement is true for the entire $T(g, n)$ when $r \to \infty$ and $X_r \to T(g, n)$.*

*Proof.* Obviously we may assume that $x = 0$. The same argumentation is applied to an arbitrary $x$ making use of quasiconformal shifts.

We invoke the infinitesimal description of the Carathéodory metric:

$$c_{X_r}^*(x, v) = \sup_h |h^*(v)|,$$

where $x$ and $v$ lie in the tangent space to $T$ at $x \in X_r$, $h$ belongs to the space $\mathrm{Hol}(X_r, U)$ of all holomorphic mappings from $X_r$ to $U$ and satisfies $h(x) = 0$, and $h^*$ is the differential of $h$.

Since $T(g, n)$ and $X_r$ are simply connected $(3g - 3 + n)$-dimensional complex analytic manifolds, there exists a holomorphic mapping $\Lambda(x)$ such that $\mathrm{Re}\, \Lambda(x) = m(x)/m(0)$, $\mathrm{Im}\, \Lambda(0) = 0$. Also, we consider the function

$$F(x) = \frac{\Lambda(x) - 1}{\Lambda(x) + 1}, \quad F(0) = 0,$$

$F \in \mathrm{Hol}(X_r, \mathbb{C})$. By an obvious inequality for the moduli of families of homotopy classes of curves, we have

$$e^{-2r} < \exp(-2\tau_T(0,x)) \leq \mathrm{Re}\, \Lambda(x) \leq \exp(2\tau_T(0,x)) < e^{2r},$$

where $\tau_T(\cdot,\cdot)$ is the Teichmüller metric on $T(g,n)$. Therefore, $F \in \mathrm{Hol}(X_r, U)$.

We put $h(t) = F \circ \Phi(t\mu_0)$, where $\mu_0 = \bar{\varphi}/|\varphi|$. Then $h(t)$ satisfies the conditions of the Schwarz lemma, whence $|h'(0)| \leq 1$. Calculating $h'(0)$ explicitly with the help of Theorem 5.2.1, we obtain

$$h'(0) = -\frac{1}{m(0)} \iint\limits_{S_0} \varphi\mu_0 \, d\sigma_\zeta = -1.$$

Hence,

$$c^*_{X_r}\left(\Phi(0), \left.\frac{\partial\Phi(t\mu_0)}{\partial t}\right|_{t=0}\right) = 1.$$

The results by E. Vezentini [172] and I. Kra [70] imply that $\Phi(t\mu_0)$, where $|t| < k = (e^{2r}-1)/(e^{2r}+1)$, is a complex geodesic in the Carathéodory metric $c_{X_r}$, i.e., $c_{X_r}(\Phi(t_1\mu_0), \Phi(t_2\mu_0)) = d(t_1, t_2)$, where $d(\cdot,\cdot)$ is the hyperbolic Poincaré metric on $U$. Now, $\Delta_\varphi$ is geodesic in the metric $\tau_T$, that proves the theorem. $\square$

**Theorem 5.5.2.** *If $\varphi$ is an extremal quadratic differential for the functional $s(x)$ at the point $x = 0$, $q$ its pull-back onto the universal covering $U$, and $s(x)$ is pluriharmonic on $X_r$, then on $\Delta_q \cap X_r$ the Teichmüller-Kobayashi metric $\tau_T$ and the Carathéodory metric $c_{X_r}$ are the same. In particular, this statement is true for the entire $T(g,n)$ when $r \to \infty$ and $X_r \to T(g,n)$.*

Thus, for the coincidence of three metrics $c_T$, $\tau_T$, $k_T$ on the Teichmüller disk $\Delta_q$ it is sufficient to check whether one of the functionals $m(x)$, $s(x)$ generated by a quadratic differential $\varphi$ is pluriharmonic on $T(g,n)$ or whether $\varphi$ has even zeros on $S_0$. Note that the coincidence of the metrics implies a progress in solution of extremal problems for compact classes of quasiconformal maps [74].

Harmonic moduli in Theorem 5.3.1 show that Theorems 5.5.1, 5.5.2 are sensible.

The modulus problem (2.12)–(2.13) reflects the Dirichlet principle. Namely, we are given a Riemann surface $S_0$ of a finite type $(g,n,l)$ possessing a complex structure, and a free family of homotopy classes of curves of the first and second types on $S_0$. The Dirichlet problem consists of finding the minimum of $L_1$-norm of a quadratic differential $\varphi$ corresponding to the complex structure of $S_0$ among all differentials of the given homotopy type from the free family. The trajectories of this differential have the lengths in the metric $\sqrt{|\varphi|}$ which are not less than preassigned positive numbers. The Dirichlet principle states that there exists a unique differential realizing this minimum. In this case the Dirichlet principle is a particular case of the modulus problem when we take only metrics $\sqrt{|\varphi|}$ from the class of all admissible metrics. The results by J. Jenkins and K. Strebel establish that the solutions of both problems involve the same metric.

A similar approach was suggested by W. Thurston and described in [39]. It also displays the Dirichlet principle. In the space of all homotopy classes of curves one can introduce measured foliations. One looks for a minimum of the $L_1$-norm in the space of differentials having reduced heights of curves from a given homotopy class in the metric $|i\sqrt{\varphi}|$ not less than the height of this class with respect to a precise measured foliation. A foliation is assigned by formal multiplication of homotopy classes by positive numbers. Each class becomes a ray of the foliation.

The norm of a measured foliation plays a similar role as the norm of the extremal quadratic differential in a modulus problem. One can derive a description of the Teichmüller metric and results about the harmonic properties of this norm in connection with invariant metrics of Kobayashi and Carathéodory similar to those for the modulus. For definition of the norm and properties of measured foliations we refer to [39], [43], [60].

Results of this chapter were obtained in [151], [152], [153], [154], [155], [160], [170].

# References

1. **W. Abikoff**, *The real analytic theory of Teichmüller spaces*, Lecture Notes in Math., Springer-Verlag, Vol. 820, 1980

2. **S. Agard**, *Distortion theorems for quasiconformal mappings*, Ann. Acad. Sci. Fenn. Ser A I, **413** (1968), 1–12

3. **N. I. Akhiezer**, *Elements of the theory of elliptic functions*, Moscow: Nauka, 1970; English transl., Trans. of Mathematical Monographs, Vol. 79. Amer. Math. Soc., Providence, RI, 1990

4. **I. A. Aleksandrov**, *Parametric continuations in the theory of univalent functions*, Nauka, Moscow, 1976 (Russian)

5. _____, **S. A. Kopanev**, *On mutual growth of the modulus of a univalent function and of the modulus of its derivative*, Sibirsk. Mat. Zh., **7** (1966), 23–30 (Russian)

6. **G. D. Anderson, M. K. Vamanamurthy, M. Vuorinen**, *Conformal invariants, inequalities, and quasiconformal mappings*, J. Wiley, 1997 (electronic version)

7. **C. Andreian Cazacu**, *On the extremal problems of the theory of quasiconformal mappings*, Modern Problems of the theory of analytic functions. Proc. Int. Conf., Moscow: Nauka, 1966, 18–27

8. **L. V. Ahlfors**, *The complex analytic structure of the space of closed Riemann surfaces*, Analytic functions, Princeton (1960), 45–66

9. _____, *Lectures on quasiconformal mappings*, Van Nostrand Math. Stud., Princeton, N.J., 1966

10. _____, *Conformal invariants. Topics in Geometric Function Theory*, McGrow-Hill, 1973

11. _____, **L. Bers**, *Spaces of Riemann surfaces and quasiconformal mappings*, Collection of papers. Moscow: Inostrannaya Literatura., 1961

12. _____, **A. Beurling**, *Conformal invariants and functiontheoretic null sets*, Acta Math., **83** (1950), no. 1-2, 101–129

13. **V. V. Aseev, B. Yu. Sultanov**, *Moduli of families of curves on a Riemannian manifold*, Sibirsk. Mat. Zh., **31** (1990), no. 5, 164–166; English transl., Siberian Math. J., **31** (1990), no. 5, 839–841

14. **A. Beardon**, *The geometry of discrete groups*, Springer, New York, 1983

15. **P. P. Belinskiĭ**, *On a distortion under quasiconformal mapping*, Dokl. Acad. Sci. USSR, **91** (1953), no. 5, 997–998 (Russian)

16. _____, *General properties of quasiconformal mappings*, Novosibirsk: Nauka., 1974 (Russian)

17. **L. Bers,** *Spaces of Riemann surfaces as bounded domains,* Bull. Amer. Math. Soc. **66** (1960), 98–103

18. _____, *Quasiconformal mappings and Teichmüller's theorem,* Analytic functions, Princeton (1960), 89–120

19. _____, *Holomorphic differentials as functions of moduli,* Bull. Amer. Math. Soc. **67** (1961), no. 2, 98–103

20. _____, **I. Kra** (editors), *A crash course on Kleinian groups,* (Lectures, Special Session, Annual Winter Meeting, Amer. Math. Soc., San Francisco, Calif., 1974), Lecture Notes in Math., Vol. 400, Springer, Berlin, 1974

21. **L. de Branges,** *The story of the verification of the Bieberbach conjecture,* The Bieberbach conjecture (West Lafayette, Ind., 1985), Math. Surveys Monographs, 21, Amer. Math. Soc., Providence, R.I., 1986, 199–203

22. **P. F. Byrd, M. D. Friedman,** *Handbook of elliptic integrals for engineers and physicists,* Springer-Verlag, Berlin, Göttingen, Heidelberg, 1954

23. **C. Carathéodory,** *Funktionentheorie,* Vol. 2, Basel, Birkhäuser, 1950

24. **S. Dëmin,** *Isoperimetric distortion problem for univalent Montel functions,* Sibirsk. Mat. Zh., **37** (1996), no. 1, 108–116; English transl., Siberian Math. J., **37** (1996), no. 1, 94–101

25. **J. Dieudonné,** *Recherches sur quelques problèmes relatifs aux polynômes et aux fonctions borneés d'une variable complexe,* Ann. Sci. École Norm. Sup., **48**(1931), 247–358

26. **B. Dittmar,** *Extremalprobleme quasikonformer Abbildungen der Ebene als Steuerungsprobleme,* Zeitschrift für Analysis und ihre Anwendungen, **5** (1986), no. 6, 563–573

27. **A. Duady, J. Hubbard,** *On the density of Strebel forms,* Invent. Math., **30** (1975), 175–179

28. **V. N. Dubinin,** *Change of harmonic measure in symmetrization,* Mat. Sb. (N.S.), **124(166)** (1984), no. 2, 272–279 (Russian)

29. _____, *Symmetrizational method in problems on non-overlapping domains,* Math. Sb. (N.S.), **128(170)** (1985), no. 1, 110–123 (Russian)

30. _____, *Symmetrization in the geometric theory of functions of a complex variable,* Uspekhi Mat. Nauk, **49** (1994), no. 1(295), 3–76 (Russian)

31. **P. Duren,** *Univalent functions,* Springer, New York, 1983

32. **C. J. Earle,** *On the Carathéodory metric in Teichmüller spaces,* Discontinuous Groups and Riemann Surfaces., Ann. Math. Stud., **79** (1974), 99–103

33. _____, **I. Kra,** *On holomorphic mappings between Teichmüller spaces,* Contributions to Analysis (ed. L. Ahlfors et al.), N.Y.: Academic Press, 1974, 107–124

34. **E. G. Emel'yanov,** *A class of functions that are univalent in an annulus,* Zap. Nauchn. Sem. Leningrad. Otdel. Mat. Inst. Steklov. (LOMI), **144** (1985), 83–93; English transl., J. Soviet Math., **38** (1988), no. 4, 2091–2100

35. _____, *Some properties of moduli of curve families,* Zap. Nauchn. Sem. Leningrad. Otdel. Mat. Inst. Steklov. (LOMI), **144** (1985), 72–82; English transl., J. Soviet Math., **38** (1988), no. 4, 2081–2090

36. _____, *On extremal partitioning problems,* Zap. Nauchn. Sem. Leningrad. Otdel. Mat. Inst. Steklov. (LOMI), **154** (1986), 76–89; English transl., J. Soviet Math., **43** (1988), no. 4, 2558–2566

37. _____, *Problems of extremal decomposition in spaces of Riemann surfaces*, Zap. Nauchn. Sem. St.-Petersb. Otdel. Mat. Inst. Steklov. (POMI), **263** (2000), 84–104 (Russian)

38. H. Farkas, I. Kra, *Riemann surfaces*, Graduate Texts in Math., Vol. 71, Springer-Verlag, 1980

39. A. Fathi, F. Laudenbach, V. Poenaru, *Travaux de Thurston sur les surfaces*, *Asterisque*, Soc. Math. Fr., Paris, 1979, 66–67

40. S. I. Fedorov, *Chebotarev's variational problem in the theory of the capacity of plane sets, and covering theorems for univalent conformal mappings*, Mat. Sb. (N.S.), **124(166)** (1984), no. 1, 121–139 (Russian)

41. _____, *Moduli of certain families of curves and the set of values of $f(\zeta_0)$ in the class of univalent functions with real coefficients*, Zap. Nauchn. Sem. Leningrad. Otdel. Mat. Inst. Steklov. (LOMI), **139** (1984), 156–167 (Russian)

42. B. Fuglede, *Extremal length and functional completion*, Acta Math., **98** (1957), 171–219

43. F. P. Gardiner, *Teichmüller theory and quadratic differentials*, Wiley Interscience, N.Y., 1987

44. _____, H. Masur, *Extremal length geometry of Teichmüller space*, Complex Variables, **16** (1991), 209–237

45. F. W. Gehring, J. Väisälä, *The coefficients of quasiconformality of domains in the space*, Acta Math., **114** (1965), 60–120

46. F. W. Gehring, E. Reich, *Area distortion under quasiconformal mappings*, Ann. Acad. Sci. Fenn., Ser. A1. Math., **388** (1966), 1–15

47. G. M. Goluzin, *On distortion theorems in the theory of conformal mappings*, Matem. Sb. (N.S.), **18(60)** 1946, no. 3, 379–390

48. _____, *Geometrische Funktionentheorie*, Deutscher Verlag, Berlin, 1957

49. A. W. Goodman, *Univalent functions*, Vol. I–II, Mariner Publishing Comp., Inc., 1983

50. V. V. Goryaĭnov, *Extremals in estimates of functionals depending on values of a univalent function and its derivative*, Theory of mappings and approximation of functions, Naukova Dumka, Kiev, 1983, 38–49 (Russian)

51. _____, V. Ya. Gutlyanskiĭ, *Extremal problems in the class $S_M$*, Mathematics collection, Izdat. Naukova Dumka, Kiev, 1976, 242–246

52. M. Gromov, *Hyperbolic cusps*, Essays in group theory, ed. S.M.Gersten. MSRI publications 8, Springer-Verlag. (1987), 75–263.

53. H. Grötzsch, *Über einige Extremalprobleme der konformen Abbildungen. I–II*, Ber. Verh.- Sächs. Akad. Wiss. Leipzig, Math.–Phys. Kl., **80** (1928), 367–376.

54. _____, *Über die Verzerrung bei nichtkonformen schlichten Abbildungen merfach zusammenhängender schlichten Bereiche*, Ber. Verh.- Sächs. Akad. Wiss. Leipzig, **82** (1930), 69–80.

55. Gung San, *Contribution to the theory of schlicht functions 1. Distortion theorems*, Acta Sci. Sinica, **4** (1955), 229–249

56. V. Ya. Gutlyanskiĭ, *Parametric representation and extremal problems in the theory of univalent functions*, Diss. Doc. Sci., Math. Inst. Acad. Sci. Ukr. SSR, Kiev, 1972

57. H. Hancock, *Elliptic integrals*, Dover Publ., NY, 1958

58. **F. Hausdorff**, *Set theory*, 3-rd Ed. (transl.), Chelsea, New York, 1957

59. **J. Hersch, A. Pfluger**, *Generalizacion du lemme de Schwarz et du principe de la mesure harmonique pour fonctions pseudo-analytiques*, C.R. Acad. Sci. Paris., **234** (1952), 43–45

60. **J. Hubbard, H. Masur**, *Quadratic differentials and foliations*, Acta Math., **142** (1979), 221–274

61. **J. Jenkins**, *Symmetrization results for some conformal invariants*, Amer. J. Math., **75** (1953), no. 3, 510–522

62. _____, *On a problem of Gronwall*, Ann. of Math., **59** (1954), no. 3, 490–504

63. _____, *On the existence of certain general extremal metrics*, Ann. of Math., **66** (1957), no. 3, 440–453

64. _____, *Univalent functions and conformal mapping*, Springer-Verlag, 1958

65. _____, *On weighted distortion in conformal mapping*, Illinois J. Math., **4** (1960), 28–37

66. _____, *On the global structure of the trajectories of a positive quadratic differential*, Illinois J. Math., **4** (1960), 405–412

67. _____, *On the existence of certain general extremal metrics. II*, Tôhoku Math. J. (2), **45** (1993), no. 2, 405–412

68. _____, **K. Oikawa**, *Conformality and semi-conformality at the boundary*, J. Reine Angew. Math., **291** (1977), 92–117

69. **S. Kerckhoff**, *The asymptotic geometry of Teichmüller space*, Topology, **19** (1980), 23–41

70. **I. Kra**, *The Carathéodory metric on Abelian Teichmüller disks*, J. d'Analyse Math., **40** (1981), 129–143

71. **S. L. Krushkal**, *Quasiconformal mappings and Riemann surfaces*, A Halsted Press Book. Scripta Series in Mathematics. With a foreword by Lipman Bers. V. H. Winston & Sons, Washington, D.C.; John Wiley & Sons, New York-Toronto, Ont.-London, 1979

72. _____, *Invariant metrics in Teichmüller spaces*, Sibisk. Mat. Zh., **22** (1981), no. 2, 209–212 (Russian)

73. _____, *Invariant metrics on Teichmüller spaces and quasiconformal extendability of analytic functions*, Ann. Acad. Sci. Fenn., **10** (1985), 299–303

74. _____, *A new method for solving variational problems in the theory of quasiconformal mappings*, Sibirsk. Mat. Zh., **29** (1988), no. 2, 105–114; English transl., Siberian Math. J., **29** (1988), no. 2, 245–252

75. _____, **R. Kühnau**, *Quasikonforme Abbildungen—neue Methoden und Anwendungen*, Teubner-Texte zur Mathematik, Vol 54. BSB B. G. Teubner Verlagsgesellschaft, Leipzig, 1983

76. **J. Krzyż**, *On the region of variability of the ratio $f(z_1)/f(z_2)$ within the class S of univalent functions*, Ann. Univ. Mariae Curie-Skłodowska. Sect A., **17** (1963), no. 8, 55–64

77. _____, **E. Złotkiewicz**, *Koebe sets for univalent functions with two pre-assigned points*, Ann. Acad. Sci. Fenn. Ser. A 1. Math., **487**, 1971

78. **G. V. Kuz'mina**, *Moduli of curve families and quadratic differentials*, Trudy Mat. Inst. Steklov., **139** (1980), 1–240; English transl., Proc. V. A. Steklov Inst. Math., 1982, no. 1

79. _____, *On a modulus problem for a curve family*, Preprint Leningrad. Otdel. Mat. Inst. Steklov. (LOMI), P-6-83, 1983

80. _____, *Modulus problem for a curve family in a circular ring*, Zap. Nauchn. Sem. Leningrad. Otdel. Mat. Inst. Steklov. (LOMI), **144** (1985), 115–127

81. _____, *On extremal properties of quadratic differentials with strip-like domains in their trajectory structure*, Zap. Nauchn. Sem. Leningrad. Otdel. Mat. Inst. Steklov. (LOMI), **154** (1986), 110–129; English transl., J. Soviet Math., **43** (1988), no. 4, 2579–2591

82. _____, *Extremal properties of quadratic differentials with trajectories that are asymptotically similar to logarithmic spirals*, Zap. Nauchn. Sem. Leningrad. Otdel. Mat. Inst. Steklov. (LOMI), **160** (1987), 121–137; English transl., J. Soviet Math., **52** (1990), no. 3, 3085–3098

83. _____, *On extremal properties of quadratic differentials with end domains in the structure of trajectories*, Zap. Nauchn. Sem. Leningrad. Otdel. Mat. Inst. Steklov. (LOMI), **168** (1988), 98–113; English transl., J. Soviet Math., **53** (1991), no. 3, 285–296

84. _____, *On the existence of quadratic differentials with poles of higher orders*, Zap. Nauchn. Sem. S.-Peterburg. Otdel. Mat. Inst. Steklov. (POMI), **212** (1994), 129–138; English transl., J. Soviet Math., **83** (1997), no. 6, 772–778

85. _____, *On the existence of quadratic differentials with prescribed properties*, Zap. Nauchn. Sem. S.-Peterburg. Otdel. Mat. Inst. Steklov. (POMI), **226** (1996), 120–137; English transl., J. Math. Sci. (New York), **89** (1998), no. 1, 996–1007

86. **M. A. Lavrentiev**, *A general problem of the theory of quasi-conformal representation of plane regions*, Mat. Sbornik (N.S), **21(63)** (1947), 85–320

87. **J. Ławrynowicz, J. Krzyż**, *Quasiconformal mappings in the plane: parametrical methods*, Lecture Notes in Math., Vol. 978, Springer-Verlag, Berlin–New York, 1983

88. **N. A. Lebedev**, *The area principle in the theory of univalent functions*, Moscow: Nauka, 1975 (Russian)

89. **J. Lehner**, *Discontinuous groups and automorphic functions*, Math. Surveys, no. 8, Amer. Math. Soc., Providence, RI, 1964

90. **O. Lehto**, *Univalent functions and Teichmüller spaces*, Graduate Texts in Mathematics, Vol. 109, Springer-Verlag, New York-Berlin, 1987

91. _____, **K. I. Virtanen**, *Quasiconformal mappings in the plane*, Second edition. Die Grundlehren der mathematischen Wissenschaften, Band 126. Springer-Verlag, New York-Heidelberg, 1973

92. **R. J. Libera, E. J. Złotkiewicz**, *Bounded Montel univalent functions*, Colloq. Math., **56** (1988), no. 1, 169–177

93. **Liu Li-Xing**, *Invariant metrics in Teichmüller spaces*, Ph.D. Thesis. Institute of Mathematics, Fudan University, Shanghai, China, 1993

94. **K. Löwner**, *Untersuchungen über schlichte konforme Abbildungen des Eiheitskreises*, J. Ann. Math., **89** (1923), 103–121

95. **W. Ma, D. Minda**, *Hyperbolically convex functions*, Ann. Polon. Math., **60** (1994), 81–100

96. **A. Marden**, *On finitely generated Fuchsian groups*, Comment. Math. Helv., **42** (1967), no. 1, 81–85

97. **A. I. Markushevich**, *Theory of functions of a complex variable*, Vol. I, II, III, Second English edition. Chelsea Publishing Co., New York, 1977

98. **J. E. McMillan**, *Boundary behaviour of a conformal mapping*, Acta Math., **123** (1969), 43–67

99. **D. Mejía, Ch. Pommerenke**, *On hyperbolically convex functions*, J. Geom. Analysis., **10** (2000), no. 2, 365–378

100. **I. M. Milin**, *Univalent functions and orthonormal systems*, Amer. Math. Soc., Providence, R.I., 1977

101. **I. P. Mityuk**, *The principle of symmetrization for multiply connected regions and certain of its applications*, Ukrain. Mat. Zh., **17** (1965), no. 4, 46–54 (Russian)

102. **A. Mori**, *On an absolute constant in the theory of quasiconformal mappings*, J. Math. Japan., **8** (1956), no. 2, 156–166

103. **S. Nag**, *Non-geodesic disks embedded in Teichmüller space*, Amer. J. Math., **104** (1982), 399–408

104. _____, *The complex analytic theory of Teichmüller spaces*, A Wiley-Interscience Publication. John Wiley & Sons, Inc., New York, 1988

105. **R. Nevanlinna**, *Uniformisierung*, Zweite Auflage. Die Grundlehren der mathematischen Wissenschaften in Einzeldarstellungen mit besonderer Bercksichtigung der Anwendungsgebiete, Band 64, Springer-Verlag, Berlin-New York, 1967

106. _____, *Analytic functions*, Translated from the second German edition by Phillip Emig. Die Grundlehren der mathematischen Wissenschaften, Band 162, Springer-Verlag, New York-Berlin, 1970

107. **M. Ohtsuka**, *Dirichlet problem, extremal length, and prime ends*, Van Nostrand, New York, 1970

108. **G. Pólya**, *Sur la simetrisation circulaire*, C.r. Acad. Sci. Paris, **230** (1950), no. 1, 25–27

109. _____, **G. Szegö**, *Über den transfiniten Darchmesser (Kapazitätskonstante) von ebenen und räumblichen Punktmengen*, J. Reine Angew. Math., **165** (1931), 4–49

110. **Ch. Pommerenke**, *Univalent functions, with a chapter on quadratic differentials by G. Jensen*, Vandenhoeck & Ruprecht, Göttingen, 1975

111. _____, *The Bieberbach conjecture*, Math. Intelligencer 7 (1985), no. 2, 23–25

112. _____, *Boundary behaviour of conformal maps*, Springer-Verlag, 1992

113. **E. A. Poletskiĭ , B. V. Shabat**, *Invariant metrics*, Current problems in mathematics. Fundamental directions, Vol. 9, Itogi Nauki i Tekhniki, Akad. Nauk SSSR, Vsesoyuz. Inst. Nauchn. i Tekhn. Inform. (VINITI), Moscow, 1986, 73–125 (Russian)

114. **V. I. Popov**, *The range of a system of functionals on the class S*, Trudy Tomsk. Gos. Univ. Ser. Meh.-Mat., **182** (1965) no. 3, 106–132 (Russian)

115. **B. Riemann**, *Gesammelte Mathematische Werke, 2*, Auf 1., Teubner, Leipzig, 1882

116. **B. Rodin, S. E. Warschawski**, *Extremal length and the boundary behaviour of conformal mappings*, Ann. Acad. Sci. Fenn. Ser. A I Math., **2** (1976), 467–500

117. **H. L. Royden**, *Automorphisms and isometrics of Teichmüller spaces*, Advances in the Theory of Riemann Surfaces, Ann. Math. Stud., **66** (1971), 369–383

118. **A. C. Schaeffer, D. C. Spenser**, *Coefficient regions for schlicht functions*, Colloquim Publ., Vol. 35, Amer. Math. Soc., NY, 1950

119. **M. Schiffer**, *On the modulus of doubly-connected domains*, Quart. J. Math., Oxford Ser., **17** (1946), 197–213

120. _____, **D. C. Spenser** *Functionals of finite Riemann surfaces*, Princeton Univ. Press., Princeton, NJ, 1954

121. _____, **G. Schober**, *A variational method for general families of quasiconformal mappings*, J. Analyse Math., **34** (1978), 240–264

122. **V. I. Semenov**, *Some dynamical systems and quasiconformal mappings*, Sibirsk. Mat. Zh., **28** (1987), no. 4, 196–206 (Russian)

123. _____, *Certain applications of the quasiconformal and quasiisometric deformations*, Rev. Roum. Math. Pures, Appl., **36** (1991), no. 9–10, 503–511

124. **B. V. Shabat**, *On the theory of quasiconformal mappings in space*, Soviet Math. Dokl., **1** (1960), 730–733

125. _____, *The modulus method in space*, Soviet Math. Dokl., **1** (1960), 165–168

126. **Shah Dao-Shing**, *Parametric representation of quasiconformal mappings*, Science Record, **3** (1959), 400–407

127. **H. Shiga, H. Tanigawa**, *Grunsky's inequality and its applications to Teichmüller spaces*, Kodai Math. J., **16** (1993), 361–378

128. **V. A. Shlyk**, *Capacity of a condenser and the modulus of a family of separating surfaces*, Zap. Nauchn. Sem. Leningrad. Otdel. Mat. Inst. Steklov. (LOMI), **185** (1990), 168–182 (Russian)

129. _____, *On the equality between p-capacity and p-modulus*, Sibirsk. Mat. Zh., **34** (1993), no. 6, 216–221; English transl., Siberian Math. J., **34** (1993), no. 6, 1196–1200

130. _____, *On the uniqueness of an extremal function for the p-capacity of a condenser*, Zap. Nauchn. Sem. S.-Peterburg. Otdel. Mat. Inst. Steklov. (POMI), **226** (1996), 228–234 (Russian)

131. **V. V. Shokurov**, *Riemann surfaces and algebraic curves*, Encyclopaedia Math. Sci., Vol. 13, Springer-Verlag, Berlin, 1994, 1–166

132. **A. Yu. Solynin**, *The dependence of a modulus problem for a family of several curve classes on parameters*, Zap. Nauchn. Semin. Leningrad. Otdel. Mat. Inst. Steklov. (LOMI), **144** (1985), 136–145; English transl., J. Soviet Math., **38** (1987), no. 4, 2131–2139

133. _____, *On harmonic measure of continua of the diameter fixed*, Zap. Nauchn. Semin. Leningrad. Otdel. Mat. Inst. Steklov. (LOMI), **144** (1985), 146–149; English transl., J. Soviet Math., **38** (1987), no. 4, 2140–2143

134. _____, *Solution of an isoperimetric problem of Pólya-Szegö*, Zap. Nauchn. Semin. Leningrad. Otdel. Mat. Inst. Steklov. (LOMI), **168** (1988), 140–153; English transl., J. Soviet Math., **53** (1991), no. 3, 311–320

135. _____, *Boundary distortion and extremal problems in some classes of univalent functions*, Zap. Nauchn. Sem. St. Petersburg. Otdel. Mat. Inst. Steklov. (POMI), **204** (1993), 115–142; English transl., J. Soviet Math., **79** (1996), no. 5, 1341–1355

136. _____, *Extremal configurations in some problems on capacity and harmonic measure*, Zap. Nauchn. Semin. S.-Peterburg. Otdel. Mat. Inst. Steklov. (POMI), **226** (1996), 170–195; English transl., J. Soviet Math., **89** (1998), no. 1, 1031–1049

137. _____, *Extremal problems on conformal moduli and estimates for harmonic measures*, Preprint S.-Peterburg. Otdel. Mat. Inst. Steklov. (POMI), PDMI Preprint–2/1996. St.-Petersburg, 1996

138. _____, *Partition into non-overlapping domains and extremal properties of conformal mappings*, Zap. Nauchn. Semin. S.-Peterburg. Otdel. Mat. Inst. Steklov. (POMI), **226** (1996), 139–163; English transl., J. Soviet Math., **89** (1998), no. 1, 1001–1027

139. _____, *Modules and extremal metric problems*, Algebra i Analiz, **11** (1999), no. 1, 1–86; English transl., St.-Petersburg Math. J., **11** (2000), no. 1, 1–70.

140. **G. Springer**, *Introduction to Riemann Surfaces*, Addison-Wesley, Reading, Mass., 1954

141. **K. Strebel**, *Quadratic differentials*, Springer-Verlag, 1984

142. **A. V. Sychëv**, *Moduli and spatial quasiconformal mappings*, Novosibirsk: Nauka, 1970 (Russian)

143. **O. Tammi**, *Extremum problems for bounded univalent functions*, Lecture Notes in Math, vol. 646, Springer, Berlin, 1978

144. _____, *Extremum problems for bounded univalent functions II*, Lecture Notes in Math, vol. 913, Springer, Berlin, 1982

145. **P. M. Tamrazov**, *Continuity of certain conformal invariants*, Ukrain. Mat. Zh., **18** (1966), no. 6, 78–84 (Russian)

146. _____, *A theorem on line integrals for extremal length*, Dopovīdī Akad. Nauk Ukraïn. RSR, **1** (1966), 51–54 (Ukranian)

147. _____, *On variational problems for the logarithmic potential*, Inst. Math. Ukr. Acad. Sci., Preprint no. 5, Kiev, 1980, 3–13

148. _____, *Capacities of condensers. A method of intermixing loads*, Mat. Sb. (N.S.), **115(157)** (1981), no. 1, 40–73 (Russian)

149. _____, *Methods for studying extremal metrics and moduli in a twisted Riemannian manifold*, Mat. Sb., **183** (1992), no. 3, 55–75; English transl., Russian Acad. Sci. Sb. Math., **75** (1993), no. 2, 333–351

150. **O. Teichmüller**, *Extremale quasikonforme Abbildungen und quadratische Differentiale*, Abhandl. Preuss. Akad. Wiss., Math.-Naturwiss. Kl., **22** (1940), 3–197

151. **A. Vasil'ev**, *Quasi-invariance of the modulus of homotopy classes of curves, and the Teichmüller problem*, Ukrain. Mat. Zh., **43** (1991), no. 3, 329–336; English transl., Ukrainian Math. J., **43** (1991), no. 3, 293–299

152. _____, *Harmonic properties of the modulus of a family of curves, and invariant metrics on a Teichmüller space*, Dokl. Russ. Akad. Nauk, **341** (1995), no. 5, 583–584

153. _____, *Homotopies of curves and mappings, and the Teichmüller metric*, Mat. Zametki, **59** (1996), no. 6, 923–926; English transl., Math. Notes, **59** (1996), no. 5-6, 668–671

154. _____, *Moduli of families of curves, and invariant metrics on the Teichmller space*, Sibirsk. Mat. Zh., **37** (1996), no. 5, 986–994; English transl., Siberian Math. J., **37** (1996), no. 5, 868–875

155. _____, *Harmonic functionals and invariant metrics on Teichmüller space*, Algebra i Analiz, **9** (1997), no. 1, 49–71; English transl., St.-Petersburg Math. J., **9** (1998), no. 1, 33–48

156. _____, *Isoperimetric growth theorems for quasiconformal automorphisms of the disk*, Izv. Vyssh. Uchebn. Zaved. Mat. (1997), no. 3, 14–22; English transl., Russian Math. (Izv. VUZ), **41** (1997), no. 3, 12–21

157. _____, *Variational-geometric methods of the solution of extremal and metric problems of the theory of conformal and quasiconformal mappings*, Dr. Sci. Hab. Thesis., Novosibirsk State Univ., Novosibirsk, Russia, 1997

158. _____, *Harmonic properties of the modulus of families of curves on Teichmüller spaces*, Math. and Applications, ed. Saratov State Univ. (1998), 19–20.

159. _____, *Two point distortion theorems for quasiconformal mappings*, International Congress of Mathematicians (ICM-98), Berlin, August 18–27, 1998, Abstracts of short communications, p.162

160. _____, *Quadratic differentials with regular trajectories and the Teichmüller metric*, Revue Roum. Math. Pures et Appl., **44** (1999), no. 4, 653–661

161. _____, *Two-point distortion theorems for quasiconformal mappings*, Preprint MAT 2000/12, Departamento de Matemática, Univ. Técnica Federico Santa María, Valparaíso, (2000), 1–19

162. _____, *On distortion under bounded univalent functions with the angular derivative fixed*, Preprint MAT 2000/12, Departamento de Matemática, Univ. Técnica Federico Santa María, Valparaíso, (2000), 1–18, to appear in Complex Variables.

163. _____, S. Fedorov, *Application of the method of moduli in an extremal problem of conformal mapping*, Izv. Vyssh. Uchebn. Zaved. Mat. 1990, no. 8, 13–22; English transl., Soviet Math. (Izv. VUZ), **34** (1990), no. 8, 13–22

164. _____, G. Kamyshova, *Moduli of strip-shaped domains in the solution of an isoperimetric problem of conformal mapping*, Sibirsk. Mat. Zh., **37** (1996), no. 1, 60–69; English transl., Siberian Math. J., **37** (1996), no. 1, 53–61

165. _____, P. Pronin, *On some extremal problems for bounded univalent functions with Montel's normalization*, Demonstratio Math., **26** (1993), no. 3-4, 703–707

166. _____, P. Pronin, *The range of a system of functionals for the Montel univalent functions*, Bol. Soc. Mat. Mexicana, **6** (2000), no. 3, 177–190

167. _____, D. Mejía, Ch. Pommerenke, *Distortion theorems for hyperbolically convex functions*, Complex Variables. **44** (2001), no. 2, 117–130

168. _____, **Ch. Pommerenke**, *On bounded univalent functions and the angular derivative*, Ann. Univ. Maria Curie-Sklodowska. Sect. A. **54** (2000), no. 8, 79–106

169. _____, **Ch. Pommerenke**, *Angular derivatives of bounded univalent functions and extremal partitions of the unit disk*, Preprint MAT 2000/14, Departamento de Matemática, Univ. Técnica Federico Santa María, Valparaíso, (2000), 1–26, to appear in Pacific J. Math.

170. _____, **R. Hidalgo**, *Harmonic moduli of families of curves on Teichmüller spaces*, Scientia. Ser. Math. Sci. **8** (2001–2002), 61–69

171. **J. Väisälä**, *Lectures on n–dimensional quasiconformal mappings*, Lecture Notes in Math. Springer-Verlag, Vol. 229, 1971

172. **E. Vezentini**, *Complex geodesics and holomorphic maps*, Sympos. Math., Vol. 26, Academic Press. London- New York, (1982), 211–230

173. **M. Vuorinen**, *Conformal invariants and quasiregular mappings*, J. d'Analyse Math., **45** (1985), 69–115

174. _____, *Conformal geometry and quasiregular mappings*, Lecture Notes in Math., Vol. 1319, Springer-Verlag, Berlin- New York, 1988

175. **Wang Chuan-Fang**, *A sharp form of Moris theorem on Q-mappings*, Kexue Jilu, **4** (1960), 334–337

176. **V. Wolontis**, *Properties of conformal invariants*, Amer. J. Math., **74** (1952), no. 3, 587–606

177. **S. Wolpert**, *Thurston's Riemannian metric for Teichmüller space*, J. Differential Geom., **23** (1986), no. 2, 143–174

178. **V. A. Zorich**, *Correspondence of boundaries under Q-quasiconformal mapping of a ball*, Dokl. Acad. Sci. USSR, **145** (1962), no. 6, 1209–1212 (Russian)

179. _____, *The global homeomorphism theorem for space quasiconformal mappings, its development and related open problems*, Quasiconformal space mappings, Lecture Notes in Math., Vol. 1508, Springer, Berlin, 1992, 132–148

180. **N. V. Zoriĭ**, *Capacities and discrete characteristics of higher-dimensional condensers*, Ukrain. Mat. Zh., **42** (1990), no. 9, 1192–1199; English transl., Ukrainian Math. J., **42** (1990), no. 9, 1059–1065

181. _____, *Capacities of a space condenser*, Bull. Soc. Sci. Lett. Łódź, Sér. Rech. Déform., **23** (1997), 127–134

# List of Symbols

| | |
|---|---|
| $\mathbb{C}$ | complex plane |
| $\overline{\mathbb{C}}$ | Riemann sphere |
| $U$ | unit disk |
| $\partial U$ | unit circle |
| $\mathbb{R}$ | real line |
| $\mathbb{R}^+$ | positive real axis |
| $\bar{D}$ | closure of $D$ |
| int $D$ | interior of $D$ |
| $f \circ g$ | superposition $f$ and $g$ |
| $k_\theta(z)$ | Koebe function |
| $p_\theta(z)$ | Pick function |
| $R(D, a)$ | conformal radius of $D$ with respect to $a$ |
| $M(D)$ | conformal modulus of a doubly connected domain or quadrilateral |
| $m(D, a)$ | reduced modulus of $D$ with respect to $a$ |
| $m(D, a, b)$ | reduced modulus of a digon $D$ with respect to its vertices $a$ and $b$ |
| $m_\Delta(D, a)$ | reduced modulus of a triangle $D$ with respect to its vertex $a$ |
| cap $C$ | capacity of a condenser $C$ |
| $S$ | normalized univalent functions |
| $S_R$ | normalized univalent functions with real coefficients |
| $B_s$ | bounded univalent functions $U \to U, 0 \to 0$ |
| $B_s(b)$ | $f \in B_s, f'(0) = b, 0 < b < 1$ |
| $M(\omega)$ | Montel's univalent functions $(f(0) = 0, f(\omega) = \omega)$ |
| $M^M(\omega)$ | bounded Montel's functions |
| $S^1(\beta)$ | functions from $B_s$ with the finite angular derivative $\beta$ at 1 |

| | |
|---|---|
| $Q_K$ | quasiconformal automorphisms of $\mathbb{C}$, $f(0) = 0$, $f(1) = 1$ |
| $U_K$ | quasiconformal automorphisms of $U$, $f(0) = 0$ |
| $\wp$ | Weierstrass elliptic function |
| **sn** | elliptic sine |
| **cn** | elliptic cosine |
| **dn** | elliptic amplitude |
| **tn** | elliptic tangent |
| $\mathbf{F}, \mathbf{E}, \Pi$ | Legendre canonical elliptic integrals of the first, second, and third kind |
| $\mathbf{K}, \mathbf{K'}$ | complete elliptic integral of the first kind and its complimentary |
| $\vartheta_{0,1,2,3}$ | Theta Functions of Jacobi |
| $\Theta, H, \Theta_1, H_1$ | old fashioned Theta Functions of Jacobi |
| $S_0 = U/G_0$ | uniformization of a Riemann surface $S_0$ by a Fuchsian group $G_0$ |
| $T(S_0) \equiv T(G_0) \equiv T(g, n)$ | Teichmüller space of a conformally finite Riemann surface $S_0$ of type $(g, n)$ |
| $\tau_T(x, y)$ | Teichmüller metric on $T(S_0)$ |
| $k_T(x, y)$ | Kobayashi metric on $T(S_0)$ |
| $c_T(x, y)$ | Carathéodory metric on $T(S_0)$ |
| $H^{p,0}$ | space of holomorphic $p$-differentials on $S_0$ |

# Index

Abelian differential, 194
admissible system of curves, 47, 50
 – domain, 48, 51
 – metric, 8, 46
angular derivative, 20,
 – limit, 19
arc, 20, 22, 34, 39, 49, 60, 68
area element, 8
area method, 1
associated domain, 48, 96
 – complete elliptic integral, 26
 – metric, 40

Beltrami differential, 176
 – equation, 141
Bieberbach Conjecture, 1, 144
boundary parameterization, 157, 173
bounded functions, 86

capacity, 17
 – , hyperbolic, 19
 – , logarithmic, 18
circular domain, 43
characteristic domain, 48
closed trajectory, 40
compact class, 60
 – surface, 184
compatibility of angles and heights, 50
complete elliptic integral, 26
condenser, 17
conformal mapping, 1, 3, 7, 57
 – invariance, 7, 8
 – structure, 37, 187
covering theorems, 57, 87, 95
 – – for Montel functions, 95
covering surface, 38
critical trajectory, 40
 – , point, 40

deck mapping, 38, 39
diameter, transfinite, 18
 – , hyperbolic transfinite, 19

differential, quadratic, 39
 – , Beltrami, 176
digon, 20
dilatation, 141
distortion theorem, 59, 88, 98
 – , two-point, 70, 151, 168
domain, doubly connected, 13, 30, 46, 47
 – , circular, 43
 – , characteristic, 48
 – , ending, 43
 – , ring, 42, 43
 – , simply connected, 20, 43, 50
 – , spiral, 43, 44

element of area, 8
 – of length, 8, 39
elliptic function, 23
 – integral, 25
 – point, 39
 – type of surfaces, 38
ending domain, 43
extremal length, 1, 3
 – function, 144, 154, 172
 – metric, 8, 9, 47
 – – , uniqueness of, 9

Fekete points, 18
finite trajectories, 40
foliation, measured, 196
free family, 47
function of Weierstrass, 24
 – of Jacobi, 26
fundamental group, 38
Fuchsian group, 38, 142, 176, 177

genus, 37
global trajectory structure, 43
growth theorem, 58, 87, 145
 – for conformal mapping, 58
 – for quasiconformal mapping, 145
Grötzsch lemmas, 13

Printing:   Strauss GmbH, Mörlenbach
Binding:   Schäffer, Grünstadt

Vol. 1701: Ti-Jun Xiao, J. Liang, The Cauchy Problem of Higher Order Abstract Differential Equations, XII, 302 pages. 1998.

Vol. 1702: J. Ma, J. Yong, Forward-Backward Stochastic Differential Equations and Their Applications. XIII, 270 pages. 1999.

Vol. 1703: R. M. Dudley, R. Norvaiša, Differentiability of Six Operators on Nonsmooth Functions and p-Variation. VIII, 272 pages. 1999.

Vol. 1704: H. Tamanoi, Elliptic Genera and Vertex Operator Super-Algebras. VI, 390 pages. 1999.

Vol. 1705: I. Nikolaev, E. Zhuzhoma, Flows in 2-dimensional Manifolds. XIX, 294 pages. 1999.

Vol. 1706: S. Yu. Pilyugin, Shadowing in Dynamical Systems. XVII, 271 pages. 1999.

Vol. 1707: R. Pytlak, Numerical Methods for Optimal Control Problems with State Constraints. XV, 215 pages. 1999.

Vol. 1708: K. Zuo, Representations of Fundamental Groups of Algebraic Varieties. VII, 139 pages. 1999.

Vol. 1709: J. Azéma, M. Émery, M. Ledoux, M. Yor (Eds), Séminaire de Probabilités XXXIII. VIII, 418 pages. 1999.

Vol. 1710: M. Koecher, The Minnesota Notes on Jordan Algebras and Their Applications. IX, 173 pages. 1999.

Vol. 1711: W. Ricker, Operator Algebras Generated by Commuting Projections: A Vector Measure Approach. XVII, 159 pages. 1999.

Vol. 1712: N. Schwartz, J. J. Madden, Semi-algebraic Function Rings and Reflectors of Partially Ordered Rings. XI, 279 pages. 1999.

Vol. 1713: F. Bethuel, G. Huisken, S. Müller, K. Steffen, Calculus of Variations and Geometric Evolution Problems. Cetraro, 1996. Editors: S. Hildebrandt, M. Struwe. VII, 293 pages. 1999.

Vol. 1714: O. Diekmann, R. Durrett, K. P. Hadeler, P. K. Maini, H. L. Smith, Mathematics Inspired by Biology. Martina Franca, 1997. Editors: V. Capasso, O. Diekmann. VII, 268 pages. 1999.

Vol. 1715: N. V. Krylov, M. Röckner, J. Zabczyk, Stochastic PDE's and Kolmogorov Equations in Infinite Dimensions. Cetraro, 1998. Editor: G. Da Prato. VIII, 239 pages. 1999.

Vol. 1716: J. Coates, R. Greenberg, K. A. Ribet, K. Rubin, Arithmetic Theory of Elliptic Curves. Cetraro, 1997. Editor: C. Viola. VIII, 260 pages. 1999.

Vol. 1717: J. Bertoin, F. Martinelli, Y. Peres, Lectures on Probability Theory and Statistics. Saint-Flour, 1997. Editor: P. Bernard. IX, 291 pages. 1999.

Vol. 1718: A. Eberle, Uniqueness and Non-Uniqueness of Semigroups Generated by Singular Diffusion Operators. VIII, 262 pages. 1999.

Vol. 1719: K. R. Meyer, Periodic Solutions of the N-Body Problem. IX, 144 pages. 1999.

Vol. 1720: D. Elworthy, Y. Le Jan, X-M. Li, On the Geometry of Diffusion Operators and Stochastic Flows. IV, 118 pages. 1999.

Vol. 1721: A. Iarrobino, V. Kanev, Power Sums, Gorenstein Algebras, and Determinantal Loci. XXVII, 345 pages. 1999.

Vol. 1722: R. McCutcheon, Elemental Methods in Ergodic Ramsey Theory. VI, 160 pages. 1999.

Vol. 1723: J. P. Croisille, C. Lebeau, Diffraction by an Immersed Elastic Wedge. VI, 134 pages. 1999.

Vol. 1724: V. N. Kolokoltsov, Semiclassical Analysis for Diffusions and Stochastic Processes. VIII, 347 pages. 2000.

Vol. 1725: D. A. Wolf-Gladrow, Lattice-Gas Cellular Automata and Lattice Boltzmann Models. IX, 308 pages. 2000.

Vol. 1726: V. Marić, Regular Variation and Differential Equations. X, 127 pages. 2000.

Vol. 1727: P. Kravanja M. Van Barel, Computing the Zeros of Analytic Functions. VII, 111 pages. 2000.

Vol. 1728: K. Gatermann Computer Algebra Methods for Equivariant Dynamical Systems. XV, 153 pages. 2000.

Vol. 1729: J. Azéma, M. Émery, M. Ledoux, M. Yor Séminaire de Probabilités XXXIV. VI, 431 pages. 2000.

Vol. 1730: S. Graf, H. Luschgy, Foundations of Quantization for Probability Distributions. X, 230 pages. 2000.

Vol. 1731: T. Hsu, Quilts: Central Extensions, Braid Actions, and Finite Groups. XII, 185 pages. 2000.

Vol. 1732: K. Keller, Invariant Factors, Julia Equivalences and the (Abstract) Mandelbrot Set. X, 206 pages. 2000.

Vol. 1733: K. Ritter, Average-Case Analysis of Numerical Problems. IX, 254 pages. 2000.

Vol. 1734: M. Espedal, A. Fasano, A. Mikelić, Filtration in Porous Media and Industrial Applications. Cetraro 1998. Editor: A. Fasano. 2000.

Vol. 1735: D. Yafaev, Scattering Theory: Some Old and New Problems. XVI, 169 pages. 2000.

Vol. 1736: B. O. Turesson, Nonlinear Potential Theory and Weighted Sobolev Spaces. XIV, 173 pages. 2000.

Vol. 1737: S. Wakabayashi, Classical Microlocal Analysis in the Space of Hyperfunctions. VIII, 367 pages. 2000.

Vol. 1738: M. Émery, A. Nemirovski, D. Voiculescu, Lectures on Probability Theory and Statistics. XI, 356 pages. 2000.

Vol. 1739: R. Burkard, P. Deuflhard, A. Jameson, J.-L. Lions, G. Strang, Computational Mathematics Driven by Industrial Problems. Martina Franca, 1999. Editors: V. Capasso, H. Engl, J. Periaux. VII, 418 pages. 2000.

Vol. 1740: B. Kawohl, O. Pironneau, L. Tartar, J.-P. Zolesio, Optimal Shape Design. Tróia, Portugal 1999. Editors: A. Cellina, A. Ornelas. IX, 388 pages. 2000.

Vol. 1741: E. Lombardi, Oscillatory Integrals and Phenomena Beyond all Algebraic Orders. XV, 413 pages. 2000.

Vol. 1742: A. Unterberger, Quantization and Non-holomorphic Modular Forms. VIII, 253 pages. 2000.

Vol. 1743: L. Habermann, Riemannian Metrics of Constant Mass and Moduli Spaces of Conformal Structures. XII, 116 pages. 2000.

Vol. 1744: M. Kunze, Non-Smooth Dynamical Systems. X, 228 pages. 2000.

Vol. 1745: V. D. Milman, G. Schechtman, Geometric Aspects of Functional Analysis. VIII, 289 pages. 2000.

Vol. 1746: A. Degtyarev, I. Itenberg, V. Kharlamov, Real Enriques Surfaces. XVI, 259 pages. 2000.

Vol. 1747: L. W. Christensen, Gorenstein Dimensions. VIII, 204 pages. 2000.

Vol. 1748: M. Ruzicka, Electrorheological Fluids: Modeling and Mathematical Theory. XV, 176 pages. 2001.

Vol. 1749: M. Fuchs, G. Seregin, Variational Methods for Problems from Plasticity Theory and for Generalized Newtonian Fluids. VI, 269 pages. 2001.

Vol. 1750: B. Conrad, Grothendieck Duality and Base Change. X, 296 pages. 2001.

Vol. 1751: N. J. Cutland, Loeb Measures in Practice: Recent Advances. XI, 111 pages. 2001.

## Recent Reprints and New Editions